Texts and
Monographs
in Physics

A. Perelomov

Generalized Coherent States and Their Applications

Springer-Verlag
Berlin Heidelberg New York
London Paris Tokyo

Professor Dr. Askold Perelomov
Institute of Theoretical and Experimental Physics
B. Cheremushkinskaya 25
SU-117259 Moscow, USSR

ISBN 3-540-15912-6 Springer-Verlag Berlin Heidelberg New York
ISBN 0-387-15912-6 Springer-Verlag New York Berlin Heidelberg

Library of Congress Cataloging-in-Publication Data. Perelomov, A. M. (Askold Mikhaĭlovich), 1935– Generalized coherent states and their applications. (Texts and monographs in physics) Bibliography: p. Includes index. 1. Lie groups. 2. Symmetric spaces. 3. Mathematical physics. I. Title. II. Title: Coherent states and their applications. III. Series. QC20.7.L54P47 1986 530.1'5255 85-30306

Monophoto typesetting, offset printing and bookbinding: Brühlsche Universitätsdruckerei, Giessen
2153/3130-543210

Preface

This monograph treats an extensively developed field in modern mathematical physics – the theory of generalized coherent states and their applications to various physical problems.

Coherent states, introduced originally by Schrödinger and von Neumann, were later employed by Glauber for a quantal description of laser light beams. The concept was generalized by the author for an arbitrary Lie group. In the last decade the formalism has been widely applied to various domains of theoretical physics and mathematics.

The area of applications of generalized coherent states is very wide, and a comprehensive exposition of the results in the field would be helpful. This monograph is the first attempt toward this aim. My purpose was to compile and expound systematically the vast amount of material dealing with the coherent states and available through numerous journal articles. The book is based on a number of undergraduate and postgraduate courses I delivered at the Moscow Physico-Technical Institute. In its present form it is intended for professional mathematicians and theoretical physicists; it may also be useful for university students of mathematics and physics.

In Part I the formalism is elaborated and explained for some of the simplest typical groups. Part II contains more sophisticated material; arbitrary Lie groups and symmetrical spaces are considered. A number of examples from various areas of theoretical and mathematical physics illustrate advantages of this approach, in Part III.

It is a pleasure for me to thank Dr. Yu. Danilov for many useful remarks.

Moscow, April 1985 *A. Perelomov*

Contents

Part III

Physical Applications

Introduction

The method which is the subject of this book originated from the early times of quantum mechanics. In 1926, *Schrödinger* [1] first introduced a system of nonorthogonal wave functions to describe nonspreading wave packets for quantum oscillators. A few years later, in the famous book by *von Neumann* [2], an important subset of these wave functions was considered, which was related to the partitioning of the phase plane of a one-dimensional dynamical system into regular cells. *Von Neumann* used this subset to investigate the coordinate and momentum measurement processes in quantum theory. For a considerable time these ideas of the eminent scientists did not attract undue attention. It was only in the early sixties that the approach was thoroughly studied [3–6]. *Glauber* [7, 8] named the states invented by Schrödinger the *coherent states* (CS) and showed that they are adequate to describe a coherent laser beam in the framework of quantum theory.

An unusual property of the CS system is its overcompleteness, yet it is just this feature which opens up new possibilities. A complete orthonormal system of basis vectors in a Hilbert space is one of the main concepts of mathematical physics and functional analysis. It was found, however, that overcomplete and nonorthogonal systems of state vectors are quite appropriate in solving some problems of quantum physics. A system of vectors is called overcomplete if at least one vector exists in the system which can be removed, while the system remains complete. So the system contains more states than necessary to decompose an arbitrary state vector. A basis like the CS system cannot be treated within the standard routine; however, it is a powerful tool when used by means of proper methods.

The CS system has a number of advantages compared with the usual orthonormal systems of states. During the past one and a half decades it has been successfully applied not only to quantum optics and radiophysics, but also in other areas of physics, for example, in the theory of superfluidity for a weakly nonideal Bose gas. The CS systems are also used to describe spin waves in the Heisenberg model of ferromagnetism, soft photon clouds around charged particles in quantum electrodynamics, and for an approximate quantum description of localized field states (solitons) in nonlinear field theories. Properties of CS systems have been considered in some detail in a number of books and reviews [9–13]. References to numerous original papers in the field can be found therein.

The standard CS system is intimately related to a group, considered first by *Weyl* [14], the so-called Heisenberg-Weyl group. The CS method is particularly effective in cases where the Heisenberg-Weyl group is the dynamical symmetry group of a considered physical system. The simplest example is a quantum oscillator under the action of a variable external driving force. In this case the Heisenberg equations of motion coincide with the corresponding equations for the classical variables. In the course of the time evolution, any coherent state remains coherent, and the motion of the phase space point representing the coherent state is described by the classical equations. This fact enables one to simplify the quantum problem significantly, reducing it to the corresponding classical problem.

The Heisenberg-Weyl group is, of course, not the universal dynamical symmetry group; other symmetry groups appear in many cases. For instance, the symmetry group for spin precession in a variable magnetic field is $SU(2)$, and for the problem of a quantum oscillator with variable frequency the symmetry group is $SU(1,1)$.

A question arises whether for other Lie groups systems of states exist having some properties similar to those of the standard CS system. The answer is positive, as shown in [15]. In that work general CS systems related to representations of an arbitrary Lie group were constructed and investigated; elaborate methods of group theory were employed to study properties of these systems.

A different generalization of the coherent state was proposed earlier by *Barut* and *Girardello* [16]. However, their approach is not applicable to all Lie groups; in particular, it is invalid for compact groups. Besides, the set of coherent states in [16] is not invariant under the action of group operators, unlike the generalized CS proposed in [15]. The particular case of the three-dimensional rotation group was considered also by *Radcliffe* [17]. The CS system constructed there coincides with the corresponding generalized CS system of [15].

Generalized coherent states, which were introduced in [15], are relevant to an arbitrary Lie group; they are parametrized by points of homogeneous spaces where the group acts. In some cases, one can consider these spaces as generalized phase spaces for classical dynamical systems. For example, the two-dimensional sphere and the Lobachevsky plane are such generalizations of the usual phase plane. If one has a situation of this type, the coherent states correspond to points in the phase space representing states of the classical system. In some cases coherent states are those quantum states closest to the corresponding classical states, as they have minimum uncertainties. Therefore the conversion of the classical system into its quantum counterpart is performed in the most natural manner in terms of the coherent states. Such generalized coherent states arise quite naturally in a number of physical problems having dynamical symmetries (relevant information has been presented in [12, 18], see also Part III of this book). Along this line I mention some nonstationary problems: the precession of spin in a variable magnetic field, a quantum oscillator under the action of a variable external force, the parametric excitation of the quantum oscillator, the creation of particle pairs in variable external fields (electric or gravitational), the relaxation of the quantum oscillator to a thermodynamical equilibrium state.

The generalized coherent states were also found to be useful in a number of purely mathematical problems, in particular, in the theory of representations of Lie groups, as well as in the investigation of special functions, automorphic functions, reproducing kernels, and in some other branches of functional analysis.

The CS formalism is also related to the so-called geometric quantization method. (Among the many works concerning this method one should mention the important ones by *Kirillov* [19] and *Kostant* [20, 21].) Geometric quantization involves the group representation space which is considered as the Hilbert space of states of a quantum dynamical system. In parallel to the quantum system, the corresponding classical system is analysed which has the same dynamical symmetry group. The essence of the quantization method [19–21] is a correspondence between a real function in the phase space of the classical dynamical system and a self-conjugate operator in the Hilbert space. The CS approach is, apparently, an adequate method in this field.

I should mention also another quantization method, developed by *Berezin* [22]. This method is applicable to a more restricted class of spaces, but it is closer to the quantization employed in physics, and enables one to get more complete results.

This book consists of three parts and several appendices.

Part I contains a study of properties of the generalized CS systems for the simplest Lie groups: the simplest nilpotent group, the Heisenberg-Weyl group; the simplest compact non-Abelian group, the three-dimensional rotation group; and the simplest noncompact non-Abelian groups, the n-dimensional Lorentz groups.

Part II considers generalized CS systems for a wider class of Lie groups: nilpotent Lie groups, compact semisimple Lie groups, and the automorphism groups of homogeneous complex symmetrical domains. This part is addressed to the reader with a higher level of knowledge in mathematics. To master the matter presented here and in the sections marked by an asterisk, the reader must be familiar with some special mathematics: first of all, the theory of Lie groups and symmetric spaces. Those who feel that these sections of the book are too difficult may skip them and proceed further.

Part III deals with applying the CS methods to solving a number of real physical problems. The subject of this part is clear from the Table of Contents.

Some calculations which are not necessary for an understanding of the main text are presented in the appendices.

As the material on coherent states is abundant, while this volume is limited, some relevant results remain beyond the scope of this book. In particular, I do not mention some properties of standard CS systems, as this subject has already been considered exhaustively in a number of books and reviews. For brevity, I also drop proofs of some statements, in the hope that the interested reader would be able to reconstruct every proof without too much effort.

Part I

Generalized Coherent States for the Simplest Lie Groups

1. Standard System of Coherent States Related to the Heisenberg-Weyl Group: One Degree of Freedom

The subject of this chapter is an overcomplete and nonorthogonal system of Hilbert-space vectors (states) – the system of the so-called standard coherent states (CS).

In quantum mechanics, the standard CS in coordinate representation describe nonspreading wave packets for the harmonic oscillator. (For a reader not familiar with quantum mechanics, the coordinate representation for a system with one degree of freedom is the realization of the Hilbert space as a space of square-integrable functions on the line.) *Schrödinger* considered them from this point of view as early as 1926 [1]. Somewhat later *von Neumann*, in his famous monograph [2], studied an important subsystem of CS, related to the regular cell partition of the phase plane for a system with one degree of freedom. This system was used by von Neumann to analyze the quantum-mechanical measurement process. Later on, after about three decades, further investigations of CS was undertaken [3–6].

Among early works in this area, the important papers by *Glauber* [7, 8] should be mentioned. There the concept of the coherent state was introduced and it was shown that CS provides an adequate means for a quantum description of coherent laser light beams.

Detailed discussion of properties of the standard CS system for a finite number of degrees of freedom, as well as references to a lot of relevant papers, may be found in [9–13]. The case of an infinite number of degrees of freedom was considered by *Segal* [5] and *Berezin* [6] and in numerous papers published in *Communications in Mathematical Physics*.

This chapter deals mainly with well-known material, but the exposition is not quite conventional. The relation between the standard CS system and the Heisenberg-Weyl group (considered originally by *Weyl* [14]) being established, the basic properties of this system are derived by group-theoretical methods. For simplicity, I restrict myself to the case of one degree of freedom. The case when the number of degrees of freedom is more than one, but still finite, is considered in Chap. 3.

1.1 The Heisenberg-Weyl Group and Its Representations

1.1.1 The Heisenberg-Weyl Group

The simplest operators used in describing a quantum-mechanical system with one degree of freedom are the *coordinate operator* q and the *momentum operator* p. They act in the standard Hilbert space $\mathscr{H}$ and satisfy the Heisenberg commutation relations

$$[q,p]=\mathrm{i}\hbar\hat{1}, \qquad [q,\hat{1}]=[p,\hat{1}]=0. \tag{1.1.1}$$

Here $\hat{1}$ is the identity operator, $\hbar$ is Planck's constant, and the bracket means the commutator $[A,B]\equiv AB-BA$.

The structure of the canonical commutation relations (1.1.1) is described by a group, the so-called Heisenberg-Weyl group [14]. The simplest properties of this group are treated in this section. (A number of more subtle mathematical aspects of this group is considered by *Cartier* [23].)

Instead of the operators q and p, another pair of operators is sometimes more suitable, the *annihilation operator* a and its conjugate, the *creation operator* a^+, defined as

$$a=\frac{q+\mathrm{i}p}{\sqrt{2\hbar}}, \qquad a^+=\frac{q-\mathrm{i}p}{\sqrt{2\hbar}}. \tag{1.1.2}$$

Here and in the following $^+$ stands for Hermitian conjugation, and the bar $^-$ means complex conjugation. The commutation relations follow immediately from (1.1.1, 2):

$$[a,a^+]=\hat{1}, \qquad [a,\hat{1}]=[a^+,\hat{1}]=0. \tag{1.1.3}$$

Relations (1.1.1 or 3) mean that the operators q, p, $\hat{1}$ (respectively, a, a^+, $\hat{1}$) are generators of a Lie algebra, which will be denoted by $\mathscr{W}_1$. This is the Heisenberg-Weyl algebra.

Introducing new quantities

$$e_1=\mathrm{i}(\hbar)^{-1/2}p, \qquad e_2=\mathrm{i}(\hbar)^{-1/2}q, \qquad e_3=\mathrm{i}\hat{1}, \tag{1.1.4}$$

and regarding them as elements of an abstract Lie algebra, not just as operators in a Hilbert space, gives the following definition.

The Heisenberg-Weyl algebra $\mathscr{W}_1$ is a real three-dimensional Lie algebra, given by the basic commutation relations

$$[e_1,e_2]=e_3, \qquad [e_1,e_3]=[e_2,e_3]=0. \tag{1.1.5}$$

In general, elements of the algebra $\mathscr{W}_1$ are written as

$$x=(s;x_1,x_2)=x_1e_1+x_2e_2+se_3 \tag{1.1.6}$$

or

$$x = \mathrm{i}s\hat{\mathrm{I}} + \frac{\mathrm{i}}{\hbar}(Pq - Qp) = \mathrm{i}s\hat{\mathrm{I}} + (\alpha a^+ - \bar{\alpha}a), \tag{1.1.7}$$

where s, x_1, and x_2 are real numbers,

$$x_1 = -(\hbar)^{-1/2}Q, \qquad x_2 = (\hbar)^{-1/2}P,$$

$$\alpha = (2\hbar)^{-1/2}(Q + \mathrm{i}P) = 2^{-1/2}(-x_1 + \mathrm{i}x_2), \qquad \bar{\alpha} = (2\hbar)^{-1/2}(Q - \mathrm{i}P).$$

The commutator of the elements $x = (s; x_1, x_2)$ and $y = (t; y_1, y_2)$ is given by

$$[x, y] = B(x, y)e_3, \qquad B(x, y) = x_1 y_2 - x_2 y_1. \tag{1.1.8}$$

Note that $B(x, y)$ is the standard symplectic form on the (x_1, x_2) plane.

Construction of the Lie group corresponding to the Lie algebra is done, as usual, by exponentiation:

$$\exp(x) = \exp(\mathrm{i}s\hat{\mathrm{I}})D(\alpha), \qquad D(\alpha) = \exp(\alpha a^+ - \bar{\alpha}a). \tag{1.1.9}$$

To find the multiplication law for the operators $D(\alpha)$ we use the operator identity (an equivalent identity was proven originally by *Weyl* [14])

$$\exp A \, \exp B = \exp\{\tfrac{1}{2}[A, B]\} \exp(A + B), \tag{1.1.10}$$

which is valid if

$$[A[A, B]] = 0, \qquad [B[A, B]] = 0. \tag{1.1.11}$$

Let us consider a simple proof of the identity (1.1.10), proposed by *Glauber* [24]. We construct an operator function

$$F(t) = \exp(tA)\exp(tB)\exp(-t(A + B)). \tag{1.1.12}$$

It satisfies

$$\frac{dF}{dt} = \exp(tA)[A, \exp(tB)]\exp(-t(A + B)) = t[A, B]F(t). \tag{1.1.13}$$

Integrating this equation up to $t = 1$, one gets (1.1.10).

Finally, substituting $A = \alpha a^+ - \bar{\alpha}a$, $B = \beta a^+ - \bar{\beta}a$ into (1.1.10), we obtain the multiplication law

$$D(\alpha)D(\beta) = \exp(\mathrm{i}\,\mathrm{Im}\{\alpha\bar{\beta}\})D(\alpha + \beta). \tag{1.1.14}$$

The corresponding formula for the product of several operators $D(\alpha)$ is

$$D(\alpha_n)D(\alpha_{n-1})\ldots D(\alpha_1) = \exp(\mathrm{i}\delta)D(\alpha_n + \alpha_{n-1} + \ldots + \alpha_1), \tag{1.1.15}$$

where

$$\delta = \operatorname{Im}\left\{\sum_{j>k} \alpha_j \bar{\alpha}_k\right\}. \tag{1.1.16}$$

The phase $\operatorname{Im}\{\alpha\bar{\beta}\}$ in (1.1.14) has a simple geometrical meaning. In fact,

$$\operatorname{Im}\{\alpha\bar{\beta}\} = 2A(0, \beta, \alpha+\beta), \tag{1.1.17}$$

where $A(\alpha, \beta, \gamma)$ is the area of the triangle with vertices at the points α, β, γ, and A is positive if the cycle $\alpha\to\beta$, $\beta\to\gamma$, $\gamma\to\alpha$ is counterclockwise, in the opposite case $A<0$. The phase angle in (1.1.15) may be interpreted similarly. Recall the relation $\alpha = (2\hbar)^{-1/2}(Q+\mathrm{i}P)$, and write the expression for δ in the form

$$\delta = \frac{1}{\hbar}\int_{\Gamma} P\,dQ, \tag{1.1.18}$$

where integration is along the boundary of the polygon with vertices at the points $0, \alpha_1, \alpha_1+\alpha_2, \ldots, \alpha_1+\alpha_2+\ldots+\alpha_n$. A reader aware of quantum mechanics would note a semiclassical form of this representation; δ is just proportional to the polygon area on the phase plane.

A consequence of (1.1.14) is

$$D(\alpha)D(\beta) = \exp(2\mathrm{i}\operatorname{Im}\{\alpha\bar{\beta}\})D(\beta)D(\alpha). \tag{1.1.19}$$

Actually, this is an integral form of the Heisenberg commutation relations. *Weyl* [14] wrote an equivalent of this relation, though in a somewhat different form:

$$\exp\left(\frac{\mathrm{i}Qp}{\hbar}\right)\exp\left(\frac{\mathrm{i}Pq}{\hbar}\right) = \exp\left(\frac{\mathrm{i}PQ}{\hbar}\right)\exp\left(\frac{\mathrm{i}Pq}{\hbar}\right)\exp\left(\frac{\mathrm{i}Qp}{\hbar}\right). \tag{1.1.19$'$}$$

An advantage of this form, (1.1.19), as compared with (1.1.3), is that unlike p, q, which are unbounded operators in the Hilbert space $\mathscr{H}$, the operators $D(\alpha)$ are bounded, so that their domain of definition is the whole space $\mathscr{H}$.

Another consequence of (1.1.14) is that the operators $\exp(\mathrm{i}t)D(\alpha)$ form a representation of the group with elements fixed by three real numbers, $g=(t; x_1, x_2)$, or by a real number t and a complex number α, $g=(t;\alpha)$. This group will be called the *Heisenberg-Weyl group*, denoted by W_1. It is not difficult to see that the multiplication law in W_1 is

$$(s; x_1, x_2)(t; y_1, y_2) = (s+t+B(x,y); x_1+y_1, x_2+y_2),$$

$$B(x,y) = x_1 y_2 - y_1 x_2. \tag{1.1.20}$$

Note that the group W_1 belongs to the class of so-called nilpotent groups; a typical example relevant to this class is the group of upper (lower) triangle matrices with unities on the main diagonal. (More general nilpotent Lie groups are considered in Part II.) In the case considered, $W_1=\{g\}$, where

$$g=\begin{pmatrix}1 & a & c\\ 0 & 1 & b\\ 0 & 0 & 1\end{pmatrix}. \tag{1.1.21}$$

These matrices form the simplest finite-dimensional nonunitary representation of the group W_1. The corresponding generators of the Lie algebra e_1, e_2, e_3 are represented by

$$\begin{pmatrix}0 & 1 & 0\\ 0 & 0 & 0\\ 0 & 0 & 0\end{pmatrix}, \quad \begin{pmatrix}0 & 0 & 0\\ 0 & 0 & 1\\ 0 & 0 & 0\end{pmatrix}, \text{ and } \begin{pmatrix}0 & 0 & 1\\ 0 & 0 & 0\\ 0 & 0 & 0\end{pmatrix}, \tag{1.1.22}$$

respectively.

1.1.2 Representations of the Heisenberg-Weyl Group

The first problem to consider is the description of all unitary irreducible representations of W_1. Note first of all that the elements $(s, 0)$ form the center of W_1, i.e., the set of all elements commuting with every element of W_1. Therefore for any unitary irreducible representation $T(g)$ of the group W_1, the operators $T((s, 0))$ form a unitary representation of the subgroup $\{(s, 0)\}$, which is determined by a real number λ:

$$T^{\lambda}((s;0))=\exp(\mathrm{i}\lambda s)\hat{I} \tag{1.1.23}$$

This problem was solved by *Stone* [25] and *von Neumann* [26]; the result is the following theorem.

Theorem. For a fixed value of λ ($\lambda\neq 0$) any two unitary irreducible representations of the group W_1 are unitarily equivalent.

In other words, for any two systems of operators $\{D(\alpha)\}$ and $\{\tilde{D}(\alpha)\}$, satisfying (1.1.19), a unitary operator U exists such that

$$\tilde{D}(\alpha)=U^{+}D(\alpha)U. \tag{1.1.24}$$

An analoguous statement is valid also for operator pairs $\tilde{a}^{+}$, $\tilde{a}$ and a^{+}, a satisfying the commutation relations (1.1.3):

$$\tilde{a}^{+}=u^{+}a^{+}u, \qquad \tilde{a}=u^{+}au. \tag{1.1.24'}$$

However, since the operators a and a^{+} are unbounded, this statement holds only if some additional conditions on the domain of operators $\tilde{a}^{+}$, $\tilde{a}$ and a^{+}, a are satisfied [23].

Thus a unitary irreducible infinite dimensional representation of the group W_1 is fixed by a single real number λ: $T(g)=T^{\lambda}(g)$, $\lambda \neq 0$. Furthermore, there are representations with $\lambda=0$. They are all one-dimensional and are fixed by a pair of real numbers, say μ and ν: $T(g)=T^{\mu\nu}(g)=\chi_{\mu\nu}(g)\hat{1}$, $\chi_{\mu\nu}(g)=\exp\{i(\mu x_1+\nu x_2)\}$. This completes the outline of the theory of representations of the Heisenberg-Weyl group. The structure of the set of representations of the group W_1 described here is explained within the theory by *Kirillov* [19], where any group representation is constructed for an orbit of the coadjoint representation of the group. More details are given in Part II. We now describe the representations $T^{\lambda}(g)$ explicitly.

1.1.3 Concrete Realization of the Representation $T^{\lambda}(g)$

The operators q, p and a^+, a act in the standard Hilbert space. Here and in the following the vectors belonging to this space are denoted by Dirac's symbol $|\psi\rangle$, the scalar product of the vector $|\varphi\rangle$ and $|\psi\rangle$ linear in $|\psi\rangle$ and antilinear in $|\varphi\rangle$ is written $\langle\varphi|\psi\rangle$, and the projection operator upon $|\psi\rangle$ is written $|\psi\rangle\langle\psi|$. The state ψ is described by a class of the vectors differing from $|\psi\rangle$ by a numerical factor.

It is known that a so-called vacuum vector $|0\rangle$ exists in $\mathscr{H}$, i.e., a normalized vector annihilated by the operator a:

$$a|0\rangle=0, \quad \langle 0|0\rangle=1. \tag{1.1.25}$$

Action of the creation operator a^+ generates a set of normalized vectors from the vacuum

$$|n\rangle=(n!)^{-1/2}(a^+)^n|0\rangle, \quad n=0,1,2,\ldots. \tag{1.1.26}$$

The vectors $\{|n\rangle\}$ form a basis in $\mathscr{H}$. The action of the operators a and a^+ in this basis is given by

$$a|n\rangle=\sqrt{n}|n-1\rangle, \quad a^+|n\rangle=\sqrt{n+1}|n+1\rangle, \quad a^+a|n\rangle=n|n\rangle. \tag{1.1.27}$$

Sometimes it is appropriate to use concrete functional realizations of the Hilbert space $\mathscr{H}$ or, using physical terminology, definite representations. In the case of the so-called coordinate representation, the vector $|\psi\rangle$ is represented by a coordinate function $\langle q|\psi\rangle=\psi(q)$, which is square-integrable

$$\int |\psi(q)|^2 dq<\infty. \tag{1.1.28}$$

The action of the coordinate operator $\hat{q}$ in the coordinate representation is just a multiplication by q, while the momentum operator $\hat{p}$ is represented by differentiation with respect to q: $\hat{p}=-i\hbar\partial/\partial q$. The basis vector $|n\rangle$ is described by the function

$$\langle q|n\rangle=\varphi_n(q)=(\pi\hbar)^{-1/4}(2^n n!)^{-1/2}H_n(\hbar^{-1/2}q)\exp[-(2\hbar)^{-1}q^2], \tag{1.1.29}$$

where $H_n(q)$ is the Hermite polynomial of degree n. In the coordinate representation, (1.1.27) become recursion relations for the Hermite polynomials:

$$\frac{d}{dq} H_n(q) = 2nH_{n-1}(q), \tag{1.1.30}$$

$$\left(2q - \frac{d}{dq}\right) H_n(q) = H_{n+1}(q). \tag{1.1.31}$$

Hence one gets at once the useful relations

$$H_n(q) = \left(2q - \frac{d}{dq}\right)^n H_0(q), \qquad H_0(q) \equiv 1,$$

$$H_n(q) = (-1)^n \exp(q^2) \frac{d^n}{dq^n} \exp(-q^2) \tag{1.1.32}$$

as well as a differential equation for the Hermite polynomials,

$$H_n'' - 2qH_n' + 2nH_n = 0. \tag{1.1.33}$$

In the coordinate representation the action of the operator $D(\alpha)$, $\alpha = (2\hbar)^{-1/2} \cdot (Q + \mathrm{i}P)$, is given by

$$D(\alpha)\varphi(q) = \exp\left(\frac{-\mathrm{i}PQ}{2\hbar}\right) \exp\left(\frac{\mathrm{i}Pq}{\hbar}\right) \varphi(q - Q). \tag{1.1.34}$$

1.2 Coherent States

This section treats some overcomplete systems of states related to the Heisenberg-Weyl group W_1, the systems of generalized coherent states. The standard system of coherent states is a particular yet very important case. The concept of the generalized coherent state is introduced here, following the presentation by the author [15].

Let $T(g)$ be a unitary irreducible representation of W_1 (described in the preceding section), and $|\psi_0\rangle$ a fixed vector in the representation space $\mathscr{H}$. It is not difficult to see that the state corresponding to the vector $|\psi_0\rangle$ is stable only under the action of the operators of the form $T((s, 0))$. (Recall that a state is represented by a set of vectors $\exp(\mathrm{i}\varphi)|\psi\rangle$, differing from the vector $|\psi\rangle$ by a phase factor only, $|\exp(\mathrm{i}\varphi)| = 1$.) In other words, the isotropy subgroup H for an arbitrary state $|\psi_0\rangle$ contains only elements of the form $(s, 0)$.

Apply now the representation operator $T(g)=T((t,\alpha))=\exp(it)D(\alpha)$ to $|\psi_0\rangle$. The result is a set of the states $\{|\alpha\rangle\}$

$$|\alpha\rangle=D(\alpha)|\psi_0\rangle, \tag{1.2.1}$$

where α is a complex number. Furthermore, since the isotropy subgroup of the state $|\psi_0\rangle$ is $H=\{h\}$, $h=(t,0)$, different α correspond to different states. The system $\{|\alpha\rangle\}$ is just a system of generalized coherent states of the type $\{T(g),|\psi_0\rangle\}$. An important particular case is the choice of the vacuum vector $|0\rangle$ as the starting vector $|\psi_0\rangle$. This is the case of standard coherent states.

The generalized CS system has a number of remarkable properties, considered below.

Note first of all that because the representation $T(g)$ is irreducible, the system is complete. However, the states are, in general, not mutually orthogonal. Actually,

$$\langle\beta|\alpha\rangle=\langle\psi_0|D^+(\beta)D(\alpha)|\psi_0\rangle=\exp(\mathrm{iIm}\{\alpha\bar{\beta}\})\langle\psi_0|D(\alpha-\beta)|\psi_0\rangle, \tag{1.2.2}$$

$$|\langle\beta|\alpha\rangle|^2=|\langle\psi_0|D(\alpha-\beta)|\psi_0\rangle|^2=\varrho(\alpha-\beta), \tag{1.2.3}$$

and the function $\varrho(\alpha)$ cannot be identically zero.

The operator $D(\alpha)$ transforms any coherent state into another coherent state,

$$D(\alpha)|\beta\rangle=\exp(\mathrm{iIm}\{\alpha\bar{\beta}\})|\alpha+\beta\rangle. \tag{1.2.4}$$

Relation (1.2.4) determines the action of group W_1 on the α plane,

$$(s;\beta)\alpha=\alpha+\beta. \tag{1.2.5}$$

This action is not effective, since the subgroup $H=\{(t,0)\}$ acts as the identity transformation in the α plane. As seen from (1.2.5), the factor group W_1/H is the group of translations of the α plane. Hence the invariant metric in the α plane is written as usual,

$$ds^2=|d\alpha|^2. \tag{1.2.6}$$

The corresponding invariant measure in the α plane is

$$d\mu(\alpha)=C\,d^2\alpha=C\,d\alpha_1 d\alpha_2, \qquad \alpha=\alpha_1+\mathrm{i}\alpha_2, \tag{1.2.7}$$

where C is a constant.

Now we turn to derivation of the so-called resolution of unity. Let $|\alpha\rangle\langle\alpha|$ be the projection operator on the state $|\alpha\rangle$. Consider the operator

$$\hat{A}=\int d\mu(\beta)|\beta\rangle\langle\beta|. \tag{1.2.8}$$

It is easily seen that $\hat{A}$ commutes with any $D(\alpha)$. Therefore, in view of Schur's lemma, this operator is the unit operator times a number,

$$\hat{A}=d^{-1}\cdot\hat{1}. \tag{1.2.9}$$

To find the constant d we calculate the average of the operator $\hat{A}$ over a coherent state $|\alpha\rangle$

$$d^{-1}=\langle\alpha|\hat{A}|\alpha\rangle=\int|\langle\alpha|\beta\rangle|^2 d\mu(\beta)=\int\varrho(\beta)d\mu(\beta). \tag{1.2.10}$$

For a bounded operator $\hat{A}$ the constant d is nonzero, so the factor C in (1.2.7) may be chosen such that $d=1$. The resulting "resolution of unity" is[1]

$$\int d\mu(\alpha)|\alpha\rangle\langle\alpha|=\hat{1}, \tag{1.2.11}$$

where $d\mu(\alpha)$ is given in (1.2.7). The constant C is determined from the condition $\int\varrho(\alpha)d\mu(\alpha)=1$.
An immediate consequence from the resolution of unity (1.2.11) is the linear dependence of the coherent states

$$\int d\mu(\alpha)|\alpha\rangle\langle\alpha|\beta\rangle=|\beta\rangle. \tag{1.2.12}$$

Clearly, the kernel $K(\alpha,\beta)=\langle\alpha|\beta\rangle$ is reproducing

$$\int K(\alpha,\beta)K(\beta,\gamma)d\mu(\beta)=K(\alpha,\gamma). \tag{1.2.13}$$

(The general theory of reproducing kernels may be found in [30, 31].)

Using the completeness condition, (1.2.11), it is not difficult to expand an arbitrary state $|\psi\rangle$ in the coherent states,

$$|\psi\rangle=\int d\mu(\alpha)\psi(\alpha)|\alpha\rangle \tag{1.2.14}$$

where the coefficient function $\psi(\alpha)$ is given by

$$\psi(\alpha)=\langle\alpha|\psi\rangle. \tag{1.2.15}$$

The function $\psi(\alpha)$ determines the state $|\psi\rangle$ completely; it is called the symbol of the state $|\psi\rangle$. Evidently,

$$\langle\psi|\psi\rangle=\int|\psi(\alpha)|^2 d\mu(\alpha). \tag{1.2.16}$$

---

[1] This identity was obtained by *Klauder* [3] for the standard system of coherent states. Overcomplete systems of states may also be considered, for which an analog of (1.2.11) is valid, with no reference to the theory of group representations. Such an approach was developed by *Klauder* and *McKenna* [27, 28], who named it "the theory of continuous representations". A number of theorems were proven by *Berezin* within this approach [29].

Up to now the state vector $|\psi_0\rangle$, which was the origin for constructing the generalized CS system, was considered as an arbitrary element of the Hilbert space $\mathscr{H}$. A question arises whether it is possible to use this arbitrariness in such a way that the resulting CS system would have some prescribed properties, for instance, the coherent states would be as close as possible to the classical states.

In the situation considered, it is natural to use the Heisenberg uncertainty relation

$$\Delta = \Delta q \cdot \Delta p \geq \hbar/2 \tag{1.2.17}$$

to determine the criterion of closeness between the classical and quantum states. Here

$$(\Delta q)^2 = \langle(\hat{q} - \langle\hat{q}\rangle)^2\rangle, \quad (\Delta p)^2 = \langle(\hat{p} - \langle\hat{p}\rangle)^2\rangle \tag{1.2.18}$$

and $\langle\hat{q}\rangle$ means the average of the operator $\hat{q}$ over the considered state $|\psi\rangle$: $\langle\hat{q}\rangle = \langle\psi|\hat{q}|\psi\rangle$.

First we show that for all the states in the system $\{|\alpha\rangle\}$ not only the uncertainty magnitude Δ, but also the dispersions Δq and Δp are universal, i.e., independent of α. To this end, we use the identity

$$D^+(\alpha) a D(\alpha) = a + \alpha, \quad \alpha = \alpha_1 + i\alpha_2 \tag{1.2.19}$$

which is easily proven, say, by expanding the operator $D(\alpha)$ in powers of the operator $(\alpha a^+ - \bar{\alpha} a)$ and using the commutation relations (1.1.3). From (1.2.19) one gets at once

$$\langle\alpha|q|\alpha\rangle = \langle\psi_0|q|\psi_0\rangle + (2\hbar)^{1/2}\alpha_1 \tag{1.2.20}$$

$$\langle\alpha|p|\alpha\rangle = \langle\psi_0|p|\psi_0\rangle + (2\hbar)^{1/2}\alpha_2 \quad \text{so that}$$

$$(\Delta q)^2_\alpha = (\Delta q)^2_0, \quad (\Delta p)^2_\alpha = (\Delta p)^2_0 \tag{1.2.21}$$

and our statement is proven.

Another consequence is that among the coherent states there is always a state for which $\langle\hat{q}\rangle = \langle\hat{p}\rangle = 0$. It is easily seen that this property is specific for the state

$$|-\alpha_0\rangle = D(-\alpha_0)|\psi_0\rangle, \tag{1.2.22}$$

where $\alpha_0 = \langle\psi_0|a|\psi_0\rangle$. Thus with no loss of generality we may assume that $\langle\psi_0|\hat{q}|\psi_0\rangle = \langle\psi_0|\hat{p}|\psi_0\rangle = 0$.

Let us now find all the states with $\langle\hat{q}\rangle = \langle\hat{p}\rangle = 0$, minimizing the Heisenberg uncertainty relation

$$\Delta q \cdot \Delta p = \hbar/2. \tag{1.2.23}$$

Consider an evident inequality

$$\langle A^+A\rangle \geq 0, \quad A = \frac{\lambda q + \mathrm{i}p}{\sqrt{2\lambda\hbar}}, \quad \lambda > 0 \quad \text{or} \tag{1.2.24}$$

$$\lambda^2(\Delta q)^2 - \lambda\hbar + (\Delta p)^2 \geq 0. \tag{1.2.25}$$

It is not difficult to see that fulfilling this inequality for all λ is equivalent to the Heisenberg uncertainty relation and that (1.2.23) may hold only if for some positive λ

$$A|\psi_0\rangle = \left(\frac{\lambda\hat{q} + \mathrm{i}\hat{p}}{\sqrt{2\lambda\hbar}}\right)|\psi_0\rangle = 0. \tag{1.2.26}$$

Specifically, such a state is the vacuum $|0\rangle$ (for $\lambda = 1$).

Note that the operators A and A^+ satisfy the commutation relation

$$[A, A^+] = \hat{\mathrm{I}} \tag{1.2.27}$$

so they may be considered as a new pair of "annihilation-creation" operators. They are obtained from the original operators via a linear canonical transformation,

$$A = ua + va^+, \quad A^+ = \bar{u}a^+ + \bar{v}a, \quad |u|^2 - |v|^2 = 1. \tag{1.2.28}$$

Note that this type of representation constitutes $SU(1,1)$, the group of linear transformations of two-dimensional complex space which leave the form $|z_1|^2 - |z_2|^2$ invariant.

Let us now look for a specific property of the state vector $|\psi_0\rangle$, satisfying (1.2.26), from the algebraic point of view. For this purpose we follow a method of a previous work [11] and consider the complex envelop $\mathscr{W}_1^{\mathrm{c}}$ of the Lie algebra $\mathscr{W}_1$, i.e., the set of all linear combinations of the basis elements $\hat{q}$, $\hat{p}$ and $\hat{\mathrm{I}}$ with complex coefficients. The isotropy subalgebra of the state $|\psi_0\rangle$ is denoted by $\mathscr{B} = \{b\}$, i.e., the set of all elements of $\mathscr{W}_1^{\mathrm{c}}$, for which $b|\psi_0\rangle = \lambda|\psi_0\rangle$. Let $\bar{\mathscr{B}} = \{\bar{b}\}$ be a subalgebra of $\mathscr{W}_1^{\mathrm{c}}$, conjugate to $\mathscr{B}$. The subalgebra $\mathscr{B}$ is called maximal here, if $\mathscr{B} + \bar{\mathscr{B}} = \mathscr{W}_1^{\mathrm{c}}$. The states for which the isotropy subalgebras are maximal are the most symmetrical, so they can be distinguished. It can be shown that in this case the coherent state is determined by a point in the coset space $W_1^{\mathrm{c}}|B = \bar{B}|D$, where $\mathscr{D} = \mathscr{B} \cap \bar{\mathscr{B}}$, and B and D are the Lie groups corresponding to the Lie algebras $\mathscr{B}$ and $\mathscr{D}$. Such coherent states may be realized naturally in a certain space of analytical functions (see the next section). In particular, the vacuum state vector has a maximal isotropy subalgebra:

$$\mathscr{B} = \{a, \mathrm{I}\}, \quad \bar{\mathscr{B}} = \{a^+, \mathrm{I}\}, \quad \mathscr{B} + \bar{\mathscr{B}} = \mathscr{W}_1^{\mathrm{c}}.$$

The state vectors satisfying the condition $(ua + va^+)|\psi\rangle = \gamma|\psi\rangle$, $|u|^2 - |v|^2 = 1$, where u, v, γ are complex numbers, also have this property. This construction is considered further in Chap. 2 in more detail.

The states $\{|\alpha\rangle\}$, $|\alpha\rangle = D(\alpha)|0\rangle$ form the standard coherent state system. Evidently, all the formulae in this section are valid also for this system. However, for this particular case a number of useful formulae also hold which are not applicable to the general case.

For instance, it is easy to see that the state $|\alpha\rangle$ is annihilated by the operator

$$D(\alpha)aD^{+}(\alpha). \tag{1.2.29}$$

As follows from (1.2.19), this fact is equivalent to

$$a|\alpha\rangle = \alpha|\alpha\rangle. \tag{1.2.30}$$

Thus the standard coherent state is an eigenstate of the annihilation operator, while any complex number α may be an eigenvalue. It is not difficult to show also that the operator a^{+} has no eigenvector in $\mathscr{H}$.

It is remarkable that because of (1.1.2) the α plane is an analog of the classical phase plane, where a point has the coordinates (Q, P). Hence the coherent states realize a mapping of the phase plane into the Hilbert space $\mathscr{H}$.

Below some other useful relations for the states $|\alpha\rangle$ are presented.

First of all, the following expressions for the operator $D(\alpha)$ may be easily derived from the identity (1.1.10):

$$D(\alpha) = \exp(-|\alpha|^2/2)\exp(\alpha a^{+})\exp(-\bar{\alpha}a) \tag{1.2.31}$$

$$D(\alpha) = \exp(|\alpha|^2/2)\exp(-\bar{\alpha}a)\exp(\alpha a^{+}). \tag{1.2.32}$$

The representation (1.2.31) is the so-called normal, or Wick, form for the operator $D(\alpha)$. In this representation all the creation operators a^{+} in the expansion stand to the left of the annihilation operators a; respectively, (1.2.32) is the antinormal, or anti-Wick, form of the operator $D(\alpha)$.

The following useful relations stem from (1.2.31, 32):

$$a^{+}D(\alpha) = \left(\frac{\partial}{\partial\alpha} + \frac{\bar{\alpha}}{2}\right)D(\alpha), \qquad D(\alpha)a^{+} = \left(\frac{\partial}{\partial\alpha} - \frac{\bar{\alpha}}{2}\right)D(\alpha) \tag{1.2.33}$$

$$aD(\alpha) = -\left(\frac{\partial}{\partial\bar{\alpha}} - \frac{\alpha}{2}\right)D(\alpha), \qquad D(\alpha)a = -\left(\frac{\partial}{\partial\bar{\alpha}} + \frac{\alpha}{2}\right)D(\alpha). \tag{1.2.34}$$

Another consequence of (1.2.31) is

$$|\alpha\rangle = \exp(-|\alpha|^2/2)\exp(\alpha a^{+})|0\rangle \tag{1.2.35}$$

which may be also rewritten in Glauber's form,

$$|\alpha\rangle = \exp(-|\alpha|^2/2)\sum_{n=0}^{\infty}\frac{\alpha^n}{\sqrt{n!}}|n\rangle. \tag{1.2.36}$$

Below explicit expressions are also given for the coherent states in the coordinate and momentum representations

$$\langle q|\alpha\rangle=(\pi\hbar)^{-1/4}\exp[\mathrm{i}(2/\hbar)^{1/2}\alpha_2 q]\exp\left(-\frac{[q-(2\hbar)^{1/2}\alpha_1]^2}{2\hbar}\right) \tag{1.2.37}$$

$$\langle p|\alpha\rangle=(\pi\hbar)^{-1/4}\exp[-\mathrm{i}(2/\hbar)^{1/2}\alpha_1 p]\exp\left(-\frac{[p-(2\hbar)^{1/2}\alpha_2]^2}{2\hbar}\right). \tag{1.2.38}$$

Further, it follows immediately from (1.2.36) that

$$\langle\alpha|\beta\rangle=\exp(-\tfrac{1}{2}|\alpha|^2-\tfrac{1}{2}|\beta|^2+\bar{\alpha}\beta), \tag{1.2.39}$$

$$\varrho(\alpha)=|\langle\alpha|0\rangle|^2=\exp(-|\alpha|^2), \qquad |\langle\alpha|\beta\rangle|^2=\exp(-|\alpha-\beta|^2). \tag{1.2.40}$$

Note that here the function $\varrho(\alpha)$ is nonzero everywhere, so that any two coherent states are nonorthogonal to each other. This property pertains also to general coherent-state systems having maximal isotropy subalgebras.

It is not difficult to obtain now the magnitude of the constant C in (1.2.7). It is π^{-1}, so that the measure is given by

$$d\mu(\alpha)=\pi^{-1}d\alpha_1\,d\alpha_2, \qquad \alpha=\alpha_1+\mathrm{i}\alpha_2. \tag{1.2.41}$$

Note the minimization of the Heisenberg uncertainty relation for the coherent state system; for these states $\Delta q\Delta p=\hbar/2$.

1.3 The Fock-Bargmann Representation

In the conventional coordinate or momentum representations no conditions of analyticity are imposed upon the functions $\varphi(q)$ and $\tilde{\varphi}(p)$, corresponding to a vector in the Hilbert space $\mathscr{H}$. However, there is a realization of the space, where any state vector is described by an entire analytical function. This realization was considered by *Fock* [32] and *Bargmann* [4], so it is called the Fock-Bargmann representation. This representation enables one to find simpler solutions for a number of problems, exploiting the theory of analytical entire functions.

Such a representation may be related to any coherent state system, having a maximal isotropy subgroup. Here we consider only the case of the usual coherent states $|\alpha\rangle=D(\alpha)|0\rangle$, $|0\rangle$ is the vacuum state vector, $a|0\rangle=0$.

Let $|\psi\rangle$ be an arbitrary normalized vector in $\mathscr{H}$. Then, as shown in the preceding section, the state $|\psi\rangle$ is completely determined by its symbol $\langle\alpha|\psi\rangle$. Let

$$|\psi\rangle=\sum_{n=0}^{\infty}c_n|n\rangle, \qquad \langle\psi|\psi\rangle=\sum_{n=0}^{\infty}|c_n|^2=1. \tag{1.3.1}$$

Then in view of (1.2.36),

$$\langle\alpha|\psi\rangle=\exp(-\tfrac{1}{2}|\alpha|^2)\psi(\bar{\alpha}), \qquad \text{where} \tag{1.3.2}$$

$$\psi(z)=\sum c_n u_n(z), \qquad u_n(z)=\frac{z^n}{\sqrt{n!}}, \tag{1.3.3}$$

The series in (1.3.3) converges uniformly in any compact domain of the z plane because of the condition $\sum_{n=0}^{\infty}|c_n|^2=1$, so $\psi(z)$ is an entire analytic function in the complex z plane, and

$$\|\psi\|^2=\langle\psi|\psi\rangle=\int \exp(-|z|^2)|\psi(z)|^2\,d\mu(z)<\infty. \tag{1.3.4}$$

The scalar product of two entire functions $\psi_1(z)$ and $\psi_2(z)$, satisfying condition (1.3.4), is defined by

$$\langle\psi_1|\psi_2\rangle=\int \exp(-|z|^2)\bar{\psi}_1(z)\psi_2(z)\,d\mu(z). \tag{1.3.5}$$

Bargmann [4] showed that this functional space is in fact a Hilbert space.

Thus we are led to a concrete realization of the Hilbert space as a space of entire analytical functions $\psi(z)$, satisfying condition (1.3.4). [A "physical" interpretation of (1.3.4) would consider it as a statistical average of the function $\psi(z)\equiv\psi(p,q)$ over the classical phase space (p,q), $z=q+\mathrm{i}p$, for a system with the classical oscillator Hamiltonian $H=(p^2+q^2)/2=|z|^2/2$, at $\beta=1/kT=2$, the distribution being $\exp(-\beta H)$.]

Originally this representation was introduced by *Fock* [32] in 1928, in a somewhat different but equivalent form. *Fock* proposed an operator solution for the Heisenberg commutation relations (1.1.3),

$$a\to d/dz, \qquad a^+\to z \tag{1.3.6}$$

in analogy to the Schrödinger solution, $\hat{p}=-\mathrm{i}\hbar d/dq$, $\hat{q}=q$. He used this representation in investigating the quantum field theory. This representation was thoroughly studied by *Bargmann* [4], and in a number of subsequent works [33–36] for a finite number of creation-annihilation operators a_j and a_k^+. (The case of an infinite number of operators was considered by *Segal* [5]. Some aspects of this case applicable to quantum field theory are discussed in [6] and in a number of papers published in *Communications in Mathematical Physics*, so it is not considered here.)

We name it the Fock-Bargmann representation, and the representation space will be denoted by $\mathscr{F}$. The scalar product in this space is given by (1.3.5). A consequence of the Schwartz inequality

$$|\langle\alpha|\psi\rangle|\leq\|\psi\|$$

is

$$|\psi(z)|\leq C\exp(\tfrac{1}{2}|z|^2) \tag{1.3.7}$$

for any $\psi(z)\in\mathscr{F}$. The inverse statement would be wrong. For example, the function $\psi(z)=\exp(z^2/2)$ satisfies (1.3.7), but has infinite norm.

As was mentioned above, (1.3.6), in the Fock-Bargmann representation the operator a^+ is the multiplication by z, while the operator a is the differentiation with respect to z, and it is easily seen that a^+ is conjugate to a for the scalar product given in (1.3.5). It is remarkable that the form of the scalar product may be derived from the requirement that the operators a and a^+ must be conjugated.

Note some properties of the Fock-Bargmann representation.

The orthonormal basis in $\mathscr{F}$ has a much simpler form than in the coordinate representation,

$$|n\rangle \to \langle z|n\rangle = u_n(z) = \frac{z^n}{\sqrt{n!}}. \tag{1.3.8}$$

The corresponding representation of the coherent state $|\alpha\rangle$ is

$$\langle z|\alpha\rangle = \exp\left(-\tfrac{1}{2}|\alpha|^2 + \alpha z\right). \tag{1.3.9}$$

The role of the δ function in the space $\mathscr{F}$ is played by

$$\delta(z, z') = \sum_{n=0}^{\infty} u_n(z)\bar{u}_n(z') = \exp(z\bar{z}'). \tag{1.3.10}$$

Actually, it is easy to see that for any analytic function $f(z)$ in $\mathscr{F}$,

$$f(z) = \int \delta(z, z')\exp(-|z'|^2) f(z')\,d\mu(z'). \tag{1.3.11}$$

Let us now consider the space L_2, the space of all functions (not necessarily analytic), satisfying the condition

$$\|f\|^2 = \int |f(z,\bar{z})|^2 \exp(-|z|^2)\,d\mu(z) < \infty. \tag{1.3.12}$$

Evidently,

$$\hat{f}(z) = \int \exp(z\bar{z}' - |z'|^2) f(z', \bar{z}')\,d\mu(z') \tag{1.3.13}$$

realizes a projection of the space L_2 onto $\mathscr{F}: L_2 \to \mathscr{F}$. Note also that if the states $|\psi\rangle$ and $|\alpha\rangle$ are orthogonal, then $\psi(\bar{\alpha})=0$.

Finally, the relation between the usual coordinate representation and the Fock-Bargmann representation is given by the kernel $\langle z|q\rangle$, satisfying

$$a\langle z|q\rangle = z\langle z|q\rangle, \tag{1.3.14}$$

where a is an operator acting in the q space. Explicitly, this equation is

$$\left[\hbar\frac{d}{dq} + q - (2\hbar)^{1/2} z\right]\langle z|q\rangle = 0, \tag{1.3.15}$$

giving

$$\langle z|q\rangle = c(z)\exp\left[-\frac{1}{2\hbar}q^2 + (2/\hbar)^{1/2}zq\right]. \tag{1.3.16}$$

In particular, the oscillator function $\varphi_0(q) = (\pi\hbar)^{-1/4}\cdot\exp(-q^2/2\hbar)$ corresponds to $f_0(z) = 1$, hence

$$[c(z)]^{-1} = (\pi\hbar)^{-1/4}\int\exp\left[-\frac{q^2}{\hbar} + (2/\hbar)^{1/2}zq\right]dq = (\pi\hbar)^{1/4}\exp(z^2/2). \tag{1.3.17}$$

Thus the kernel is

$$K(z,q) = \langle z|q\rangle = (\pi\hbar)^{-1/4}\exp\left[-\frac{z^2}{2} + (2/\hbar)^{1/2}zq - \frac{q^2}{2\hbar}\right] \tag{1.3.18}$$

and the formulae producing the relation are

$$f(z) = \int K(z,q)\varphi(q)dq, \tag{1.3.19}$$

$$\varphi(q) = \lim_{r\to\infty}\int_{|z|<r}\overline{K(z,q)}f(z)\exp(-|z|^2)\,d\mu(z) \tag{1.3.20}$$

$$K(z,q) = \sum\frac{z^n}{\sqrt{n!}}\varphi_n(q) = (\pi\hbar)^{-1/4}\sum\frac{z^n}{n!}2^{-n/2}H_n((\hbar)^{-1/2}q)\exp\left(-\frac{q^2}{2\hbar}\right). \tag{1.3.21}$$

(They were obtained by *Bargmann* [4].) Comparing this expression with (1.3.18) gives the generating function for the Hermite polynomials:

$$\exp(-z^2 + 2zq) = \sum_{n=0}^{\infty}\frac{z^n}{n!}H_n(q). \tag{1.3.22}$$

In particular, an integral representation for the Hermite polynomials is obtained from (1.3.20)

$$H_n(q) = \frac{2^{n/2}}{\pi}\int\exp\left[-\frac{z^2}{2} + \sqrt{2}zq - |z|^2\right]\bar{z}^n d^2z,$$

$$d^2z = dx\,dy, \qquad z = x + \mathrm{i}y. \tag{1.3.23}$$

In conclusion, I present the formulae describing the action of the operators $D(\alpha)$ in coordinate representation

$$D(\alpha)=\exp\left[\frac{\mathrm{i}}{\hbar}(Pq-Qp)\right]$$

$$=\exp\left(-\frac{\mathrm{i}}{2\hbar}PQ\right)\exp\left(\mathrm{i}\frac{Pq}{\hbar}\right)\exp\left(-\mathrm{i}\frac{Qp}{\hbar}\right), \tag{1.3.24}$$

$$D(\alpha)\psi(q)=\exp\left(-\frac{\mathrm{i}}{2\hbar}PQ\right)\exp\left(\mathrm{i}\frac{Pq}{\hbar}\right)\psi(q-Q), \tag{1.3.25}$$

and in the Fock-Bargmann representation

$$D(\alpha)f(z)=\exp(-\tfrac{1}{2}|\alpha|^2)\exp(\alpha z)f(z-\bar{\alpha}). \tag{1.3.26}$$

The Fock-Bargmann representation will be exploited in solving the problems considered in the subsequent sections.

1.4 Completeness of Coherent-State Subsystems

In Sect. 1.2 it was mentioned that the coherent-state system $\{|\alpha\rangle\}$ is over-complete. Hence subsystems of CS must exist which are complete. Here I indicate criteria of completeness for a subsystem $\{|\alpha_k\rangle\}$ corresponding to a set of points $\{\alpha_k\}$ in the complex α plane.

First suppose that the subsystem $\{|\alpha_k\rangle\}$ is not complete. Then a vector $|\psi\rangle \neq 0$, belonging to the Hilbert space $\mathscr{H}$, exists orthogonal to any $|\alpha_k\rangle$: $\langle\psi|\alpha_k\rangle=0$. Hence it is clear that the function

$$\psi(\alpha)=\exp(|\alpha|^2/2)\langle\psi|\alpha\rangle \tag{1.4.1}$$

vanishes at every point of the set $\{\alpha_k\}$. Meanwhile, as shown above, the function $\psi(\alpha)$ is an entire analytic function of the complex variable α satisfying the condition

$$\mathrm{I}=\int|\psi(\alpha)|^2\exp(-|\alpha|^2)d\mu(\alpha)<\infty. \tag{1.4.2}$$

(Only the standard CS system is considered in this section.) In other words, $\psi(\alpha)$ belongs to the space $\mathscr{F}$. If, however, the system $\{|\alpha_k\rangle\}$ is complete, no such function exists.

Thus we have proven the following proposition.

Proposition 1. The subsystem of states $\{|\alpha_k\rangle\}$ is complete if and only if no function $\psi(\alpha)\in\mathscr{F}$, $\psi\neq 0$, vanishing at all points of the set $\{\alpha_k\}$, exists.

A number of examples of complete subsystems within the CS system $\{|\alpha\rangle\}$ are taken from [4].

i) Any set $\{\alpha_k\}$ with a limiting point in the finite part of the α plane remains a complete subsystem even after a finite number of states is removed.

ii) Any infinite set $\{\alpha_k\}$ which does not contain the origin $\alpha=0$ and satisfies the condition

$$\sum_k |\alpha_k|^{-2-\varepsilon}=\infty \tag{1.4.3}$$

at some $\varepsilon>0$ is a complete subsystem. This condition is a consequence of general theorems relating the *order of increase* of an entire function at $|\alpha|\to\infty$ to the distribution of its zeros (App. A).

iii) Of special interest is the case where the points α_k form a regular lattice L on the α plane, $\alpha_k\equiv\alpha_{mn}=m\omega_1+n\omega_2$, where the periods of the lattice ω_1 and ω_2 are linearly independent, $\mathrm{Im}\{\bar{\omega}_2\omega_1\}\neq 0$, and m, n are arbitrary integers. The simplest such CS subsystem corresponds to a square lattice with the given area of the lattice's elementary cell, $S=\pi$. *Von Neumann* [2] considered this subsystem long ago in view of the problem of the most accurate simultaneous measurement of both coordinate and momentum. Such a system was used in an analysis of the problem of a nonexponential decay law of an unstable particle [37]. For the results obtained to be valid, the system $\{|\alpha_{mn}\rangle\}$ should be complete. However, no proof that such a system is complete was published by *von Neumann*.

A complete solution of the completeness problem for the system $\{|\alpha_{mn}\rangle\}$ corresponding to a lattice was given in a theorem proven in [38], and also, in part, in [39]. (Over-completeness of the system at $S=\pi$ was not considered in [39]. The theorem in view was also proven in [40], while some mathematical aspects relevant to it were considered in [41–43].)

Theorem. Let $\{|\alpha_{mn}\rangle\}$ be a subsystem of coherent states, corresponding to a lattice whose elementary cell has an area S, the area element being $d\alpha_1 d\alpha_2$. Then

i) The system is overcomplete for $S<\pi$, and remains overcomplete if a finite number of states are removed;
ii) the subsystem $\{|\alpha_{mn}\rangle\}$ is not complete for $S>\pi$;
iii) the subsystem is complete for $S=\pi$. It remains complete if one state is removed, but becomes incomplete when any two states are removed.

This theorem is proved in Appendix A.

Note that the α plane is an analog of the phase plane for the classical system, and the cell of the α plane of the area π corresponds to the phase plane cell of the area $2\pi\hbar$. Hence a physical interpretation of the obtained result is clear: the CS system corresponding to a lattice on the phase plane with a density equal to one state per Planck cell, or with a higher density, is complete; if the density is less, the system is not complete. This fact provides more evidence in favor of the fundamental importance of partitioning the phase plane into Planck cells.

Thus at $S=\pi$, i.e., if one coherent state is chosen for any Planck cell, the system $\{|\alpha'_{mn}\rangle\}$ of all the states excluding the vacuum state $|0\rangle$ may be taken as a complete and minimal system of states. Expanding the vacuum state $|0\rangle$ over the basis $\{|\alpha'_{mn}\rangle\}$ gives a linear relation between all states of the system $\{|\alpha_{mn}\rangle\}$. The calculations lead to a formal relation [38]

$$\sum_{m,n} (-1)^{mn+m+n}|\alpha_{mn}\rangle \sim 0. \tag{1.4.4}$$

Let us now look at this relation in the Fock-Bargmann representation. The state $|\alpha\rangle$ is represented by $\psi_\alpha(z)=\exp(-\frac{1}{2}|\alpha|^2)\exp(\alpha z)$, and (1.4.4) is rewritten as

$$f(z)=\sum_{m,n} (-1)^{mn+m+n}\exp(-\tfrac{1}{2}|\alpha_{mn}|^2)\exp(\alpha_{mn}z)\equiv 0. \tag{1.4.5}$$

Note that the series in (1.4.5) converges uniformly for any compact subset of the α plane, so $f(z)$ is an entire function. The fact that $f(z)\equiv 0$ may be also confirmed directly [38].

In particular, setting $z=0$ in (1.4.5),

$$\sum_{m,n} (-1)^{mn+m+n}\exp\left[-\frac{\pi}{2}(am^2+2bmn+cn^2)\right]\equiv 0, \tag{1.4.6}$$

where $\pi a=|\omega_1|^2$, $\pi b=\operatorname{Re}\{\omega_1\bar{\omega}_2\}$, $\pi c=|\omega_2|^2$, $ac-b^2=1$, $a>0$, $c>0$. For a rectangular lattice, $b=0$, and the above identity is equivalent to a bilinear relation for the theta functions

$$\theta_3(\tau_1)\theta_3(\tau_2)-\theta_3(\tau_1)\theta_4(\tau_2)-\theta_4(\tau_1)\theta_3(\tau_2)-\theta_4(\tau_1)\theta_4(\tau_2)=0. \tag{1.4.7}$$

Here $\tau_1=ic/2$, $\tau_2=ic^{-1}/2$, and the theta functions $\theta_3(\tau)$ and $\theta_4(\tau)$ are defined by the series [47]

$$\theta_3(\tau)=\sum_{m=-\infty}^{\infty}\exp(i\pi\tau m^2),\qquad \theta_4(\tau)=\sum_{m=-\infty}^{\infty}(-1)^m\exp(i\pi\tau m^2). \tag{1.4.8}$$

Note in conclusion that since the subsystem $|\alpha_{mn}\rangle$ is complete, any subsystem of the form $|\psi_{mn}\rangle=U|\alpha_{mn}\rangle$, where U is an arbitrary unitary operator, is also complete. Suppose we choose the operator U so that

$$UaU^+=ua+va^+,\qquad |u|^2-|v|^2=1. \tag{1.4.9}$$

(The *Stone* [25]–*von Neumann* [26] theorem proves that such an operator exists.) Then

$$UD(\alpha)U^+=D(\beta),\qquad \beta=\bar{u}\alpha-v\bar{\alpha} \tag{1.4.10}$$

and the linear mapping $\alpha\to\beta$ is area preserving. Hence the subsystem $\{|\psi_{mn}\rangle\}$, $|\psi_{mn}\rangle=D(\beta_{mn})|\psi_0\rangle$, of the CS system of type $(T,|\psi_0\rangle)$, $|\psi_0\rangle=U|0\rangle$, where the operator U satisfies (1.4.9), is also complete.

1.5 Coherent States and Theta Functions

The preceding section showed that the theta functions arise during the analysis of completeness of the CS subsystem, corresponding to a regular lattice in the α plane. An investigation of the theta-function properties based on the group-theoretical arguments related to the Heisenberg-Weyl group was begun by *Cartier* [23]. The present section considers the relation between coherent states and theta functions in more detail.

Let L be a regular lattice in the α plane, that is the set of vectors of the form $\alpha_n = \sum_j n_j \omega_j$, $j=1,2$, where n_j are integers and the vectors ω_1, ω_2 (the lattice periods) are linearly independent over the field of real numbers. Consider also the set of operators $\{D(\alpha_n)\}$ and let us try to find their eigenvector $|\theta\rangle$ (such a vector does not exist in the space $\mathscr{H}$, but in an extended space, see below). A common eigenvector exists only if all the operators $D(\alpha_n)$ commute, which is the case when two operators $D(\omega_j)$ commute. According to (1.1.14), this condition is equivalent to

$$\frac{1}{\pi}\,\mathrm{Im}\,\{\omega_1\bar{\omega}_2\} = k, \tag{1.5.1}$$

where k is an integer. In other words, the area of the parallelogram based on the vectors ω_1 and ω_2 must be a multiple of π.

The lattice L satisfying condition (1.5.1) is called admissible. The desired eigenvector is characterized by two real numbers ε_1 and ε_2, $|\theta\rangle = |\theta_\varepsilon\rangle$, and it must satisfy

$$D(\omega_j)|\theta_\varepsilon\rangle = \exp(\mathrm{i}\pi\varepsilon_j)|\theta_\varepsilon\rangle, \quad j=1,2. \tag{1.5.2}$$

Thus the quantities ε_1 and ε_2 lie in the interval $0 \le \varepsilon_j < 2$, so that the vector $|\theta_\varepsilon\rangle$ corresponds to a point on the two-dimensional torus.

It is not difficult to see, however, that the vector $|\theta_\varepsilon\rangle$ cannot belong to the Hilbert space $\mathscr{H}$, where the operators $D(\alpha)$ act. Let us describe in brief the extended space $\mathscr{H}_{-\infty}$ containing the vector $|\theta_\varepsilon\rangle$. A more detailed discussion of this construction was given in [23].

Denote by $\mathscr{H}_\infty$ a subspace of $\mathscr{H}$ containing only such vectors $|\psi\rangle$ that the function

$$\langle\varphi|T(g)|\psi\rangle = \exp(\mathrm{i}t)\langle\varphi|D(\alpha)|\psi\rangle$$

is infinitely differentiable for any fixed $\varphi \in \mathscr{H}$. Elements of the space will be called C^∞ vectors. Note that some C^∞ vectors have the form

$$|\psi_f\rangle = \int d\mu(\alpha) f(\alpha) D(\alpha)|\psi\rangle,$$

where $|\psi\rangle$ is an arbitrary vector in $\mathscr{H}$, and $f(\alpha)$ is an infinitely differentiable function with a compact support. As shown by *Gårding* [44], such vectors form a dense subset in $\mathscr{H}$. The space $\mathscr{H}_{-\infty}$ is defined as a set of all continuous antilinear forms on $\mathscr{H}_{\infty}$, while $\mathscr{H}$ is identified with a subspace of $\mathscr{H}_{-\infty}$ mapping any vector $|\psi\rangle\in\mathscr{H}$ onto the antilinear form $\langle\psi|\varphi\rangle$ on $\mathscr{H}_{\infty}$. Then the representation $T(g)$, defined originally in $\mathscr{H}$, may be extended to a representation in the space $\mathscr{H}_{-\infty}$.

Among the states $|\theta_\varepsilon\rangle$ we select now a state, say $|\theta_0\rangle$, and let it be subject to the action of all the operators $D(\alpha)$. The result is a system of generalized coherent states

$$|\theta_\alpha\rangle = D(\alpha)|\theta_0\rangle. \tag{1.5.3}$$

It is not difficult to see that the system of states obtained is just $\{|\theta_\varepsilon\rangle\}$, defined by (1.5.2). This is quite natural because the isotropy subgroup H of the state $|\theta_\varepsilon\rangle$ contains elements (t, α_n), so the coset space G/H is a two-dimensional torus.

It is also remarkable that the possibility to parametrize the states by means of a complex number α, see (1.5.3), arises from the well-known fact that any two-dimensional torus is a complex manifold [45].

Thus the system $\{|\theta_\varepsilon\rangle\}$ is a CS system related to an admissible lattice L. To establish its relevance to the theta functions consider the state $|\theta_\varepsilon\rangle$ in the Fock-Bargmann representation, $|\theta_\varepsilon\rangle\to\theta_\varepsilon(z)$ [for simplicity we restrict ourselves to the so-called principal lattice corresponding to $k=1$ in (1.5.1)]. In this case it follows from (1.5.2) that

$$D(\alpha_m)|\theta_\varepsilon\rangle = \mathrm{e}^{\mathrm{i}\pi F_\varepsilon(m)}|\theta_\varepsilon\rangle, \tag{1.5.4}$$

where $\alpha_m = m_1\omega_1 + m_2\omega_2$ is an arbitrary lattice vector, and

$$F_\varepsilon(m) = m_1 m_2 + \varepsilon_1 m_1 + \varepsilon_2 m_2. \tag{1.5.5}$$

Using (1.3.26), the equality (1.5.4) may be rewritten as

$$\theta_\varepsilon(z+\beta_m) = \mathrm{e}^{\mathrm{i}\pi F_\varepsilon(-m)}\mathrm{e}^{|\beta_m|^2/2}\mathrm{e}^{\bar{\beta}_m z}\theta_\varepsilon(z), \tag{1.5.6}$$

where $\beta_m = \bar{\alpha}_m$ is the conjugated lattice vector. Equation (1.5.6) is just the familiar functional equation for the theta functions [46, 47].

Besides $\theta_\varepsilon(z)$, the function

$$\tilde{\theta}_\varepsilon(z,\bar{z}) = \mathrm{e}^{-|z|^2/2}\theta_\varepsilon(z) \tag{1.5.7}$$

is also useful. The following functional equation arises from (1.5.6) for this function

$$\tilde{\theta}_\varepsilon(z+\beta_m, \bar{z}+\bar{\beta}_m) = \mathrm{e}^{\mathrm{i}\pi F_\varepsilon(-m)}\mathrm{e}^{\mathrm{i}\,\mathrm{Im}(z\bar{\beta}_m)}\tilde{\theta}_\varepsilon(z,\bar{z}). \tag{1.5.8}$$

Thus $|\tilde{\theta}_\varepsilon(z,\bar{z})|^2 = \varrho(z,\bar{z})$, where $\varrho(z,\bar{z})$ is a nonnegative function periodical with respect to $\bar{L}$, the lattice conjugate to L. So the usual norm of the function $\tilde{\theta}_\varepsilon(z)$ is infinite.

As for the solution of (1.5.6), it is not difficult to verify that the function

$$h_\varepsilon(z)=\sum_n \mathrm{e}^{-\mathrm{i}\pi F_\varepsilon(n)} D(-\bar{\beta}_n)h(z) \tag{1.5.9}$$

is a solution for an arbitrary entire function $h(z)$ such that the series (1.5.9) is converging. A more explicit form is

$$h_\varepsilon(z)=\sum_n \mathrm{e}^{-\mathrm{i}\pi F_\varepsilon(n)} \exp(-\tfrac{1}{2}|\beta_n|^2)\exp(-\bar{\beta}_n z)h(z+\beta_n). \tag{1.5.10}$$

In particular, for $h(z)\equiv 1$ one gets the following solution

$$f_\varepsilon(z)=c_\varepsilon\theta_\varepsilon(z)=\sum_n \mathrm{e}^{-\mathrm{i}\pi F_\varepsilon(n)}\mathrm{e}^{-|\alpha_n|^2/2}\mathrm{e}^{\bar{\alpha}_n z} \tag{1.5.11}$$

written as a superposition of coherent states. Note, however, that for certain values of ε, e.g., for $\varepsilon=(1,1)$, the function $f_\varepsilon(z)$ is identically zero.

To conclude this section, consider the relation between the present approach and that proposed by *Cartier* [23]. He showed that the theta functions arise in representations of the Heisenberg-Weyl group $\mathscr{W}_1$, which are induced by representations of its discontinuous subgroup Γ related to a lattice L. (A number of general problems related to discontinuous subgroups of continuous groups is considered in [48].)

Let us consider the set of the operators $D_{mn}=D(m\omega_1+n\omega_2)$, where $\alpha_{mn}=m\omega_1+n\omega_2$ is a point of an admissible lattice L with the elementary cells of area $S=\pi$. These operators form a discontinuous commutative group with the multiplication law

$$D_{k,l}D_{m,n}=(-1)^{B(k,l,m,n)}D_{k+m,l+n}; \qquad B(k,l;m,n)=kn-lm. \tag{1.5.12}$$

Let Γ be the discontinuous subgroup of $\mathscr{W}_1$ consisting of the elements $g=(k\pi,\alpha_{mn})$. The operators $\{\pm D_{mn}\}$ form a representation of this subgroup, while the set of the states $\{|\alpha_{mn}\rangle\}$ is a basis for a representation of the group Γ. Now we try to do without the sign factor in (1.5.12); in other words, our purpose is to get a representation of the factor group Γ/Γ_0, where $\Gamma_0=\{(k\pi,0)\}$. Introduce some new operators,

$$D_{k,l}=(-1)^{F(k,l)}\tilde{D}_{k,l} \tag{1.5.13}$$

and impose the requirement

$$\tilde{D}_{k,l}\tilde{D}_{m,n}=\tilde{D}_{k+m,l+n}. \tag{1.5.14}$$

Hence we get a functional equation for $F(k,l)$:

$$F(k+m,l+n)=F(k,l)+F(m,n)+B(k,l;m,n)\,(\mathrm{mod}\,2), \tag{1.5.15}$$

where B is defined in (1.5.12). At $m=1, n=1$ this equation coincides with Eq. (71)

of [23]. The solution of (1.5.15) is

$$F(k,l)=kl+k+l. \tag{1.5.16}$$

Thus a new set of states arises

$$|\tilde{\alpha}_{kl}\rangle=\tilde{D}_{kl}|0\rangle=(-1)^{kl+k+l}|\alpha_{kl}\rangle, \tag{1.5.17}$$

where the operator $\tilde{D}_{kl}$ acts as follows:

$$\tilde{D}_{kl}|\tilde{\alpha}_{mn}\rangle=|\tilde{\alpha}_{k+m,l+n}\rangle. \tag{1.5.18}$$

Hence a relation between the states in the system $|\tilde{\alpha}_{kl}\rangle$ may be obtained at once. As explained above, this relation looks like $\sum C_{kl}|\tilde{\alpha}_{kl}\rangle\sim 0$ and is unique. Consequently, it is invariant under the action of any operator $\tilde{D}_{kl}$. It is not difficult to see that the only relation having this property is

$$\sum_{m,n}|\tilde{\alpha}_{m,n}\rangle=\sum_{m,n}(-1)^{mn+m+n}|\alpha_{m,n}\rangle\sim 0 \tag{1.5.19}$$

and coincides with (1.4.4).

It is also remarkable that the uniqueness of the solution of the functional equation (1.5.6) arises from the fact that the representation of the group W_1 induced by that of the discontinuous subgroup Γ_2 is irreducible [23, 48]. In particular, at $\varepsilon_1=\varepsilon_2=1$ the state $|\theta_{11}\rangle$ corresponds to the function

$$f_{11}(z)=c\sigma(z)\exp(-vz^2) \tag{1.5.20}$$

where $v=\mathrm{i}/4\pi(\eta_1\omega_2-\eta_2\omega_1)$, $\eta_j=\zeta(\bar{\omega}_j/2)$, $\sigma(z)$ and $\zeta(z)$ are the well-known Weierstrass functions [38].

In conclusion, I mention an interesting work [49], where the CS related to the lattice were used to describe the motion of an electron in a periodical magnetic field.

1.6 Operators and Their Symbols

As shown in Sect. 1.2, any vector $|\psi\rangle$ of the Hilbert space $\mathscr{H}$ is determined completely by a function $\langle\alpha|\psi\rangle$ which may be called a symbol of the vector (however, not every function of α determines a vector of Hilbert space). Thus one gets a functional realization of the Hilbert space. Similarly, using the CS system, one may represent operators acting in the Hilbert space and belonging to a certain class by functions. These determine the operators completely and are called their symbols.

Such a correspondence between operators and their symbols is rather useful. For instance, questions relevant to the operators can be formulated in terms of the functions. The inverse correspondence is also valuable. Actually, operator symbols may sometimes be considered as functions on the phase space of a classical dynamical system. Constructing operators for these functions is in fact the quantization of the system. This is the reasoning underlying the works by *Kirillov* [19] and *Kostant* [20, 21], in which quantization is treated as a general procedure to construct unitary irreducible representations of the Lie groups. The coherent states provide the most natural means for this correspondence: in a number of cases they are quantum states whose properties are as close as possible to those of the classical states. Therefore the construction of a CS system is in a sense the completion of the quantization procedure, so it may be called the postquantization.

Let us consider for simplicity the standard CS system. Suppose $\hat{A}$ is an operator. We represent it by the functions $A(\bar{\alpha}, \beta)$ and $\tilde{A}(\bar{\alpha}, \beta)$ defined by

$$A(\bar{\alpha}, \beta) = \langle \alpha | \hat{A} | \beta \rangle, \tag{1.6.1}$$

$$\tilde{A}(\bar{\alpha}, \beta) = \exp\left(\frac{|\alpha|^2 + |\beta|^2}{2}\right) A(\bar{\alpha}, \beta). \tag{1.6.2}$$

It is not difficult to see that the function $\tilde{A}(\bar{\alpha}, \beta)$ determines the operator $\hat{A}$ completely and is an analytical function of the complex variables $\bar{\alpha}$ and β. In fact, as shown in [36], this function is determined completely by its "diagonal" values, i.e., by the function $\tilde{A}(\bar{\alpha}, \alpha)$. To verify this fact, it is appropriate to introduce new variables $u = (\bar{\alpha} + \beta)/2$, $v = \mathrm{i}(\bar{\alpha} - \beta)/2$, so that $\beta = u + \mathrm{i}v$, $\bar{\alpha} = u - \mathrm{i}v$. Then $F(u, v) = \tilde{A}(\bar{\alpha}, \beta)$ is an entire function of u and v. At the diagonal, $\alpha = \beta$, u and v are real variables, so the above statement stems from the well-known fact that any entire function of variables u, v is determined completely by its values at real u and v.

Thus the function $A(\bar{\alpha}, \beta)$ is also determined completely by its diagonal values, i.e., by the function

$$Q_A(\alpha) = \langle \alpha | \hat{A} | \alpha \rangle = A(\bar{\alpha}, \alpha) = \exp(-|\alpha|^2) \tilde{A}(\bar{\alpha}, \alpha). \tag{1.6.3}$$

On the other hand, in a number of cases the operator $\hat{A}$ may be represented by

$$\hat{A} = \int P_A(\alpha) |\alpha\rangle \langle \alpha| d\mu(\alpha), \tag{1.6.4}$$

where $|\alpha\rangle \langle \alpha|$ is projection operator on state $|\alpha\rangle$. This integral representation was considered originally by *Glauber* [7, 8] and *Sudarshan* [50].

A reasonable question arises, namely which properties must have a function of the variables α, $\bar{\alpha}$ in order to be the P (or Q) symbol of an operator; and for which operators such symbols exist. It was found that if an operator $\hat{A}$ is bounded, then it always has a symbol $Q_A(\alpha)$, which is a value of an entire function $A(\bar{\alpha}, \beta)$ at $\beta = \alpha$, provided that the inequality

$$|A(\bar{\alpha}, \beta)| \leq \|A\| \exp\left(\frac{|\alpha|^2 + |\beta|^2}{2}\right) \tag{1.6.5}$$

is satisfied. Here $\|\hat{A}\|$ is, as usual, the upper bound of the expectation values $\langle\psi|\hat{A}|\psi\rangle$ over all the normalized states ψ, $\langle\psi|\psi\rangle=1$. The Q symbols also exist for some unbounded operators, say, for operators which are polynomials of the annihilation-creation operators a and a^+.

The following result is known for the P symbols. If $P(\alpha, \bar{\alpha})$ is a function, such that

$$\int |P(\alpha, \bar{\alpha})|^r \exp(-|\alpha|^2) d\mu(\alpha) < \infty \quad \text{at } r>2, \tag{1.6.6}$$

then it is the P symbol of an operator with a dense domain of definition.

The Q and P symbols of operators were investigated in some detail by *Berezin* [29, 51], who called them covariant (or Wick) and contravariant (or anti-Wick) symbols, respectively. A number of properties of these symbols are, in a sense, dual. Here I present some of them without proof.

1. For an operator $\hat{A}$ let its average values $\langle\psi|\hat{A}|\psi\rangle$ over states $|\psi\rangle$, belonging to a dense set in the Hilbert space, form a point set $D(\hat{A})$ in the complex plane, the values of $Q_A(\alpha, \bar{\alpha})$ form a set $D(Q)$, and the convex hull of the set of values of $P_A(\alpha, \bar{\alpha})$ form a set $D(P)$. It is established that

$$D(Q) \subset D(\hat{A}) \subset D(P). \tag{1.6.7}$$

In particular, for an operator $\hat{A}$ to be bounded, its Q symbol must be bounded, and it is sufficient that its P symbol be bounded. Besides, an estimate holds:

$$\sup |Q_A(\bar{\alpha}, \alpha)| \leq \|\hat{A}\| \leq \sup |P_A(\bar{\alpha}, \alpha)|. \tag{1.6.8}$$

2. If an operator $\hat{A}$ is defined over finite vectors (a vector $|\psi\rangle$ is called finite if its symbol $\exp(\frac{1}{2}|\alpha|^2)\langle\alpha|\psi\rangle$ is a polynomial in $\bar{\alpha}$), its closure is self-adjoint and it has the symbol $P(\bar{\alpha}, \alpha)$. If $\int |P|^k \exp(-|\alpha|^2) d\mu(\alpha) < \infty$ with $k>1$, then it has also the Q symbol $Q(\alpha)$, and

$$D(Q) \subset \sigma(\hat{A}) \subset D(P), \tag{1.6.9}$$

where $\sigma(\hat{A})$ is the convex hull of the spectrum of $\hat{A}$.

3. For an operator A be nuclear, its symbol $Q(\alpha, \bar{\alpha})$ must be integrable, and it is sufficient that its symbol $P(\alpha, \bar{\alpha})$ be integrable. In this situation the following relation is valid

$$\operatorname{tr}\{\hat{A}\} = \int Q(\alpha, \bar{\alpha}) d\mu(\alpha) = \int P(\alpha, \bar{\alpha}) d\mu(\alpha). \tag{1.6.10}$$

(Operator $\hat{A}$ is called nuclear if $\sum_n |\langle\psi_n|\hat{A}|\psi_n\rangle| < \infty$ for any orthonormal basis $\{|\psi_n\rangle\}$ in the Hilbert space. For the nuclear operator the sum $\sum_n \langle\psi_n|A|\psi_n\rangle$ does not depend on choice of the basis and is called the trace of the operator $\hat{A}$.)

4. For an operator $\hat{A}$ be completely continuous, it is necessary that its symbol $Q(\alpha, \bar{\alpha})$ vanishes at $|\alpha| \to \infty$, and it is sufficient that its symbol $P(\alpha, \bar{\alpha})$ vanishes at $|\alpha| \to \infty$.

5. If $\hat{H}$ is a self-adjoint operator having both Q and P symbols, then

$$\int e^{-\tau Q(\alpha, \bar{\alpha})} d\mu(\alpha) \leq \operatorname{tr} e^{-\tau H} \leq \int e^{-\tau P(\alpha, \bar{\alpha})} d\mu(\alpha) \tag{1.6.11}$$

(note that in statistical mechanics $\tau = T^{-1}$, T is temperature) and the fact that the operator $\exp(-\tau H)$ is nuclear is a consequence of the existence of the integral $\int \exp[-\tau P(\alpha, \bar{\alpha})] d\mu(\alpha)$. The left inequality in (1.6.11) follows from (1.6.10) and from an equality for the Q symbol $U(\alpha, \bar{\alpha}|\tau)$ of the operator $\exp(-\tau H)$,

$$U(\alpha, \bar{\alpha}|\tau) \geq \exp[-\tau H(\alpha, \bar{\alpha})]. \tag{1.6.12}$$

The right inequality in (1.6.11) is an analog of the familiar Feynman inequality for an operator of the type $\hat{H} = \frac{1}{2}\hat{p}^2 + V(q)$, i.e., of the inequality

$$\operatorname{tr}\{e^{-\tau \hat{H}}\} \leq \frac{1}{2\pi\hbar} \int \exp\left\{-\tau \left[\frac{p^2}{2} + V(q)\right]\right\} dp\, dq. \tag{1.6.13}$$

The proof of inequality (1.6.13) was based on representing the operator $\exp(-\tau H)$ as an integral over Wiener's measure, and it cannot be extended to operators which are not written as $\frac{1}{2}p^2 + V(q)$.

Under quite general assumptions, inequality (1.6.11) enables one to obtain the $E \to \infty$ asymptotic behavior of $N(E)$, the number of eigenvalues of the operator $\hat{H}$ which are less than E:

$$\underset{E \to \infty}{N(E)} = [1 + O(1)] \int_{P(\alpha) < E} d\mu(\alpha) = [1 + O(1)] \int_{Q(\alpha) < E} d\mu(\alpha). \tag{1.6.14}$$

The proof of (1.6.14), given in [29], assumed that both P and Q symbols of the operator H are regular and satisfy the condition

$$\int_{P(\alpha) < E} d\mu(\alpha) = AE^{\gamma}[1 + O(1)], \qquad \gamma > 0. \tag{1.6.15}$$

It is notable also that the exponential in inequality (1.6.11) may be substituted for any concave function, $\exp(x) \to \varphi(x)$.

The CS system may be used to strengthen the so-called *Golden–Thompson* inequality [52, 53]

$$\operatorname{tr}\{\exp[-\tau(\hat{A} + \hat{B})]\} \leq \operatorname{tr}\{\exp(-\tau\hat{A})\exp(-\tau\hat{B})\}. \tag{1.6.16}$$

The following inequality may be also proven [54]

$$\operatorname{tr}\{\exp[-\tau(\hat{A} + \hat{B})]\} \leq \int d\mu(x) \langle x|e^{-\tau\hat{A}}|x\rangle e^{-\tau P_B(x)}, \quad \text{where} \tag{1.6.17}$$

$$\hat{B} = \int P_B(x)|x\rangle\langle x| d\mu(x). \tag{1.6.18}$$

The Q and P symbols are related by a simple formula

$$Q_A(\alpha)=\int |\langle\alpha|\beta\rangle|^2 P_A(\beta)d\mu(\beta)$$

$$=\pi^{-1}\int \exp(-|\alpha-\beta|^2)P_A(\beta)d^2\beta. \qquad (1.6.19)$$

The kernel in this relation is smoothing. Therefore $Q(\alpha)$ exists for any given $P(\alpha)$. The inverse proposition would be wrong, in general; operators exist having Q symbols but no P symbols.

The Q and P symbols of an operator are closely related to its so-called normal (Wick) and antinormal (anti-Wick) form. Recall that the Wick form of an operator is its representation by means of the series

$$\hat{A}=\sum_{m,n} A_{mn}(a^+)^m a^n, \qquad (1.6.20)$$

where the powers of the operator a^+ are posed to the left of the powers of a. Note that any bounded operator may be written in such a form. It is easily seen that

$$Q_A(\bar{\alpha},\alpha)=\langle\alpha|\hat{A}|\alpha\rangle=\sum_{m,n} A_{mn}\bar{\alpha}^m\alpha^n. \qquad (1.6.21)$$

Hence the expansion of the function $Q_A(\bar{\alpha},\alpha)$ in powers of α and $\bar{\alpha}$ produce the coefficients A_{mn} in the normal form of the operator $\hat{A}$.

Suppose now that the operator $\hat{A}$ is written in the antinormal (anti-Wick) form

$$\hat{A}=\sum_{m,n} A'_{mn}a^m(a^+)^n. \qquad (1.6.22)$$

Rewrite it equivalently

$$\hat{A}=\sum_{m,n} A'_{mn}a^m\hat{1}(a^+)^n,$$

where $\hat{1}$ is the unit operator. Substituting $\hat{1}$ by (1.2.11) yields for the P symbol

$$P_A(\bar{\alpha},\alpha)=\sum_{m,n} A'_{mn}\alpha^m\bar{\alpha}^n. \qquad (1.6.23)$$

Thus the P representation is intimately related to the antinormal ordering of the operator.

Consider an expression for the symbol of a product of operators. Suppose $\hat{C}=\hat{A}\hat{B}$, then

$$C(\bar{\alpha},\gamma)=\int A(\bar{\alpha},\beta)B(\bar{\beta},\gamma)d\mu(\beta), \qquad \text{or} \qquad (1.6.24)$$

$$\tilde{C}(\bar{\alpha},\gamma)=\int \tilde{A}(\bar{\alpha},\beta)\tilde{B}(\bar{\beta},\gamma)\exp(-|\beta|^2)d\mu(\beta). \qquad (1.6.25)$$

Note also that the action of an operator upon a state

$$|\psi\rangle = A|\varphi\rangle \tag{1.6.26}$$

in terms of the symbol of the state vector

$$\psi(\bar{\alpha}) = \langle\alpha|\psi\rangle, \qquad \varphi(\bar{\alpha}) = \langle\alpha|\varphi\rangle \tag{1.6.27}$$

is written as

$$\psi(\bar{\beta}) = \int \langle\beta|\alpha\rangle\, P_A(\alpha)\,\varphi(\bar{\alpha})\,d\mu(\alpha), \tag{1.6.28}$$

$$\psi(\bar{\beta}) = \int A(\bar{\beta}, \alpha)\,\varphi(\bar{\alpha})\,d\mu(\alpha). \tag{1.6.29}$$

The symbols are rather useful; let us now look at some examples where their use is quite suitable.

a) Schur's Lemma [36]

Lemma. Any bounded linear operator $\hat{B}$, commuting with every operator $D(\alpha)$, is the unit operator times a number.

Proof. The symbol corresponding to the operator is

$$\tilde{B}(\bar{z}, z) = \exp(|z|^2)\,\langle 0|D^+(z)\,\hat{B}D(z)|0\rangle.$$

From the fact of the commutation one gets at once $\tilde{B}(\bar{z}, z) = \exp(\bar{z}z)\,B_{00}$. Hence $\tilde{B}(\bar{z}, w) = \exp(\bar{z}w)\,B_{00}$, and the corresponding operator is $B_{00}\hat{1}$.

b) Evaluation of tr$\{D(\gamma)\}$

Another example is the calculation of the trace of $D(\gamma)$, in other words, the character of the representation $T(g)$, using the symbols. It is not difficult to see that the symbols corresponding to the operator $D(\gamma)$ are

$$P(\alpha) = \exp(\tfrac{1}{2}|\gamma|^2)\exp(\gamma\bar{\alpha} - \bar{\gamma}\alpha), \quad Q(\alpha) = \exp(-\tfrac{1}{2}|\gamma|^2)\exp(\gamma\bar{\alpha} - \bar{\gamma}\alpha) \tag{1.6.30}$$

Hence

$$\mathrm{tr}\{D(\gamma)\} = \exp(-\tfrac{1}{2}|\gamma|^2)\int \exp[2\mathrm{i}\,\mathrm{Im}(\gamma\bar{\alpha})]\,d\mu(\alpha) = \pi\delta^2(\gamma), \tag{1.6.31}$$

where

$$\delta^2(\gamma) = \delta(\gamma_1)\,\delta(\gamma_2), \qquad \gamma = \gamma_1 + \mathrm{i}\gamma_2.$$

From (1.6.31) one gets the important consequences

$$\pi^{-1}\,\mathrm{tr}\{D(\alpha)D^{-1}(\beta)\} = \delta^2(\alpha - \beta) \tag{1.6.32}$$

$$\hat{A} = \int d\mu(\eta)\,\mathrm{tr}\{D(\eta)\hat{A}\}\,D^{-1}(\eta). \tag{1.6.33}$$

By means of the symbols, it is not difficult to obtain also an explicit expression for matrix elements of the operator $D(\gamma)$. Let us consider a function related to the matrix element in view,

$$G(\bar{\alpha}, \beta; \gamma) = \exp\left[\tfrac{1}{2}(|\alpha|^2 + |\beta|^2)\right] \langle \alpha | D(\gamma) | \beta \rangle. \tag{1.6.34}$$

Clearly, using (1.6.30), one gets

$$G(\bar{\alpha}, \beta; \gamma) = \exp(-\tfrac{1}{2}|\gamma|^2) \exp(\bar{\alpha}\beta + \bar{\alpha}\gamma - \beta\bar{\gamma}). \tag{1.6.34'}$$

On the other hand,

$$G(\bar{\alpha}, \beta; \gamma) = \sum_{m,n} \bar{u}_m(\alpha) D_{mn}(\gamma) u_n(\beta); \quad u_m(\alpha) = \frac{\alpha^m}{\sqrt{m!}} \tag{1.6.35}$$

so that $G(\bar{\alpha}, \beta; \gamma)$ is just the generating function for the matrix elements. Now (1.6.34) must be expanded in powers of $\bar{\alpha}$ and β. The result is *Schwinger's* formula [55]:

$$D_{mn}(\gamma) = \begin{cases} \sqrt{\dfrac{n!}{m!}} \exp(-|\gamma|^2/2) \gamma^{m-n} \mathrm{L}_n^{m-n}(|\gamma|^2), & m \geq n \\[2ex] \sqrt{\dfrac{m!}{n!}} \exp(-|\gamma|^2/2) (-\bar{\gamma})^{n-m} \mathrm{L}_m^{n-m}(|\gamma|^2), & m \leq n, \end{cases} \tag{1.6.36}$$

where $\mathrm{L}_n^k(x)$ is the Laguerre polynomial. In particular, at $m = n$

$$\langle n | \mathrm{e}^{\alpha a^+} \mathrm{e}^{-\bar{\alpha} a} | n \rangle = \mathrm{L}_n(|\alpha|^2). \tag{1.6.36'}$$

A derivation of these formulae, as well as a number of the resulting properties of the Laguerre polynomials, is given in App. B. Note that even before Schwinger, the matrix elements (1.6.36) were obtained by *Feynman* [56], though in a somewhat different form.

Beside the symbols $Q_A(\alpha)$ and $P_A(\alpha)$, which are related to the Wick and anti-Wick ordering of the operator $\hat{A}$, another function is sometimes suitable, determining the operator completely. Consider the symmetrical ordering of the operator,

$$\hat{A} = \sum_{m,n} A_{mn} \{(a^+)^m a^n\}, \tag{1.6.37}$$

where the symmetrical monomial $\{(a^+)^m a^n\}$ is defined by

$$\{(a^+)^m a^n\} = \frac{1}{(m+n)!} P[(a^+)^m a^n], \tag{1.6.38}$$

and the symmetrization operator P is just a sum of $(m+n)!$ permutations of the factors. The function in view, corresponding to the operator, is written by means of the series (1.6.37)

$$W_A(\alpha)=\sum_{m,n} A_{mn}\bar{\alpha}^m\alpha^n. \tag{1.6.39}$$

To find a relation between the operator $\hat{A}$ and its symbol W_A, let us write the Fourier integral representation for W_A:

$$W_A(\alpha)=\int \exp(\bar{\alpha}\eta-\alpha\bar{\eta})\chi_A(\eta)d\mu(\eta). \tag{1.6.40}$$

It is not difficult to verify that the operator $\hat{A}$ is given by

$$\hat{A}=\int D(\eta)\chi_A(\eta)d\mu(\eta)=\int \exp(a^+\eta-a\bar{\eta})\chi_A(\eta)d\mu(\eta). \tag{1.6.41}$$

The function $\chi_A(\eta)$ here determines the operator completely. It is called the characteristic function and is considered in more detail in Sect. 1.7. Thus the operator $\hat{A}$ is constructed from the function W_A, substituting the variables α, $\bar{\alpha}$ in the integrand in (1.6.40) by the operators a, a^+. This operator-function correspondence was established by *Weyl* [14, 57], so the function $W_A(\alpha)$ is called the Weyl symbol of the operator.

The operator $\hat{A}$ is easily written in terms of $W_A(\alpha)$ [59–61]. To this end one must just extract the function $\chi_A(\eta)$ from (1.6.40) and insert it into (1.6.41):

$$\hat{A}=\int W_A(\alpha)\hat{T}(\alpha)d\mu(\alpha), \quad \text{where} \tag{1.6.42}$$

$$\hat{T}(\alpha)=\int d\mu(\eta)\exp(\alpha\bar{\eta}-\bar{\alpha}\eta)D(\eta) \tag{1.6.43}$$

$$\hat{T}(\alpha)=2\,D(\alpha)\hat{J}D^{-1}(\alpha)\equiv 2\,D(2\alpha)\hat{J}=2\hat{J}D(-2\alpha). \tag{1.6.44}$$

Here $\hat{J}$ is the operator of inversion in the α plane,

$$\hat{J}D(\alpha)\hat{J}=D(-\alpha), \qquad \hat{J}^2=\hat{1}, \qquad \hat{J}_{mn}=(-1)^m\delta_{mn}. \tag{1.6.45}$$

The operators $\hat{T}(\alpha)$, as the operators $D(\alpha)$, form a complete orthogonal system

$$\frac{1}{\pi}\operatorname{tr}\{[T(\alpha),T(\beta)]\}=\delta^2(\alpha-\beta), \tag{1.6.46}$$

$$\int d\mu(\alpha)T_{ki}(\alpha)T_{mn}(\alpha)=\delta_{im}\delta_{kn}. \tag{1.6.47}$$

Therefore any operator may be decomposed over the system $T(\alpha)$:

$$\hat{A}=\int d\mu(\alpha)(\operatorname{tr}\{T(\alpha)\hat{A}\})T(\alpha). \tag{1.6.48}$$

It is not difficult to find relations between the functions Q, P, and W. The formulae result from (1.6.40, 41):

$$W_A(\alpha)=\int P_A(\beta)\exp(-2|\alpha-\beta|^2)d\mu(\beta) \tag{1.6.49}$$

$$\begin{aligned} Q_A(\alpha)&=\int W_A(\beta)\exp(-2|\alpha-\beta|^2)d\mu(\beta)\\ &=\int P_A(\beta)\exp(-|\alpha-\beta|^2)d\mu(\beta). \end{aligned} \tag{1.6.50}$$

Thus the functions $W_A(\alpha)$ and $Q_A(\alpha)$ are smoothed versions of $P_A(\alpha)$. In the classical limit all three functions variate substantially only at large distances, $\Delta\alpha \gg 1$, so in this limit all the functions coincide.

1.7 Characteristic Functions

This section treats the basic properties of the characteristic functions [58–62].

Let $W(q,p)$ be a probability distribution on the phase space of a classical dynamical system. In classical statistics, the characteristic function $\chi(\xi,\pi)$ is defined as the expectation of the quantity $\exp[\mathrm{i}(-\pi q+\xi p)]$, i.e., as the Fourier transform of the distribution $W(q,p)$. Introduce complex variables

$$\alpha=\frac{1}{\sqrt{2}}(q+\mathrm{i}p), \qquad \eta=\frac{1}{\sqrt{2}}(\xi+\mathrm{i}\pi). \tag{1.7.1}$$

Then

$$\chi(\eta)=\int W(\alpha)\exp(-\eta\bar{\alpha}+\bar{\eta}\alpha)d\mu(\alpha), \qquad d\mu(\alpha)=\frac{1}{\pi}\,d\alpha_1\,d\alpha_2 \tag{1.7.2}$$

$$W(\alpha)=\int \chi(\eta)\exp(\eta\bar{\alpha}-\bar{\eta}\alpha)d\mu(\eta). \tag{1.7.3}$$

Turning to the quantum case, let us first look at the most important case where $W(q,p)$ is the symbol of the density matrix: $W(\bar{\alpha},\alpha)=\langle\alpha|\varrho|\alpha\rangle=Q(\alpha)$, i.e., $W(\bar{\alpha},\alpha)$ is a normalized real and nonnegative function. Hence the characteristic function $\chi(\eta)$ must satisfy the following conditions

a) $\chi(0)=1,$ (1.7.4a)

b) $\bar{\chi}(\eta)=\chi(-\eta),$ (1.7.4b)

c) $\sum \bar{z}_i z_j \chi(\eta_i-\eta_j)\geq 0,$ (1.7.4c)

and the last inequality must be valid for any sets of the complex numbers $z_1,\ldots,z_n$ and $\eta_1,\ldots,\eta_n$. This is the familiar positive definiteness or the Bochner-Khinchin condition [63].

The characteristic function $\chi(\eta)$ is suitable for calculating moments of random quantities

$$M_{mn}=\langle\bar{\alpha}^m\alpha^n\rangle=\int d\mu(\alpha)\,W(\alpha)\bar{\alpha}^m\alpha^n=\left(-\frac{\partial}{\partial\eta}\right)^m\left(\frac{\partial}{\partial\bar{\eta}}\right)^n\chi(\eta)|_{\eta=0}. \tag{1.7.5}$$

Suppose α is a sum of several independent random quantities, $\alpha=\alpha_1+\ldots+\alpha_n$. Because of the multiplicative property of the exponential, $\exp(x+y)=\exp(x)\exp(y)$, the characteristic function here is a product of the partial characteristic functions $\chi_j(\eta)$, and this is the main advantage of using the characteristic functions. The corresponding probability distribution $W(\alpha)$ is found as a convolution,

$$\chi(\eta)=\prod_{j=1}^{n}\chi_j(\eta) \tag{1.7.6}$$

$$W(\alpha)=\int\ldots\int\delta^2(\alpha-\alpha_1-\ldots-\alpha_n)\prod_{j=1}^{n}W_j(\alpha_j)d^2\alpha_j. \tag{1.7.6$'$}$$

It is clear from (1.7.6′) that, in particular, the distribution $W(\alpha)$ is surely nonnegative for nonnegative partial distributions $W_j(\alpha)$. The property (1.7.6) is widely exploited in many applications of the characteristic functions in the theory of probability, specifically in the problems of random walks and Brownian motions [58, 63].

Two novel aspects appear in applications to quantum theory. First, a state of the system is represented by a density matrix $\hat{\varrho}$, i.e., by an Hermitian, positive-definite operator, with the trace equal to 1. For any operator F the expectation value is given by

$$\langle\hat{F}\rangle=\mathrm{tr}\,\{\varrho\hat{F}\}. \tag{1.7.7}$$

Second, the variables $\hat{q}$, $\hat{p}$ are now operators and do not commute (the operators $\hat{a}$, $\hat{a}^+$ do not commute), so the order of the factors $\exp(\eta a^+)$ and $\exp(-\bar{\eta}a)$ is essential. Usually, three variants of the characteristic function are used:

$$\chi_{\mathrm{N}}(-\eta)=\mathrm{tr}\,\{\varrho e^{\eta a^+}e^{-\bar{\eta}a}\}, \tag{1.7.8a}$$

$$\chi_{\mathrm{O}}(-\eta)=\mathrm{tr}\,\{\varrho e^{\eta a^+-\bar{\eta}a}\}, \tag{1.7.8b}$$

$$\chi_{\mathrm{A}}(-\eta)=\mathrm{tr}\,\{\varrho e^{-\bar{\eta}a}e^{\eta a^+}\}. \tag{1.7.8c}$$

The subscripts N, O, A indicate the normal, symmetrical and antinormal ordering of $\hat{a}$ and $\hat{a}^+$, respectively; the subscript O is often omitted.

These characteristic functions are simply related through (1.2.31, 32)

$$\chi_k(\eta)=\exp(-\sigma_k|\eta|^2)\chi_{\mathrm{N}}(\eta) \tag{1.7.9}$$

$$\sigma_k=\begin{cases}0 & \text{for}\quad k=\mathrm{N},\\ \tfrac{1}{2} & \text{for}\quad k=\mathrm{O},\\ 1 & \text{for}\quad k=\mathrm{A}.\end{cases} \tag{1.7.10}$$

Any variant is sufficient to determine the density matrix.

Suppose the density matrix is diagonal in the occupation-number representation, $\varrho_{mn} = w_n \delta_{mn}$. Then the characteristic function is

$$\chi_k(\eta) = e^{-\sigma_k|\eta|^2} \sum_n w_n L_n(|\eta|^2). \tag{1.7.11}$$

Using (1.6.33, 7.9), clearly

$$\varrho = \int \chi_O(\eta) D(\eta) d\mu(\eta) \tag{1.7.12a}$$

$$\varrho = \int \chi_N(\eta) e^{-\bar{\eta} a} e^{\eta a^+} d\mu(\eta) \tag{1.7.12b}$$

$$\varrho = \int \chi_A(\eta) e^{\eta a^+} e^{-\bar{\eta} a} d\mu(\eta), \tag{1.7.12c}$$

(note the order of the operators $\hat{a}$ and $\hat{a}^+$ in the integrands). Thus each characteristic function $\chi_k(\eta)$ may be used to determine the density matrix ϱ.

Just as in the classical theory, for any k

$$\chi_k(0) = 1, \quad \bar{\chi}_k(\eta) = \chi_k(-\eta) \tag{1.7.13}$$

because the operator $\hat{\varrho}$ is Hermitian and normalized. The nonnegativity of the density matrix,

$$\langle \psi | \varrho | \psi \rangle \geq 0 \tag{1.7.14}$$

for any state vector $|\psi\rangle$, leads to the following condition for the characteristic function $\chi_k(\eta)$ [64]:

$$\sum z_i \bar{z}_j \exp(\bar{\eta}_i \eta_j + \sigma_k |\eta_i - \eta_j|^2) \chi_k(\eta_i - \eta_j) \geq 0 \tag{1.7.15}$$

for any sets of complex numbers $z_1, \ldots, z_n$ and $\eta_1, \ldots, \eta_n$.

As in the classical case, the quantum characteristic functions are suitable for calculating the moments,

$$M_{mn}^{(k)} = \left(-\frac{\partial}{\partial \eta}\right)^m \left(\frac{\partial}{\partial \bar{\eta}}\right)^n \chi_k(\eta)\bigg|_{\eta=0}. \tag{1.7.16}$$

Here $M_{mn}^{(k)}$ are the mean values of the normal, antinormal, and symmetrical products,

$$M_{mn}^{(N)} = \langle (a^+)^m a^n \rangle, \quad M_{mn}^{(A)} = \langle a^n (a^+)^m \rangle, \quad M_{mn}^{(O)} = \langle \{a^n (a^+)^m\} \rangle. \tag{1.7.17}$$

The characteristic functions provide a very appropriate approach to solving a number of problems, for example, relaxation of a quantum oscillator to thermodynamic equilibrium (Part III).

2. Coherent States for Arbitrary Lie Groups

The constructions in the preceding chapter can be generalized to any Lie group. This is the purpose of the present chapter, where, following [15], the concept of the coherent state is introduced for arbitrary Lie groups and some of their properties are investigated. A reader whose interests lie mostly in CS for the simplest Lie groups may skip this chapter.

Note that though the generalized CS may be defined for linear representations of an arbitrary group (for instance for a finite group), we restrict ourselves to unitary irreducible representations of the Lie groups: this case enables a substantial theory of such states to be constructed.

2.1 Definition of the Generalized Coherent State

Let G be an arbitrary Lie group and $T(g)$ its unitary irreducible representation, acting in the Hilbert space $\mathscr{H}$.

Take a fixed vector $|\psi_0\rangle$ in the Hilbert space $\mathscr{H}$ and consider a set $\{|\psi_g\rangle\}$, where $|\psi_g\rangle = T(g)|\psi_0\rangle$ and g is any element of the group G. It is not difficult to see that two vectors $|\psi_{g^1}\rangle$ and $|\psi_{g_2}\rangle$ correspond to the same state, i.e., differ by a phase factor ($|\psi_{g^1}\rangle = \exp(\mathrm{i}\alpha)|\psi_{g_2}\rangle$, $|\exp(\mathrm{i}\alpha)| = 1$), only if $T(g_2^{-1} g_1)|\psi_0\rangle = \exp(\mathrm{i}\alpha)|\psi_0\rangle$. Suppose $H = \{h\}$ is a subgroup of the group G, such that its elements have the property

$$T(h)|\psi_0\rangle = \exp[\mathrm{i}\alpha(h)\}|\psi_0\rangle. \qquad (2.1.1)$$

When the subgroup H is maximal, it will be called the isotropy subgroup for the state $|\psi_0\rangle$.

This construction shows that the vectors $|\psi_g\rangle$ for all the group elements g, belonging to a left coset class of G with respect to the subgroup H, differ only in a phase factor and so determine the same state. Choosing a representative $g(x)$ in any equivalence class x, one gets a set of states $\{|\psi_{g(x)}\rangle\}$, where $x \in X = G/H$. A more concise form $\{|x\rangle\}$, $|x\rangle \in \mathscr{H}$, will be used for this set here. (The group G may be considered as a fiber bundle with base $X = G/H$, and fiber H. A choice of $g(x)$ is a cross section in the fiber bundle.)

Now we are able to present a definition of the generalized coherent state.

Definition. The system of states $\{|\psi_g\rangle\}$, $|\psi_g\rangle = T(g)|\psi_0\rangle$, where g are elements of the group G (T is a representation of the group G, acting in the Hilbert space $\mathscr{H}$, and $|\psi_0\rangle$ is a fixed vector in this space) is called the coherent-state system $\{T, |\psi_0\rangle\}$.

Let H be the isotropy subgroup for the state $|\psi_0\rangle$. Then a coherent state $|\psi_g\rangle$ is determined by a point $x = x(g)$ in the coset space G/H, corresponding to the element g: $|\psi_g\rangle = \exp(i\alpha)|x\rangle$, $|\psi_0\rangle = |0\rangle$.

Remark 1. The state corresponding to the vector $|x\rangle$ may also be considered as a one-dimensional subspace in $\mathscr{H}$, or as a projector $P_x = |x\rangle\langle x|$, $\dim P_x = 1$, in $\mathscr{H}$. Thus the system of generalized CS determines a set of one-dimensional subspaces in $\mathscr{H}$, parametrized by points of the homogeneous space $X = G/H$.

Even more general systems may be considered, where any point of the homogeneous space X is mapped into a subspace P_x in $\mathscr{H}$, which is not one-dimensional, while $T(g)P_xT(g^{-1}) = P_{gx}$. Such systems were investigated in [65].

Remark 2. In a number of cases it is useful to consider not the Hilbert space $\mathscr{H}$, but a wider space $\tilde{\mathscr{H}} \supset \mathscr{H}$, the so-called rigged Hilbert space. Then the action of the representation $T(g)$ is extended to $\tilde{\mathscr{H}}$. *If an element of* $\tilde{\mathscr{H}}$ *is taken as the starting state vector* $|\psi_0\rangle$, then a more general CS system is constructed. Such systems appear when representations belonging to the principal series are considered for a semisimple Lie group (Chaps. 6–8), as well as when the isotropy subgroup H is a discrete subgroup of G (see Sect. 1.5 concerning theta functions and Sect. 3.2).

An important class of CS systems corresponds to the coset space $X = G/H$ is a homogeneous sympletic manifold. (Definition and properties of sympletic manifolds may be found in [66]. Homogeneous sympletic manifolds coincide with orbits of coadjoint representations [19].) Then X may be considered as the phase space of a classical dynamical system, and the mapping $x \to P_x$ is the "quantization" for this system.

Constructing unitary irreducible representations of the group G, based on a classical dynamical system with the phase space X, was developed by *Kirillov* [19] and *Kostant* [20], who, however, did not consider the correspondence principle and the analog of Planck's constant. When, in addition, analyticity is assumed (in other words, when X is not only sympletic but also a complex manifold, i.e., a homogeneous Kählerian manifold) a somewhat different approach to the quantization was developed by *Berezin* [22]. He could then obtain more advanced results (though for a narrower class of groups). It is remarkable that in this case the quantization is performed via the coherent states. Moreover, unlike the general case, it is possible to verify here also the correspondence principle: classical dynamics is a limit of quantum dynamics when a parameter, which is an analog of Planck's constant, tends to zero.

Further below in this chapter, the generalized coherent state is called just the coherent state, for brevity.

2.2 General Properties of Coherent-State Systems

Note first of all that a direct consequence of (2.1.1) is $\exp[i\alpha(h_2h_1)] = \exp[i\alpha(h_2)]\exp[i\alpha(h_1)]$, i.e., $\exp[i\alpha(h)]$ is the one-dimensional unitary representation (character) of the subgroup H. If this representation is not identical, i.e., $\alpha(h)\not\equiv 0$, then the quotient A of the group H by its commutant H' is not trivial, i.e., it contains elements other than unity, and the character of A determines the representation of H completely. Recall that the commutant H' of a group H contains elements of the form $h'=h_1h_2h_1^{-1}h_2^{-1}$. The commutant is an invariant subgroup in H, and the coset space H/H' is an Abelian group.

If, however, $\alpha(h)\equiv 0$, then H is the usual isotropy subgroup for the vector $|\psi_0\rangle$. In both cases, the representation T of the group G, when restricted to the subgroup H, must contain the one-dimensional (identity) representation of H. Note that if the subgroup H is simply connected, the vector $|\psi_0\rangle$ is an eigenvector of the generators of the representation of H. [Useful information on possible representations $T(g)$ may often be obtained from the Frobenius duality theorem, namely, if T^α is a representation of G, induced by the character $\exp(i\alpha)$ of the subgroup H, then the representation T must be present in the decomposition of T^α over the irreducible representations [19].]

Suppose H_0 is a subgroup of H, containing elements h_0 such that $T(h_0)=I$. Evidently, H_0 is an invariant subgroup of H. Let us consider the quotient group H/H_0. If it is compact, the vector $|\psi_0\rangle$ which is an eigenvector for the operators $T(h)$, belongs to the Hilbert space $\mathscr{H}$. If the quotient group H/H_0 is not compact, the vector $|\psi_0\rangle$ is not contained in $\mathscr{H}$ but belongs to the space $\mathscr{H}_{-\infty}$, Sect. 1.5.

Let us consider the action of the operator $T(g)$ upon the state $|\psi_0\rangle=|0\rangle$,

$$T(g)|0\rangle=\exp[i\alpha(g)]|x(g)\rangle. \tag{2.2.1}$$

This defines a mapping $\pi\colon G\to\tilde{M}$, where $\tilde{M}$ is a fiber bundle whose base is $X=G/H$ and the fiber is a circle. Here the function $\alpha(g)$ is defined for any element g of the group G, while for $g\in H$ it coincides with the function $\alpha(h)$ considered above. Replacing g in (2.2.1) by gh gives

$$\alpha(gh)=\alpha(g)+\alpha(h). \tag{2.2.2}$$

Apply now the operator $T(g_1)$ to an arbitrary coherent state,

$$T(g_1)|x\rangle=e^{-i\alpha(g)}T(g_1)T(g)|0\rangle=e^{i\beta(g^1,g)}|g_1x\rangle,$$

$$\beta(g_1,g)=\alpha(g_1g)-\alpha(g). \tag{2.2.3}$$

Here $x=x(g)$, $g_1x=x_1\in X$, $x(g)$ is determined by the action of the group G in the homogeneous space $X=G/H$. Note that in view of (2.2.2) the equality (2.2.3) is correct, since the rhs does not depend on the element g, but on the equivalence class $x(g)$: $\beta(g_1,g)=\tilde{\beta}(g_1,x)$.

It is not difficult to see that the scalar product of two coherent states $|x_1\rangle = |x(g_1)\rangle$ and $|x_2\rangle = |x(g_2)\rangle$ is given by

$$\langle x_1|x_2\rangle = \exp\{\mathrm{i}[\alpha(g_1) - \alpha(g_2)]\} \langle 0|T(g_1^{-1}g_2)|0\rangle \tag{2.2.4}$$

and is independent of the choice of the representatives g_1 and g_2, because of (2.2.2). As the representation is unitary, $|\langle x_1|x_2\rangle < 1$ for $x_1 \neq x_2$, and

$$\overline{\langle x_1|x_2\rangle} = \langle x_2|x_1\rangle, \tag{2.2.5}$$

$$\langle gx_1|gx_2\rangle = \exp\{\mathrm{i}[\tilde{\beta}(g, x_1) - \tilde{\beta}(g, x_2)]\} \langle x_1|x_2\rangle. \tag{2.2.6}$$

2.3 Completeness and Expansion in States of the CS System

The first point to be mentioned in discussing completeness is that it is a direct consequence of the irreducibility of the representation $T(g)$. Suppose that a measure $dy(g)$, which is invariant under left and right shifts, exists for the group G. It induces an invariant measure dx in the homogeneous space $X = G/H$. Assuming that convergence conditions are satisfied, let us consider the operator

$$\hat{B} = \int dx|x\rangle \langle x|, \tag{2.3.1}$$

where $|x\rangle\langle x|$ is the projector for the state $|x\rangle$. Because of the definition of $\hat{B}$, the invariance of the measure dx, and in view of (2.2.3), one has at once

$$T(g)\hat{B}[T(g)]^{-1} = \hat{B}. \tag{2.3.2}$$

Thus $\hat{B}$ commutes with all the operators $T(g)$ and must be equal to the unity operator times a numerical factor, because the representation $T(g)$ is irreducible,

$$\hat{B} = d\hat{1}. \tag{2.3.3}$$

To fix the constant d, it is appropriate to calculate the expectation value of $\hat{B}$ for a state $|y\rangle$ (recall the normalization $\langle y|y\rangle = 1$)

$$\langle y|\hat{B}|y\rangle = \int |\langle y|x\rangle|^2 dx = \int |\langle 0|x\rangle|^2 dx = d. \tag{2.3.4}$$

Hence it is seen, by the way, that a necessary condition for $\hat{B}$ to exist is the convergence of the integral in (2.3.4). In this case the CS system is called square-integrable. (The CS are square-integrable for a number of cases: all representations of compact semisimple groups, representations of discrete series for real semisimple groups, and some representations of solvable Lie groups.) Introducing the factor into the measure in X, $d\mu(x) = d^{-1}dx$, one gets an important identity (the resolution of unity)

$$\int d\mu(x)|x\rangle \langle x| = \hat{1}. \tag{2.3.5}$$

Thus any state $|\psi\rangle$ may be expanded in the CS system,

$$|\psi\rangle=\int d\mu(x)c(x)|x\rangle, \tag{2.3.6}$$

where $c(x)=\langle x|\psi\rangle$, and

$$\langle\psi|\psi\rangle=\int d\mu(x)|c(x)|^2. \tag{2.3.7}$$

The function $c(x)$, the symbol of the state $|\psi\rangle$, is not arbitrary, but satisfies

$$c(x)=\int \langle x|y\rangle c(y)d\mu(y). \tag{2.3.8}$$

Hence the kernel $K(y,y)=\langle x|y\rangle$ is reproducing,

$$K(x,z)=\int d\mu(y)K(x,y)K(y,z), \tag{2.3.9}$$

while the function $c(x)=\int K(x,y)f(y)d\mu(y)$ satisfies (2.3.8) for an arbitrary function $f(x)$.

Clearly, some "linear dependences" exist for the coherent states. A consequence of (2.3.6) is

$$|x\rangle=\int \langle y|x\rangle|y\rangle d\mu(y). \tag{2.3.10}$$

Thus the CS system is overcomplete, i.e., it contains some subsets of coherent states which are complete systems. An important class of such subsets may be constructed by means of discrete subgroups of G. Let Γ be a discrete subgroup of G, such that the volume of the coset space $\Gamma\backslash X=\Gamma\backslash G/H$ is finite. Consider the subset of the coherent states

$$\{|x_l\rangle\}=\{|x(\gamma_l)\rangle\}, \quad \gamma_l\in\Gamma. \tag{2.3.11}$$

The problem to consider is whether such subsets are complete. The solution of the problem for the simplest nilpotent group, the Heisenberg-Weyl group, is presented in Sect. 1.4. Below is given the solution for a number of noncompact semisimple Lie groups with discrete representation series.

2.4 Selection of Generalized CS Systems with States Closest to Classical

The above construction shows that the CS system depends essentially on the choice of the original state $|\psi_0\rangle$. We consider now the problem of which state vector $|\psi_0\rangle$ must be taken to generate states as close as possible to the classical states. This principle of selection was proposed in [11] and developed further in [67], and is also considered in this section.

The idea is to extend the Lie algebra $\mathscr{G}$ of the group G up to the complex algebra $\mathscr{G}^c$, and to consider in $\mathscr{G}^c$ the isotropy subalgebra $\mathscr{B}$ for the state $|\psi_0\rangle$. Those vectors for which the subalgebra is maximal are of special interest; as shown below, the corresponding states are the closest to the classical states. Actually, for a compact simple Lie group these states have the least uncertainty, in the sense of an invariant definition proposed in [68, 69]. However, the method presented here is more general and is applicable to other groups, for instance, the Heisenberg–Weyl group, and some nilpotent and solvable groups, as well as to noncompact semisimple Lie groups with discrete representation series.

Let us now consider the method.

Let $\mathscr{G}$ be the Lie algebra for the group G, and $\mathscr{G}^c$ its complex hull, i.e., the set of all linear combinations of elements of $\mathscr{G}$ with complex coefficients. Let $T(g)$ be the considered unitary irreducible representation of the group G, T_g the corresponding representation of the algebra $\mathscr{G}$, and $|\psi_0\rangle$ a fixed vector in the representation space. Denote by $\mathscr{B}=\{b\}$ the isotropy subalgebra for the state $|\psi_0\rangle$, i.e., the set of elements of $\mathscr{G}^c$ such that

$$T_b|\psi_0\rangle=\lambda_b|\psi_0\rangle. \tag{2.4.1}$$

Denote by $\bar{\mathscr{B}}$ the subalgebra of $\mathscr{G}^c$, conjugate to $\mathscr{B}$. The subalgebra $\mathscr{B}$ is called maximal, if $\mathscr{B}\oplus\bar{\mathscr{B}}=\mathscr{G}^c$. The states for which their isotropy subalgebras are maximal are most symmetrical, and, as shown below, they are the closest to the classical states.

It can be shown that when the subalgebra $\mathscr{B}$ is maximal, the coherent state is determined by a point in the coset space $G^c/B=\bar{B}/D$, where $B=\exp\mathscr{B}$, $D=\exp\mathscr{D}$, $\mathscr{D}=\mathscr{B}\cap\bar{\mathscr{B}}$. Note that the concept of maximal isotropy subalgebra is intimately related to the concept of completely polarization, introduced by *Kostant* [21].

Several properties of this construction are remarkable. First, $\mathscr{B}\cap\mathscr{G}=\mathscr{H}$ is the isotropy subalgebra of the state $|\psi_0\rangle$ within the Lie algebra $\mathscr{G}$. It is not difficult to see also that $\mathscr{D}=\mathscr{B}\cap\bar{\mathscr{B}}$ is a complex subalgebra of $\mathscr{G}^c$, which may be written as $\mathscr{H}^c=\mathscr{H}\oplus\mathrm{i}\mathscr{H}$.

Further, a consequence of (2.4.1) is $(T_{b^1}T_{b_2}-T_{b_2}T_{b^1})|\psi_0\rangle=0$, so $[\mathscr{B},\mathscr{B}]\subset\mathscr{B}_0$, where $\mathscr{B}_0$ is the subalgebra of $\mathscr{B}$ containing elements annihilating $|\psi_0\rangle$. Note that the adjoint representation $\mathrm{Ad}(h)$ for the subgroup H acting in $\mathscr{G}$ may be continued to $\mathscr{G}^c$ by means of the linearity. Evidently, the representation $\mathrm{Ad}(h)$ conserves the subalgebra $\mathscr{B}$,

$$\mathrm{Ad}(h)\,T_b|\psi_0\rangle=\lambda_b|\psi_0\rangle.$$

Here the coset space $X=G/H$ can have a natural complex homogeneous structure [Ref. 19, p. 224], that appears when X is identified with $G^c/B=\bar{B}/D$. The elements of G act in the space X as holomorphic transformations.

Let us consider some examples.

I) Suppose G is the Heisenberg-Weyl group, $G=W_n$. Its Lie algebra $\mathscr{G}$ contains $(2n+1)$ basic elements $p_1,\ldots,p_n,q_1,\ldots,q_n,\hat{\mathrm{I}}$, satisfying the Heisenberg commutation relations

$$[q_j,p_k]=\mathrm{i}\delta_{jk}\hat{\mathrm{I}},\qquad [q_j,\hat{\mathrm{I}}]=[p_k,\hat{\mathrm{I}}]=[q_j,q_k]=[p_j,p_k]=0. \tag{2.4.2}$$

Within the complex Lie algebra $\mathscr{G}^c$, the following elements may be considered (creation–annihilation operators)

$$a_j^+ = \frac{\omega_j q_j - \mathrm{i} p_j}{\sqrt{2\omega_j}}, \qquad a_k = \frac{\omega_k q_k + \mathrm{i} p_k}{\sqrt{2\omega_k}}, \tag{2.4.3}$$

for any positive ω_j. In this case the "vacuum" vector $|\psi_0\rangle$ defined by $a_k|\psi_0\rangle = 0$ has maximal isotropy subalgebra:

$$\mathscr{B} = \{a_j, \hat{\mathrm{I}}\}, \qquad \bar{\mathscr{B}} = \{a_j^+, \hat{\mathrm{I}}\}, \qquad \mathscr{B} \oplus \bar{\mathscr{B}} = \mathscr{G}^c.$$

On the other hand, it is well known that the CS constructed with $|\psi_0\rangle$ have the least uncertainty (for them $\Delta p_j \Delta q_j = \frac{1}{2}\hbar$), so they are closest to the classical states.

II) An example of a solvable group is the so-called oscillator group. Consider the Lie algebra $\mathscr{G}$ with $2n+2$ basic elements $p_1, \ldots, p_n$, $q_1, \ldots, q_n$, A, $\hat{\mathrm{I}}$, satisfying the following commutation relations

$$\begin{aligned} &[q_j, p_k] = \mathrm{i}\delta_{jk}\hat{\mathrm{I}}, \quad [p_j, \hat{\mathrm{I}}] = [q_k, \hat{\mathrm{I}}] = [q_j, q_k] = [p_j, p_k] = [A, \hat{\mathrm{I}}] = 0 \\ &[q_j, A] = \mathrm{i} p_j, \qquad [p_k, A] = -\mathrm{i} q_k. \end{aligned} \tag{2.4.4}$$

I use the name 'oscillator algebra', because the element A may be considered as a Hamiltonian for the quantum n-dimensional oscillator

$$A = \tfrac{1}{2} \sum_j (p_j^2 + q_j^2).$$

Again we introduce the operators a_j^+, a_k instead of q_j, p_k. As in the preceding example, the "vacuum" vector has maximal isotropy subalgebra. Now $\mathscr{B} = \{a_j, \hat{\mathrm{I}}, A\}$, $\bar{\mathscr{B}} = \{a_j^+, \mathrm{I}, A\}$, $\mathscr{B} \oplus \bar{\mathscr{B}} = \mathscr{G}^c$, and the coherent states constructed based on the "vacuum" are closest to the classical states.

III) Suppose G is a compact simple Lie group, H is its Cartan subgroup, and $\mathscr{H}$ is the corresponding Lie algebra.[1] Let $T^\Lambda(g)$ be a representation of G with the highest weight Λ, and $|\psi_m\rangle$ be the representation vector corresponding to a certain weight m. Take the vector $|\psi_m\rangle$ as a starting point for constructing the coherent states. Evidently, to get $\mathscr{B} \oplus \bar{\mathscr{B}} = \mathscr{G}^c$ it is necessary to have $\dim \mathscr{B} \geq (n+r)/2$, where n is the group dimensionality and r is its rank. It is not difficult to see that this property may be realized only for the state vector $|\psi_m\rangle$, corresponding to a dominant weight, i.e., to the highest weight Λ, or to a weight equivalent to the highest one with respect to a transformation of the Weyl group. Note also that for general representations $\dim \mathscr{B} = (n+r)/2$ always, while the inequality arises only for degenerate representations. Clearly, the above statement is true also for a general vector in the representation space

[1] The objects we deal with are defined in [70]. The construction of quantum (coherent) states closest to the classical states for the three-dimensional rotation group was considered in [71]. In this section, $\mathscr{H}$ denotes the Hilbert space, or the Cartan subalgebra; both notations are standard, and should not lead to a confusion here.

which may be written as $|\psi_0\rangle = \sum_m c_m |\psi_m\rangle$. Therefore, with no loss of generality, one may assume that $|\psi_0\rangle = |\psi_\Lambda\rangle$.

Such a state, as well as all the CS constructed on its basis [69], have least uncertainty if the latter is defined invariantly. Namely, these states minimize the dispersion ΔC_2,

$$\begin{aligned} &C_2 = g^{jk} X_j X_k, \\ &\Delta C_2 = \langle\psi_0|C_2|\psi_0\rangle - g^{jk}\langle\psi_0|X_j|\psi_0\rangle\langle\psi_0|X_k|\psi_0\rangle, \end{aligned} \tag{2.4.5}$$

where C_2 is the quadratic Casimir operator, X_j is the generator of the representation of the Lie algebra $\mathscr{G}$, and g^{jk} is the Killing–Cartan metric tensor.

Thus we have considered another example of CS closest to the classical states.

IV) Let us consider the simplest real semisimple Lie group, $G = SU(1,1)$, having a discrete series of representations $T^k(g)$. Let $\mathscr{G}$ be the Lie algebra for G, with the standard basis K_0, K_1, K_2. The representation is characterized by a nonnegative number k, and the basis vectors $|\psi_m\rangle$ correspond to the weights $m = k + n$, $n = 0, 1, 2, \ldots$; they are eigenvectors of the operator K_0: $K_0|\psi_m\rangle = m|\psi_m\rangle$. The Casimir operator here is $C_2 = -K_0^2 + K_1^2 + K_2^2$; for a given representation T^k it equals the unity times a number, $C_2|\psi\rangle = \lambda|\psi\rangle$, $\lambda = k(1-k)$. The dispersion

$$\begin{aligned} \Delta C_2 &= \langle K_1^2 + K_2^2 - K_0^2\rangle - \langle K_1\rangle^2 - \langle K_2\rangle^2 + \langle K_0\rangle^2 \\ &= k - k^2 + (k+n)^2 = k + 2kn + n^2 \end{aligned}$$

is minimal, $\Delta C_2 = k$ for $m = k$, $n = 0$, i.e., for the vector corresponding to the lowest weight, $|\psi_0\rangle = |\psi_k\rangle$. Thus in this case, where $\mathscr{B} = \{K_0, K_+ = K_1 + \mathrm{i}K_2\}$, $\bar{\mathscr{B}} = \{K_0, K_- = K_1 - \mathrm{i}K_2\}$, the CS system constructed based on the vector $|\psi_0\rangle = |\psi_k\rangle$ is the closest to the classical system.

Note that this property is specific also for real simple Lie groups with discrete series of representations. (It is known that the Lie group has a discrete series of representations if it contains a compact Cartan subgroup.) Furthermore, their maximal compact subgroups contain one-dimensional centers. In this case these representations are induced by representations of the maximal compact subgroup, which are nontrivial only on the center.

It is known also that in the considered case $X = G/H$ may be represented as a complex homogeneous bounded domain, and the Hilbert space may be realized as a certain space of analytic functions (a generalization of the Fock-Bargmann representation, Sect. 1.3). The CS systems arising in this case were investigated in [72, 73].

Note that this statement is true also for a certain class of solvable Lie groups considered in [74].

Thus the principle for selecting generalized CS systems containing states closest to the classical states, proposed in [11] and investigated further in [67], is applicable to a wide class of Lie groups.

3. The Standard System of Coherent States; Several Degrees of Freedom

The case of several degrees of freedom is essentially similar to that of one degree of freedom considered in Chap. 1. Many formulae from Chapt. 1 are extended to several degrees of freedom in a trivial manner. Here we present only the most important results.

A reader mainly interested in applications of the CS method in theoretical physics should turn to Chaps. 8, 9.

3.1 General Properties

Recall that the basic operators used to describe a quantum system with a finite number of degrees of freedom are the coordinate operators $\hat{q}_j$ and the momentum operators $\hat{p}_k$ $(j, k=1, \ldots, N)$. They satisfy the Heisenberg commutation relations:

$$[\hat{q}_j, \hat{p}_k] = \mathrm{i}\hbar\delta_{jk}\hat{\mathrm{I}}, \quad [\hat{q}_j, \hat{q}_k] = [\hat{q}_j, \hat{\mathrm{I}}] = 0, \quad [\hat{p}_j, \hat{p}_k] = [\hat{p}_j, \hat{\mathrm{I}}] = 0. \tag{3.1.1}$$

Here $\hat{\mathrm{I}}$ is the unity operator and $\hbar$ is Planck's constant.

The group-theoretical structure of the commutation relations is given by the corresponding group, called the Heisenberg–Weyl group W_N. As in Chap. 1, another operator basis is suitable:

$$a_j = \frac{\hat{q}_j + \mathrm{i}\hat{p}_j}{\sqrt{2\hbar}}, \qquad a_k^+ = \frac{\hat{q}_k - \mathrm{i}\hat{p}_k}{\sqrt{2\hbar}}. \tag{3.1.2}$$

The commutation relations are obtained from (3.1.1, 2),

$$[a_j, a_k^+] = \delta_{jk}\hat{\mathrm{I}}, \quad [a_j, a_k] = [a_j, \hat{\mathrm{I}}] = 0, \quad [a_j^+, a_k^+] = [a_j^+, \hat{\mathrm{I}}] = 0. \tag{3.1.3}$$

The operators $\hat{q}_j, \hat{p}_k$, as well as a_j, a_k^+, act in the standard Hilbert space $\mathscr{H}$. It is known that in $\mathscr{H}$ a so-called vacuum vector $|0\rangle$ exists which is a normalized vector vanishing under the action of any a_j:

$$a_j|0\rangle = 0, \quad \langle 0|0\rangle = 1. \tag{3.1.4}$$

Products of the operators a_k^+ generate a set of the normalized vectors

$$|n_1,\ldots,n_N\rangle=\frac{(a_1^+)^{n^1}\ldots(a_N^+)^{n_N}}{\sqrt{n_1!\ldots n_N!}}\,|0\rangle. \tag{3.1.5}$$

A concise notations proposed in [33] are useful

$$[n]=(n_1,\ldots,n_N),\qquad [n]!=n_1!\ldots n_N!$$
$$[a^+]^{[n]}=(a_1^+)^{n^1}\ldots(a_N^+)^{n_N}, \tag{3.1.6}$$

so that (3.1.5) can be written as

$$|[n]\rangle=\frac{(a^+)^{[n]}}{\sqrt{[n]!}}\,|0\rangle. \tag{3.1.5'}$$

The set of the vectors $\{|[n]\rangle\}$ is a basis in the Hilbert space $\mathscr{H}$. The action of the operators a_j and a_k^+ in the space $\mathscr{H}$ is written as

$$a_j|[n]\rangle=\sqrt{n_j}\,|[n']\rangle,\qquad a_k^+|[n]\rangle=\sqrt{n_{k+1}}\,|[n'']\rangle,$$
$$[n']=(n_1,\ldots,n_{j-1},n_j-1,\ldots,n_N),\quad [n'']=(n_1,\ldots,n_k+1,\ldots,n_N). \tag{3.1.7}$$

Note also the coordinate representation of the basis vectors

$$\langle q|[n]\rangle=\varphi_{[n]}(q)=\prod_{j=1}^{N}\varphi_{n_j}(q_j),$$
$$\varphi_n(q)=(2^n n!)^{-1/2}(\pi\hbar)^{-1/4}\mathrm{H}_n(q/\sqrt{\hbar})\exp(-q^2/2\hbar), \tag{3.1.8}$$

where H_n is the Hermite polynomial of degree n.

Let us now turn to (3.1.3). These relations mean that the operators a_j, a_k^+ and $\hat{\mathrm{I}}$ form a basis of the Lie algebra of W_N, which is just the Heisenberg-Weyl algebra. Any element of the algebra is written as

$$\hat{X}=s\hat{\mathrm{I}}+\frac{1}{\hbar}(P\hat{q}-Q\hat{p})=s\hat{\mathrm{I}}-\mathrm{i}(\alpha a^+-\bar{\alpha}a), \tag{3.1.9}$$

where s is a real number, $\alpha=(Q+\mathrm{i}P)/\sqrt{2\hbar}$, $\bar{\alpha}=(Q-\mathrm{i}P)/\sqrt{2\hbar}$. Here and below the following notations are used: $P=(P_1,\ldots,P_N)$ and $Q=(Q_1,\ldots,Q_N)$ are N-dimensional real vectors, $\alpha=(\alpha_1,\ldots,\alpha_N)$, $a=(a_1,\ldots,a_N)$, $\alpha a^+=\sum_{j=1}^{N}\alpha_j a_j^+$.

The operator $\hat{X}$ is self-adjoint, and the corresponding operator $\exp(\mathrm{i}\hat{X})$ is unitary,

$$\exp(\mathrm{i}\hat{X})=\exp(\mathrm{i}s\hat{\mathrm{I}})D(\alpha),\qquad D(\alpha)=\exp(\alpha a^+-\bar{\alpha}a). \tag{3.1.10}$$

It describes a finite element of the group W_N with Lie algebra $\mathscr{W}_N$.

With the above notations, most formulae of Chap. 1 are also valid for this case, or must be modified in a trivial way. For example, an expression for the CS, analogous to that in (1.2.36), is

$$|\alpha\rangle=\exp(-|\alpha|^2/2)\sum_{[n]}\frac{\alpha^{[n]}}{\sqrt{[n]!}}|[n]\rangle, \tag{3.1.11}$$

while the integration measure is

$$d\mu(\alpha)=\pi^{-N}\prod_{j=1}^{N}\prod_{k=1}^{N} d\,\mathrm{Re}\{\alpha_j\}\, d\,\mathrm{Im}\{\alpha_k\}. \tag{3.1.12}$$

3.2 Coherent States and Theta Functions for Several Degrees of Freedom

The relation between coherent states and theta functions is somewhat more complicated than in the one-dimensional case (the relevant aspects were considered in [23, 75]).

First of all, one should extend the representation $T(g)$ of the Heisenberg-Weyl group, given originally in the space $\mathscr{H}$, up to a representation in the space $\mathscr{H}_{-\infty}$, as described in Sect. 1.5. Consider a regular lattice L in the N-dimensional complex space $\mathbb{C}^N$, i.e., the set of vectors written as $\alpha_n=\sum_1^{2N} n_j\omega_j$, where n_j are integers, and $2N$ vectors ω_j in $\mathbb{C}^N$ (periods of the lattice) are linearly independent when taken with real coefficients. Consider also the set of operators $\{D(\alpha_n)\}$ and try to find their eigenvector $|\theta\rangle$ in the space $\mathscr{H}_{-\infty}$. For such a state vector to exist, it is necessary that the operators $D(\alpha_n)$ commute, while it is sufficient that the operators $D(\omega_j)$ commute. According to (1.1.14), this requirement is equivalent to the requirement that a matrix B has only integer elements

$$B_{ij}=\pi^{-1}\,\mathrm{Im}\{\omega_i\bar{\omega}_j\}\equiv 0 (\mathrm{mod}\, 1). \tag{3.2.1}$$

The conditions (3.2.1) are a particular case of the so-called Riemann-Frobenius conditions, imposed on the periods ω_j to make the quotient space $\mathbb{C}/L$ an Abelian manifold [45].

A lattice L, satisfying (3.2.1), will be called admissible. Then the eigenvector $|\theta\rangle$ must satisfy the set of equations

$$D(\omega_j)|\theta_\varepsilon\rangle=\mathrm{e}^{\mathrm{i}\pi\varepsilon_j}|\theta_\varepsilon\rangle, \quad j=1,2,\ldots,2N \tag{3.2.2}$$

while the state corresponding to the vector $|\theta_\varepsilon\rangle$ is determined by the real numbers $\varepsilon_1,\ldots,\varepsilon_{2N}$ lying in the interval $0\le\varepsilon_j<2$; in other words, by a point in the $2N$-dimensional torus. The isotropy subgroup H for the vector $|\theta_\varepsilon\rangle$ consists of elements of the form (t,α_n).

Applying an operator $D(\alpha)$, say, to the vector $|\theta_0\rangle$ yields a system of generalized coherent states corresponding to the lattice L,

$$|\theta_\alpha\rangle = D(\alpha)|\theta_0\rangle. \tag{3.2.3}$$

Here $\alpha = (\alpha_1, \ldots, \alpha_N)$ is an N-dimensional complex vector; α runs through the coset space G/H, which is a complex torus and here (for an admissible lattice), even an Abelian manifold.

Thus a CS system is constructed corresponding to the admissible lattice L. The theta functions may also be related to the lattice L. Actually, the theta functions may be defined as automorphic forms with respect to the lattice L [46], i.e., as entire functions, satisfying the functional equation $f(z+\alpha_n) = \exp[v(z,\alpha_n)] f(z)$, where $v(z,\alpha_n)$ is a first-order polynomial with respect to z. (This equation has a solution only when the periods ω_j satisfy the Riemann-Frobenius conditions [46]). As shown below, also [75], the states $|\theta_\varepsilon\rangle$ written in the Fock-Bargmann representation ($|\theta_\varepsilon\rangle \to \theta_\varepsilon(z)$) are given just by the theta functions.

In the following we discuss only the principal lattices, for which $\det B = 1$, so we deal with admissible lattices of the least possible volume of the elementary cell. A special basis ω_j', ω_k'' ($j,k = 1, \ldots, N$) can be introduced [46], for which

$$\operatorname{Im}\{\omega_j' \bar{\omega}_k'\} = 0, \qquad \operatorname{Im}\{\omega_j'', \bar{\omega}_k''\} = 0, \qquad \pi^{-1} \operatorname{Im}(\omega_j' \bar{\omega}_k'') = \delta_{jk}. \tag{3.2.4}$$

In other words, in this case the matrix may be reduced to the special form $\begin{pmatrix} 0 & \mathrm{I} \\ -\mathrm{I} & 0 \end{pmatrix}$, where 0 and I are the zero and unit $N \times N$ matrices, respectively. Supposing that this condition is satisfied, (3.2.2) can be written in z representation

$$D(\alpha_m)\theta_\varepsilon(z) = \exp[\mathrm{i}\pi F_\varepsilon(m)]\theta_\varepsilon(z), \tag{3.2.5}$$

where $\alpha_n = \sum (m_j' \omega_j' + m_j'' \omega_j'')$ is an arbitrary vector belonging to the lattice, and

$$F_\varepsilon(m) = \sum_{j=1}^{N} (m_j' m_j'' + \varepsilon_j' m_j' + \varepsilon_j'' m_j'). \tag{3.2.6}$$

By means of (1.3.25), the equality (3.2.5) may be rewritten as

$$\theta_\varepsilon(z+\beta_m) = \mathrm{e}^{\mathrm{i}\pi F_\varepsilon(-m)} \mathrm{e}^{|\beta_m|^2/2} \mathrm{e}^{\bar{\beta}_m z} \theta_\varepsilon(z), \tag{3.2.7}$$

where $\beta_m = \bar{\alpha}_m$ is a vector of the conjugate lattice.

Now we show that the norm of the function $\theta_\varepsilon(z)$ is infinite. Actually, the norm is given by the integral

$$\int |\tilde{\theta}_\varepsilon(z)|^2 d\mu(z), \qquad \text{where} \tag{3.2.8}$$

$$\tilde{\theta}_\varepsilon(z) = \exp(-|z|^2/2)\theta_\varepsilon(z).$$

The functional equation for $\tilde{\theta}_\varepsilon(z)$ is obtained from (3.2.7) immediately:

$$\tilde{\theta}_\varepsilon(z+\beta_m, \bar{z}+\bar{\beta}_m)=\exp[\mathrm{i}\pi F_\varepsilon(-m)] \exp[\mathrm{i}\,\mathrm{Im}(z\bar{\beta}_m)]\theta_\varepsilon(z,\bar{z}). \tag{3.2.9}$$

Thus $|\tilde{\theta}_\varepsilon(z,\bar{z})|^2=\varrho(z,\bar{z})$, where $\varrho(z,\bar{z})$ is a nonnegative function, periodical with respect to the lattice $\bar{L}$, conjugate to L. Therefore the norm of $\theta_\varepsilon(z)$ is infinite.

Our next task is to consider the solution of (3.2.5). We shall look for a solution of the form

$$h_\varepsilon(z)=\sum_n C_n^\varepsilon D(\alpha_n)h(z), \tag{3.2.10}$$

where $n=(n', n'')$, n' and n'' are integer N-dimensional vectors, and $h(z)$ is an unknown function, rising at infinity not too rapidly, so that the series in (3.2.10) is convergent. It is not difficult to verify that $h_\varepsilon(z)$ is a solution of (3.2.5) for $C_n^\varepsilon=\exp[-\mathrm{i}\pi F_\varepsilon(n)]$ at an arbitrary $h(z)$, provided that the series in (3.2.10) is convergent. This solution is appropriately written as

$$h_\varepsilon(z)=T_\varepsilon h(z), \tag{3.2.11}$$

where the operator T_ε is

$$T_\varepsilon=\sum_n \exp[-\mathrm{i}\pi F_\varepsilon(-n)]D(\alpha_n). \tag{3.2.12}$$

A more explicit form of this solution is

$$h_\varepsilon(z)=\sum_n \exp[-\mathrm{i}\pi F_\varepsilon(-n)] \exp(-\tfrac{1}{2}|\beta_n|^2) \exp[-\bar{\beta}_n z)h(z+\beta_n). \tag{3.2.13}$$

It is known, however, that in the case of a principal lattice considered, the functional equation (3.2.7) has a unique solution [46], so the function $h_\varepsilon(z)$ is the standard theta function times a constant,

$$h_\varepsilon(z)=T_\varepsilon h(z)=C_n^\varepsilon\theta_\varepsilon(z). \tag{3.2.14}$$

To determine the constant C_h^ε, it is sufficient to calculate $h_\varepsilon(z)$ at a single point, where $\theta_\varepsilon(z) \neq 0$. Taking $h(z)\equiv 1$, one gets the simplest solution

$$\tilde{f}_\varepsilon(z)=\sum_n \exp[-\mathrm{i}\pi F_\varepsilon(n)] \exp(-|\alpha_n|^2/2) \exp(\alpha_n z) \tag{3.2.15}$$

written as a superposition of the coherent states.

A consequence of the functional equation (3.2.7) is $\theta_\varepsilon(-z)=c\theta_\varepsilon(z)$, so for cases of integer characteristics (all components of the $2N$-dimensional vector ε are integers equal to 0 or 1), the function $\theta_\varepsilon(z)$ has a definite parity,

$$\theta_\varepsilon(-z)=P_\varepsilon\theta_\varepsilon(z). \tag{3.2.16}$$

It is not difficult to show [75] that the parity of the theta function is given by

$$P_\varepsilon=(-1)^{\sum \varepsilon_j' \varepsilon_j''}=(-1)^{F_0(\varepsilon)}. \tag{3.2.17}$$

Thus among all 2^{2N} theta functions with integer characteristics, $2^{N-1}(2^N-1)$ functions are odd, and $2^{N-1}(2^N+1)$ are even.

Note that it follows from (3.2.12, 13) that for an integer ε the operator T_ε does not change the parity of the function $h(z)$, so $h_\varepsilon(z)$ has the same parity as $h(z)$. Since the solution of the functional equation (3.2.7) is unique, then $h_\varepsilon(z) = C_h^\varepsilon \theta_\varepsilon(z)$, so the parity of the function $h_\varepsilon(z)$ is the same as the parity of P_ε. Thus the following theorem follows.

Theorem [75]. If the parities of the characteristics ε and function $h(z)$ are different, i.e., $P_\varepsilon P=-1$, then $h_\varepsilon(z)$ is zero identically,

$$h_\varepsilon(z)\equiv 0. \tag{3.2.18}$$

If the parities coincide, $P_\varepsilon P=+1$, then $h_\varepsilon(z)$ is proportional to the theta function of characteristics of ε,

$$h_\varepsilon(z)=C_h^\varepsilon \theta_\varepsilon(z). \tag{3.2.19}$$

The identities (3.2.18, 19) generalize those obtained in [38] for the one-dimensional case.

A consequence of (3.2.18) is obtained for the odd characteristics of $\varepsilon=(\varepsilon', \varepsilon'')$, i.e., $(\varepsilon' \cdot \varepsilon'')\equiv 1 \pmod 2$. Taking $h(z)\equiv 1$, one gets $2^{N-1}(2^N-1)$ identities:

$$f_\varepsilon(z)=\sum_n (-1)^{F_\varepsilon(n)} \exp(-\tfrac{1}{2}|\beta_n|^2)\exp(\bar{\beta}_n z)\equiv 0. \tag{3.2.20}$$

4. Coherent States for the Rotation Group of Three-Dimensional Space

The CS system for the group of rotations of three-dimensional Euclidean space, group $SO(3)$, was originally considered in [17], where such states were called spin CS. Properties of this system were investigated in [15, 76]. Here we follow the general pattern presented in [15]. Some applications of the spin CS will be considered in Part III of the book.

4.1 Structure of the Groups $SO(3)$ and $SU(2)$

The three-dimensional rotation group $SO(3)$ is the simplest and most thoroughly investigated of the compact, non-Abelian Lie groups. (An exhaustive exposition of the theory of the group $SO(3)$ and its representation may be found in [70, 77–79].) It is locally isomorphic to the group $SU(2)$, the group of unitary 2×2 matrices with unit determinant. To be more precise, $SO(3)$ is the quotient of $SU(2)$ by its center $\mathbb{Z}_2 = \{\mathrm{I}, -\mathrm{I}\}$ (the center of a group is the set of its elements commuting with every element of the group), where I is unit 2×2 matrix, $SO(3) = SU(2)/\mathbb{Z}_2$. The difference between $SO(3)$ and $SU(2)$ is inessential for our purpose here, and the group in view is actually $G = SU(2)$.

We shall consider the group $G = \{g\}$, where

$$g = \begin{pmatrix} \alpha & \beta \\ -\bar{\beta} & \bar{\alpha} \end{pmatrix}, \quad |\alpha|^2 + |\beta|^2 = 1, \quad \alpha = \alpha_1 + \mathrm{i}\alpha_2, \quad \beta = \beta_1 + \mathrm{i}\beta_2. \tag{4.1.1}$$

Here α_j and β_j are real numbers, and the bar means complex conjugation. Hence one can see, in particular, that G is homeomorphic (topologically equivalent) to the three-dimensional sphere $S^3 = \{\alpha_1, \alpha_2, \beta_1, \beta_2 : \alpha_1^2 + \alpha_2^2 + \beta_1^2 + \beta_2^2 = 1\}$, so it is simply connected. This means that any closed contour in G may be continuously deformed into a point.

The group G is not complex, but it naturally embeds into the complex group $G^c = SL(2, \mathbb{C})$, which is the group of all complex 2×2 matrices with the unit determinant,

$$g = \begin{pmatrix} \alpha & \beta \\ \gamma & \delta \end{pmatrix}, \quad \alpha\delta - \beta\gamma = 1. \tag{4.1.2}$$

The group G^c has a number of subgroup classes. Among the subgroups the following are of special interest here.

i) The groups of upper (and lower) triangular matrices, $B_{\pm}=\{b_{\pm}\}$,

$$b_{+}=\begin{pmatrix} b_{11} & b_{12} \\ 0 & b_{22} \end{pmatrix}, \quad b_{-}=\begin{pmatrix} b_{11} & 0 \\ b_{21} & b_{22} \end{pmatrix}, \quad b_{11}b_{22}=1. \tag{4.1.3}$$

These subgroups are maximal solvable groups in G^{c}.

ii) The groups of upper (and lower) triangular matrices, having unit diagonal elements, $Z_{\pm}=\{z_{\pm}\}$,

$$z_{+}=\begin{pmatrix} 1 & z_{12} \\ 0 & 1 \end{pmatrix}, \quad z_{-}=\begin{pmatrix} 1 & 0 \\ z_{21} & 1 \end{pmatrix}. \tag{4.1.4}$$

These are maximal nilpotent subgroups in G^{c}.

iii) The subgroup of complex diagonal matrices, $H^{c}=\{h\}$,

$$h=h(\varepsilon)=\begin{pmatrix} \varepsilon^{-1} & 0 \\ 0 & \varepsilon \end{pmatrix}. \tag{4.1.5}$$

The important Gaussian decomposition is valid [70] for any element of G^{c} (however, it is not realized within G),

$$g=\begin{pmatrix} \alpha & \beta \\ \gamma & \delta \end{pmatrix}=z_{+}hz_{-}=b_{+}z_{-}=z_{+}b_{-}, \quad \text{where} \tag{4.1.6}$$

$$b_{+}=z_{+}h, \quad b_{-}=hz_{-}, \quad z_{+}=\begin{pmatrix} 1 & \zeta(g) \\ 0 & 1 \end{pmatrix},$$

$$z_{-}=\begin{pmatrix} 1 & 0 \\ z(g) & 1 \end{pmatrix}, \quad h=\begin{pmatrix} \varepsilon^{-1}(g) & 0 \\ 0 & \varepsilon(g) \end{pmatrix}. \tag{4.1.7}$$

It is not difficult to get from (4.1.6)

$$\zeta=\zeta(g)=\beta\delta^{-1}, \quad z=z(g)=\gamma\delta^{-1}, \quad \varepsilon(g)=\delta. \tag{4.1.8}$$

Thus the Gaussian decomposition is unique, and almost every element of G^{c} (respectively, G), except elements of the form $g=\begin{pmatrix} \alpha & \beta \\ -\beta^{-1} & 0 \end{pmatrix}$, may be decomposed accordingly.

Note that for elements of G the parameters ζ, z and ε are related simply,

$$|\varepsilon(g)|^{2}=(1+|z(g)|^{2})^{-1}=(1+|\zeta(g)|^{2})^{-1}. \tag{4.1.9}$$

The quotients of the group G^{c} by its subgroups $B_{\pm}$ are interesting homogeneous spaces (with respect to the group G, as well as G^{c}), which are isomorphic to the complex plane $\mathbb{C}$,

$$X_{+}=G^{c}/B_{-}\sim Z_{+}, \quad X_{-}=B_{+}\backslash G^{c}\sim Z_{-}. \tag{4.1.10}$$

The action of group G^c in these spaces is obtained easily by means of the Gaussian decomposition. For instance, for X_+,

$$gz_+ = \begin{pmatrix} \alpha & \beta \\ \gamma & \delta \end{pmatrix} \begin{pmatrix} 1 & \zeta \\ 0 & 1 \end{pmatrix} = z'_+ h' z'_- \quad \text{and} \tag{4.1.11}$$

$$g: \zeta \to \zeta' = (\alpha\zeta + \beta)/(\gamma\zeta + \delta). \tag{4.1.12}$$

Respectively, for the space X_-

$$z_- g = \begin{pmatrix} 1 & 0 \\ z & 1 \end{pmatrix} \begin{pmatrix} \alpha & \beta \\ \gamma & \delta \end{pmatrix} = z''_+ h'' z''_- \quad \text{and} \tag{4.1.13}$$

$$g: z \to z'' = (\alpha z + \gamma)/(\beta z + \delta). \tag{4.1.14}$$

Let us now turn to the group $G = SU(2)$. It contains a subgroup of diagonal matrices

$$H = \left\{ \begin{pmatrix} \alpha & 0 \\ 0 & \bar{\alpha} \end{pmatrix} \right\}, \qquad \alpha = \exp(\mathrm{i}\psi/2).$$

It is not difficult to see that the quotient space $X = G/H$ is isomorphic to the set of elements of G of the form

$$\left\{ \begin{pmatrix} \alpha & \beta \\ -\bar{\beta} & \alpha \end{pmatrix} \right\}, \qquad \beta = \beta_1 + \mathrm{i}\beta_2, \qquad \alpha^2 + \beta_1^2 + \beta_2^2 = 1. \tag{4.1.15}$$

With the parametrization

$$\alpha = \cos\frac{\theta}{2}, \qquad \beta = -\sin\frac{\theta}{2}\,\mathrm{e}^{-\mathrm{i}\varphi}, \qquad 0 \le \theta < 2\pi, \qquad 0 \le \varphi < 2\pi, \tag{4.1.15$'$}$$

clearly space X is just the unit two-dimensional sphere S^2, i.e., the set of unit three-dimensional vectors $\boldsymbol{n} = (\sin\theta\cos\varphi, \sin\theta\sin\varphi, \cos\theta)$. Any element of the space X may also be written as

$$g_{\boldsymbol{n}} = \exp\left[\mathrm{i}\frac{\theta}{2}(m_1\sigma_1 + m_2\sigma_2)\right], \tag{4.1.15$''$}$$

where $m_1 = \sin\varphi$, $m_2 = -\cos\varphi$, σ_1, σ_2 are the Pauli matrices:

$$\sigma_1 = \begin{pmatrix} 0 & 1 \\ 1 & 0 \end{pmatrix}, \qquad \sigma_2 = \begin{pmatrix} 0 & -\mathrm{i} \\ \mathrm{i} & 0 \end{pmatrix}.$$

Thus this matrix describes a rotation by the angle θ around the vector $\boldsymbol{m}$, belonging to the equatorial plane of the sphere S^2, and perpendicular to vector $\boldsymbol{n}$.

The Gaussian decomposition for the element g_n indicates the isomorphism between the coset spaces X and X_+,

$$g_n \to g_\zeta = (1+|\zeta|^2)^{-1/2} \begin{pmatrix} 1 & \zeta \\ -\bar{\zeta} & 1 \end{pmatrix}, \qquad \zeta = -\tan\frac{\theta}{2}\, \mathrm{e}^{-\mathrm{i}\varphi}. \tag{4.1.16}$$

Evidently, the isomorphism is just the familiar stereographical projection of the sphere from its south pole to the complex ζ plane. To be more precise, to establish the isomorphism one has to make the ζ plane compact, adding to it the point at infinity, $\{\infty\}$, corresponding to the south pole of the sphere,

$$S^2 = \mathbb{C} \cup \{\infty\}. \tag{4.1.17}$$

The isomorphism between X and X_- is established similarly.

Actually, the above construction introduces a complex coordinate for the sphere S^2, or, more accurately, for the sphere with the south pole removed. Note that it is impossible to introduce such a coordinate system on the whole sphere, since the topological structure of the sphere is not that of the plane. Nevertheless, the sphere S^2, as well as any orbit of the adjoint representation for a compact Lie group, is a compact complex manifold, so it may be described with a combination of several coordinate systems. In the case considered, it is sufficient to take the second coordinate system, obtained through stereographical projection from the north pole of the sphere.

The expression for the G-invariant metrics on the sphere is

$$ds^2 = d\boldsymbol{n} \cdot d\boldsymbol{n} = \frac{4\, d\zeta d\bar{\zeta}}{(1+|\zeta|^2)^2} \tag{4.1.18}$$

and the closed 2-form is written as

$$\omega = 2\mathrm{i}\, \frac{d\zeta \wedge d\bar{\zeta}}{(1+|\zeta|^2)^2} = (\boldsymbol{n}, [d\boldsymbol{x}, d\boldsymbol{y}]), \tag{4.1.19}$$

which is evidently G invariant and here coincides with the element of area on the sphere. The symbol $\wedge$ here means the external product, $d\zeta \wedge d\bar{\zeta} = -d\bar{\zeta} \wedge d\zeta$, and $d\boldsymbol{x}$ and $d\boldsymbol{y}$ are vectors tangent to S^2 at point $\boldsymbol{n}$. It is remarkable also that the expressions in (4.1.18, 19) may be represented as

$$ds^2 = 4\, \frac{\partial^2 F}{\partial\zeta \partial\bar{\zeta}}\, d\zeta \cdot d\bar{\zeta} \tag{4.1.18'}$$

$$\omega = 2\mathrm{i}\, \frac{\partial^2 F}{\partial\zeta \partial\bar{\zeta}}\, d\zeta \wedge d\bar{\zeta}, \quad \text{where} \tag{4.1.19'}$$

$$F = \log(1+|\zeta|^2). \tag{4.1.20}$$

This is possible since the coset space is not only a complex manifold, but also the so-called Kählerian manifold; the function F in (4.1.20) is the Kählerian potential (Sect. 11.2).

4.2 Representation of $SU(2)$

Any unitary irreducible representation $T(g)$ of the group $SU(2)$ is given by a nonnegative integer or half-integer j: $T(g)=T^j(g)$, $\dim T^j=2j+1$. In the representation space $\mathscr{H}^j$ the canonical basis $|j,\mu\rangle$ exists, where μ runs from $-j$ to j with the unity steps $(-j\leq\mu\leq j)$. The infinitesimal operators $J_\pm=J_1\pm \mathrm{i}J_2$, $J_0=J_3$ of the group representation $T^j(g)$ satisfy the standard commutation relations

$$[J_0, J_\pm]=\pm J_\pm, \qquad [J_-, J_+]=-2J_0. \tag{4.2.1}$$

The operator J_k $(k=1,2,3)$ is related to the infinitesimal rotation around the kth axis. The representation space vectors $|j,\mu\rangle$ are eigenvectors for the operators J_0 and $\boldsymbol{J}^2=J_1^2+J_2^2+J_3^2$,

$$J_0|j,\mu\rangle=\mu|j,\mu\rangle, \qquad \boldsymbol{J}^2|j,\mu\rangle=j(j+1)|j,\mu\rangle. \tag{4.2.2}$$

Respectively, the operator $\exp[\mathrm{i}\omega(\boldsymbol{m}\boldsymbol{J})]$, $\boldsymbol{m}^2=1$, describes the rotation by the angle ω around the axis directed along $\boldsymbol{m}$. The action of the operators $J_\pm$ in the canonical basis is given by

$$\begin{aligned} J_+|j,\mu\rangle&=\sqrt{(j-\mu)(j+\mu+1)}\,|j,\mu+1\rangle,\\ J_-|j,\mu\rangle&=\sqrt{(j+\mu)(j-\mu+1)}\,|j,\mu-1\rangle. \end{aligned} \tag{4.2.3}$$

Hence

$$J_-|j,-j\rangle=0, \qquad |j,\mu\rangle=\sqrt{\frac{(j-\mu)!}{(j+\mu)!(2j)!}}\,(J_+)^{j+\mu}|j,-j\rangle. \tag{4.2.4}$$

Now we describe a standard realization of the representation $T^j(g)$ that is appropriate for our purpose. It is a consequence of (4.1.14) that a representation of the group $SU(2)$ acts in the space $\mathscr{F}^j$ of the polynomials of degree less than $(2j+1)$ on the group Z_-. By linearity, this representation is extended to that of the group $G^c=\mathrm{SL}(2,\mathbb{C})$. It is given by

$$T^j(g)f(z)=(\beta z+\delta)^{2j}f\left(\frac{\alpha z z_g+\gamma}{\beta z+\delta}\right), \qquad z_g=\frac{\alpha z+\gamma}{\beta z+\delta}. \tag{4.2.5}$$

The corresponding scalar product is written as

$$\langle f_1|f_2\rangle=\frac{2j+1}{\pi}\int\frac{\overline{f_1(z)}\,f_2(z)}{(1+|z|^2)^{2j+2}}\,d^2z, \qquad d^2z=dx\,dy,$$

$$z=x+\mathrm{i}y, \qquad f_0\equiv 1, \qquad \langle f_0|f_0\rangle=1 \tag{4.2.6}$$

and the infinitesimal operators are represented by the first-order differential operators

$$\tilde{J}_+ = -z^2 \frac{d}{dz} + 2jz, \quad \tilde{J}_- = \frac{d}{dz}, \quad \tilde{J}_0 = z\frac{d}{dz} - j. \tag{4.2.7}$$

The operators J_+ and J_- are conjugated with respect to the norm (4.2.6), while J_0 is a self-conjugate operator. The basis vectors $|j,\mu\rangle$ are represented by the monomials $f_\mu = c_\mu z^{j+\mu}$, $c_\mu = [2j!/(j-\mu)!(j+\mu)!]^{1/2}$.

In this representation $T^j(g)$ has a simple "semiclassical" meaning and may be rewritten as follows (cf. [80], where the general case of a simple compact Lie group is considered)

$$T^j(g)f(z) = \exp[\mathrm{i}\,S(g,z)]f(z_g), \quad z_g = \frac{\alpha z - \bar{\beta}}{\beta z + \bar{\alpha}}. \tag{4.2.8}$$

Here $S(g,z)$ is just the classical action integral for free motion on the sphere,

$$S(g,z) = \int_0^z (\theta - g\theta) + S(g,0), \quad \theta = \frac{\partial F(z,\bar{z})}{\partial z}\,dz, \quad F = j\ln(1+z\bar{z}).$$

The 1-form θ is an analog of the form $\theta = \bar{z}\,dz$ for the plane case.

4.3 Coherent States

Applying operators in the representation $T^j(g)$ to a fixed vector $|\psi_0\rangle$ according to the general scheme described in Chap. 2 yields a CS system.

As seen from (4.1.15″), the operator $T^j(g)$ may be written as

$$T^j(g) = T^j(g_{\boldsymbol{n}})T^j(h), \quad h \in H = U(1). \tag{4.3.1}$$

Hence, if a vector $|j,\mu\rangle$ is taken as the fundamental vector $|\psi_0\rangle$, the coherent state is determined by a unit vector $\boldsymbol{n} = (\sin\theta\cos\varphi, \sin\theta\sin\varphi, \cos\theta)$:

$$|\boldsymbol{n}\rangle = \mathrm{e}^{\mathrm{i}\alpha(\boldsymbol{n})}T(g_{\boldsymbol{n}})|\psi_0\rangle. \tag{4.3.2}$$

Thus, in accordance with the general reasoning, the CS corresponds to a point of the two-dimensional sphere $S^2 = SO(3)/SO(2) = SU(2)/U(1)$, which is the orbit of the coadjoint representation and therefore may be considered as the phase space of a classical dynamical system, "the classical spin".

The phase factor $\exp(\mathrm{i}\alpha(\boldsymbol{n}))$ may be chosen equal to unity, so

$$|\boldsymbol{n}\rangle = D(\boldsymbol{n})|\psi_0\rangle, \quad \text{where} \tag{4.3.3}$$

$$D(\boldsymbol{n}) = T^j(g_{\boldsymbol{n}}) = \exp[\mathrm{i}\theta(\boldsymbol{mJ})], \quad 0 \le \theta < \pi \tag{4.3.4}$$

and $\boldsymbol{m}$ is a unit vector, orthogonal both to $\boldsymbol{n}$ and $\boldsymbol{n}_0=(0,0,1)$: $\boldsymbol{m}=(\sin\varphi, -\cos\varphi, 0)$. Note that this definition of $\boldsymbol{m}$ is valid for any $\boldsymbol{n}$, excluding that corresponding to the south pole, $\boldsymbol{n}=(0,0,-1)$.

Here we present another form for the operator $D(\boldsymbol{n})$, analogous to (1.1.9),

$$D(\boldsymbol{n})=D(\xi)=\exp(\xi J_+ - \bar{\xi} J_-). \tag{4.3.4'}$$

Note that though the operators $D(\boldsymbol{n})$ do not form a group, their multiplication law may be written within $SU(2)$ as

$$D(\boldsymbol{n}_1)D(\boldsymbol{n}_2)=D(\boldsymbol{n}_3)\exp(\mathrm{i}\Phi(\boldsymbol{n}_1,\boldsymbol{n}_2)J_0). \tag{4.3.5}$$

Straightforward calculations show that the quantity $\Phi(\boldsymbol{n}_1,\boldsymbol{n}_2)$ here is just the area of the geodesical triangle on the sphere, with the vertices at the points $\boldsymbol{n}_0, \boldsymbol{n}_1, \boldsymbol{n}_2$:

$$\Phi(\boldsymbol{n}_1,\boldsymbol{n}_2)=A(\boldsymbol{n}_0,\boldsymbol{n}_1,\boldsymbol{n}_2). \tag{4.3.6}$$

As in Sect. 1.1, this fact reveals a semiclassical meaning of the constructed CS system.

For the systems in view, any vector $|j,\mu\rangle$ with an arbitrary μ may be taken as the fundamental vector $|\psi_0\rangle$. At first sight, all such CS systems are equivalent. However, as mentioned in Sect. 2.4, for the vectors $|j,\pm j\rangle$ the dispersion of the quadratic Casimir operator $\boldsymbol{J}^2=J_1^2+J_2^2+J_3^2$ is minimal. Then the dispersion is

$$\Delta\boldsymbol{J}^2=(\Delta\boldsymbol{J}^2)_{\min}=j \tag{4.3.7}$$

so the states $|j,\pm j\rangle$ determine the system of CS, which are the closest to the classical states. Either one of two states may be taken as $|\psi_0\rangle$, since both lead to the same CS system. Within our formalism the state $|\psi_0\rangle=|j,-j\rangle$ is more suitable. Another possibility would be to take $|\psi_0\rangle=|l,0\rangle$ for integer $j=l$. We will not consider this system now, since a similar system is considered in Chap. 6 for the group $SO(2,1)\sim SU(1,1)$.

The CS system constructed on this basis has a larger symmetry than one constructed with $|\psi_0\rangle=|j,\mu\rangle$, $\mu\neq\pm j$. To verify this statement, let us consider the complexification $\mathscr{G}^c$ of the Lie algebra $\mathscr{G}$, i.e., the set of linear combinations of the basis operators J_1, J_2, J_3 with complex coefficients. This is the Lie algebra for the Lie group $G^c=SL(2,\mathbb{C})$, the group of complex 2×2 matrices with unit determinant.

It is not difficult to see that the isotropy subalgebra $\mathscr{B}$ in $\mathscr{G}^c$ for the state $|\psi_0\rangle=|j,-j\rangle$ is generated by the elements J_0, J_-, and the corresponding group B is that of the lower triangular matrices,

$$B_-=\{b\}, \quad b=\begin{pmatrix}\delta^{-1} & 0\\ \gamma & \delta\end{pmatrix}. \tag{4.3.8}$$

Therefore such a coherent state may be parametrized by a complex number ζ.

Actually, from the Gaussian decomposition (4.1.6)

$$T(g)|\psi_0\rangle = T(z_+)T(h)T(z_-)|\psi_0\rangle = e^{i\varphi} N T(z_+)|\psi_0\rangle,$$

$$z_+ = \begin{pmatrix} 1 & \zeta \\ 0 & 1 \end{pmatrix}, \qquad |\psi_0\rangle = |j, -j\rangle.$$

Hence

$$N^{-2} = \langle\psi_0|T((z_+)^+)T(z_+)|\psi_0\rangle, \qquad N = (1+|\zeta|^2)^{-j}. \tag{4.3.9}$$

So the phase factor $\exp(i\Phi)=1$ for the operators $T(g)$ which are $D(\boldsymbol{n})$, (4.3.4), and then

$$|\boldsymbol{n}\rangle \to |\zeta\rangle = (1+|\zeta|^2)^{-j} T(z_+)|j, -j\rangle,$$

$$|\zeta\rangle = (1+|\zeta|^2)^{-j} \exp(\zeta J_+)|j, -j\rangle \tag{4.3.10}$$

$$\zeta = -\tan\frac{\theta}{2} e^{-i\varphi}, \qquad \boldsymbol{n} = (\sin\theta\cos\varphi, \sin\theta\sin\varphi, \cos\theta).$$

Thus, as mentioned above, the relation between $\boldsymbol{n}$ and ζ is given by the stereographical projection. Another form of the operator $D(\boldsymbol{n})$ is

$$D(\xi) = \exp(\xi J_+ - \bar{\xi} J_-), \qquad \xi = i\frac{\theta}{2}(m_1 - im_2) = -|\xi| e^{-i\varphi}. \tag{4.3.11}$$

By means of the Gaussian decomposition (4.1.6) this operator may be rewritten in the "normal form"

$$D(\xi) = \exp(\zeta J_+)\exp(\eta J_0)\exp(\zeta' J_-), \qquad \text{where} \tag{4.3.12}$$

$$\zeta = -\tan\frac{\theta}{2} e^{-i\varphi}, \qquad \eta = -2\ln\cos|\xi| = \ln(1+|\zeta|^2), \qquad \zeta' = -\bar{\zeta}. \tag{4.3.13}$$

The "antinormal form" of the operator is

$$D(\xi) = \exp(\zeta' J_-)\exp(-\eta J_0)\exp(\zeta J_+). \tag{4.3.14}$$

The parameters arc given in (4.3.13).

Since the parameters ζ, η, ζ' involved in (4.3.12, 14) are independent of j, it is sufficient to verify these formulae at $j=1/2$, when $\boldsymbol{J} = 1/2\boldsymbol{\sigma}$, and $\boldsymbol{\sigma}$ are the Pauli matrices.

Expanding the exponential in (4.3.10) and using (4.2.4), one gets the decomposition of the CS over the orthonormalized basis, cf. (1.2.36),

$$|\zeta\rangle = \sum_{\mu=-j}^{j} u_\mu(\zeta)|j, \mu\rangle, \; u_\mu(\zeta) = \left[\frac{(2j)!}{(j+\mu)!(j-\mu)!}\right]^{1/2} (1+|\zeta|^2)^{-j}\zeta^{j+\mu} \tag{4.3.15}$$

or, in angular variables,

$$|\zeta\rangle = \sum_{\mu=-j}^{j} u_\mu(\theta,\varphi)|j,\mu\rangle,$$

$$u_\mu(\theta,\varphi)=\left(\frac{(2j)!}{(j+\mu)!(j-\mu)!}\right)^{1/2}\left(-\sin\frac{\theta}{2}\right)^{j+\mu}\left(\cos\frac{\theta}{2}\right)^{j-\mu}$$
$$\cdot\exp[-\mathrm{i}(j+\mu)\varphi]. \tag{4.3.16}$$

The expression for the CS in z representation is

$$\langle z|\zeta\rangle=\psi_\zeta(z)=(1+|\zeta|^2)^{-j}(1+\zeta z)^{2j}. \tag{4.3.17}$$

Note that the spin CS is an eigenvector for the operator $(\boldsymbol{nJ})$

$$(\boldsymbol{nJ})|\boldsymbol{n}\rangle=-j|\boldsymbol{n}\rangle, \tag{4.3.18}$$

and besides

$$(J_-+2\zeta J_0-\zeta^2 J_+)|\zeta\rangle=0. \tag{4.3.19}$$

Formula (4.3.18) is a direct consequence of

$$J_0|\boldsymbol{n}_0\rangle=-j|\boldsymbol{n}_0\rangle, \qquad \boldsymbol{n}_0=(0,0,1)$$

because of

$$D(\boldsymbol{n})\,J_0 D^{-1}(\boldsymbol{n})=(\boldsymbol{nJ}). \tag{4.3.20}$$

The proof of (4.3.19) is analogous. Formulae (4.3.18, 19) determine the spin CS up to a phase factor $\exp(\mathrm{i}\alpha)$. The spin CS system has all the properties of the standard CS system, described in Sect. 1.2. Here I present a list of the properties, omitting the evident proofs.

1. An operator $T^j(g)$ transforms any CS into another state of the system,

$$T^j(g)|\boldsymbol{n}\rangle=\exp[\mathrm{i}\Phi(\boldsymbol{n},g)]|\boldsymbol{n}_g\rangle, \qquad \text{where} \tag{4.3.21}$$

$$\Phi(\boldsymbol{n},g)=jA(\boldsymbol{n}_0,\boldsymbol{n},\boldsymbol{n}_g). \tag{4.3.22}$$

For an infinitesimal transformation here

$$T^j(\mathrm{I}+\delta g)|\zeta\rangle=\exp(\mathrm{i}\delta\Phi)|\zeta+\delta\zeta\rangle \qquad \text{and} \tag{4.3.21$'$}$$

$$\mathrm{i}\delta\Phi=j\left(\frac{\partial F(\zeta,\bar\zeta)}{\partial\zeta}\,\delta\zeta-\frac{\partial F}{\partial\bar\zeta}\,\delta\bar\zeta\right), \qquad F=\ln(1+|\zeta|^2). \tag{4.3.22$'$}$$

Thus $\Theta=\delta\Phi$ is the connection form for the one-dimensional fiber bundle over the sphere S^2, and the fiber is a circle [81]. Note that $\Theta=\int\omega$, where

$$\omega=d\Theta \tag{4.3.22''}$$

is the area 2-form. Hence (4.3.22) follows.

2. The system of spin CS is complete.
3. The coherent states are not mutually orthogonal,

$$\langle \boldsymbol{n}_1|\boldsymbol{n}_2\rangle=\exp\left[\mathrm{i}\Phi(\boldsymbol{n}_1,\boldsymbol{n}_2)\right]\left(\frac{1+\boldsymbol{n}_1\boldsymbol{n}_2}{2}\right)^j,$$

$$|\langle \boldsymbol{n}_1|\boldsymbol{n}_2\rangle|^2=\left(\frac{1+\boldsymbol{n}_1\boldsymbol{n}_2}{2}\right)^{2j} \tag{4.3.23}$$

$$\langle\xi|\eta\rangle=\left[(1+|\xi|^2)(1+|\eta|^2)\right]^{-j}(1+\bar{\xi}\eta)^{2j},$$

$$|\langle\xi|\eta\rangle|^2=\left(\frac{(1+\bar{\xi}\eta)(1+\xi\bar{\eta})}{(1+|\xi|^2)(1+|\eta|^2)}\right)^{2j}\quad \text{where} \tag{4.3.24}$$

$$\Phi(\boldsymbol{n}_1,\boldsymbol{n}_2)=jA(\boldsymbol{n}_0,\boldsymbol{n}_1,\boldsymbol{n}_2),\qquad \Phi(\xi,\eta)=\frac{j}{2\mathrm{i}}\ln\left(\frac{1+\bar{\xi}\eta}{1+\xi\bar{\eta}}\right).$$

It is quite natural that Φ is given by the area of the spherical triangle. Actually, the sphere S^2 can be considered as the phase manifold for spin, and the spin CS corresponds to semiclassical states. Note that the area 2-form coincides with the symplectic 2-form, cf. (4.2.8).

4. The spin CS minimize the Heisenberg uncertainty relation; the inequality

$$\langle J_1^2\rangle\langle J_2^2\rangle\geq\tfrac{1}{4}\langle J_3\rangle^2 \tag{4.3.25}$$

is saturated for the state $|\boldsymbol{n}_0\rangle$. Respectively, for the transformed angular momentum operator components

$$\tilde{J}_k=D(\boldsymbol{n})\,J_k\,D^{-1}(\boldsymbol{n}), \tag{4.3.26}$$

the uncertainty relation

$$\langle\tilde{J}_1^2\rangle\langle\tilde{J}_2^2\rangle\geq\tfrac{1}{4}\langle\tilde{J}_3\rangle^2 \tag{4.3.27}$$

is minimized for the state $|\boldsymbol{n}\rangle$.

5. "Resolution of unity" may be written as

$$\frac{2j+1}{4\pi}\int d\boldsymbol{n}|\boldsymbol{n}\rangle\langle\boldsymbol{n}|=\hat{1}. \tag{4.3.28}$$

If the points of the complex ζ plane are used to parametrize the CS system, then

$$\int d\mu_j(\zeta)|\zeta\rangle\langle\zeta|=\hat{1}, \quad \text{where} \tag{4.3.29}$$

$$d\mu_j(\zeta)=\frac{2j+1}{\pi}\,\frac{d^2\zeta}{(1+|\zeta|^2)^2}\,. \tag{4.3.30}$$

6. Using the above formulae, one is able to decompose an arbitrary state over the coherent states,

$$|\psi\rangle=\sum_\mu c_\mu|j,\mu\rangle, \qquad |\psi\rangle=\int d\mu_j(\zeta)\psi(\bar{\zeta})|\zeta\rangle, \tag{4.3.31}$$

where

$$\psi(\bar{\zeta})=\langle\zeta|\psi\rangle=(1+|\zeta|^2)^{-j}\tilde{\psi}(\bar{\zeta})$$

$$\tilde{\psi}(\bar{\zeta})=\sum_{\mu=-j}^{j} c_\mu u_\mu(\bar{\zeta}), \qquad u_\mu(\bar{\zeta})=\left(\frac{(2j)!}{(j+\mu)!(j-\mu)!}\right)^{1/2}\bar{\zeta}^{j+\mu}. \tag{4.3.32}$$

Here $\tilde{\psi}(\bar{\zeta})$ is a polynomial in $\bar{\zeta}$ of an order $m\leq 2j$.

A consequence of these formulae is that for any function $f(\zeta)=P_m(\zeta)/(1+|\zeta|^2)^j$, where $P_m(\zeta)$ is an arbitrary polynomial of degree $m\leq 2j$, there is a state vector $|\psi\rangle$ given by

$$f(\zeta)=\langle\psi|\zeta\rangle, \qquad \int d\mu_j(\eta)\langle\xi|\eta\rangle f(\eta)=f(\xi). \tag{4.3.33}$$

Recall that the set of such functions is just the Hilbert space $\mathscr{F}^j$ describing states of a spin-j system. Using (4.3.31, 32), it is not difficult to prove that the subsystem $\zeta_1,\ldots,\zeta_{2j+1}$ is complete for any choice of noncoinciding points $\zeta_1,\ldots,\zeta_{2j+1}$. It is evident, since any polynomial of degree $2j$ is determined completely by its values at $2j+1$ points.

7. The infinitesimal operators in the CS representation are

$$\langle\zeta|J_0|\zeta\rangle=-j\,\frac{1-|\zeta|^2}{1+|\zeta|^2}, \qquad \langle\zeta|J_+|\zeta\rangle=2j\,\frac{\bar{\zeta}}{1+|\zeta|^2} \tag{4.3.34}$$

or, equivalently,

$$\langle\boldsymbol{n}|\boldsymbol{J}|\boldsymbol{n}\rangle=-j\boldsymbol{n}. \tag{4.3.35}$$

8. It is remarkable that the spin CS approach the standard CS system in the high spin limit, and all the formulae valid for the $SU(2)$ group are transformed to the corresponding formulae of the W_1 group. To verify this fact one has to perform the substitution

$$J_+=(2j)^{1/2}a^+, \qquad \zeta=(2j)^{-1/2}\alpha, \qquad J_-=(2j)^{1/2}a, \tag{4.3.36}$$

and let j go to infinity. For example,

$$|\zeta\rangle \Rightarrow \lim_{j\to\infty} (1+|\alpha|^2/2j)^{-j} \exp(\alpha a^+)|\psi_0\rangle = |\alpha\rangle. \tag{4.3.37}$$

9. The CS representation is appropriate for describing operators, in particular the spin density matrix. In fact, the density matrix is determined completely by its symbols $P(\boldsymbol{n})$, or $Q(\boldsymbol{n})$, defined by

$$\varrho = \int d\mu_j(\boldsymbol{n}) P(\boldsymbol{n}) |\boldsymbol{n}\rangle\langle \boldsymbol{n}|, \qquad d\mu_j(\boldsymbol{n}) = \frac{2j+1}{4\pi}\, d\boldsymbol{n}, \tag{4.3.38}$$

$$Q(\boldsymbol{n}) = \langle \boldsymbol{n}|\varrho|\boldsymbol{n}\rangle. \tag{4.3.39}$$

Unlike the case of W_N considered above, here $P(\boldsymbol{n})$ and $Q(\boldsymbol{n})$ exist for any operator A since the representations of $SU(2)$ are finite-dimensional. The decomposition of these functions over the spherical harmonics $Y_{lm}(\boldsymbol{n})$ contains only those terms with $l \leq 2j$. For example,

$$P(\boldsymbol{n}) = \sum_{l,m} C_{lm} Y_{lm}(\boldsymbol{n}). \tag{4.3.40}$$

Hence we get the decomposition of the density matrix,

$$\varrho = \sum_{l,m} C_{lm} \hat{P}_{lm}, \qquad \text{where} \tag{4.3.41}$$

$$\hat{P}_{lm} = \int d\mu_j(\boldsymbol{n})\, Y_{lm}(\boldsymbol{n}) |\boldsymbol{n}\rangle\langle \boldsymbol{n}|.$$

Calculating the involved integral, we get

$$\langle j, \nu'|\hat{P}_{lm}|j, \nu\rangle = \sqrt{\frac{2l+1}{4\pi}}\, \langle j, \nu'; l, m|j, \nu\rangle \langle j, -j; l, 0|j, -j\rangle, \tag{4.3.42}$$

where $\langle j, \nu'; l, m|j, \nu\rangle$ are the Clebsch-Gordan coefficients.

10. Here is a list of the simplest examples, taken from [82]

Operator	$Q(\boldsymbol{n})$	$P(\boldsymbol{n})$
J_3	$-j\cos\theta$	$-(j+1)\cos\theta$
J_1	$-j\sin\theta\cos\varphi$	$-(j+1)\sin\theta\cos\varphi$
J_2	$-j\sin\theta\sin\varphi$	$-(j+1)\sin\theta\sin\varphi$
J_3^2	$j(j-\frac{1}{2})\cos^2\theta+\frac{1}{2}j$	$(j+1)(j+\frac{3}{2})\cos^2\theta-\frac{1}{2}(j+1)$
J_1^2	$j(j-\frac{1}{2})(\sin\theta\cos\varphi)^2+\frac{1}{2}j$	$(j+1)(j+\frac{3}{2})(\sin\theta\cos\varphi)^2-\frac{1}{2}(j+1)$
J_2^2	$j(j-\frac{1}{2})(\sin\theta\sin\varphi)^2+\frac{1}{2}j$	$(j+1)(j+\frac{3}{2})(\sin\theta\sin\varphi)^2-\frac{1}{2}(j+1)$

For the operator $T(\tau, \boldsymbol{v}) = \exp[\mathrm{i}\tau(\boldsymbol{vJ})]$, $\boldsymbol{v}^2 = 1$, the function $Q(\boldsymbol{n})$ is

$$Q(\boldsymbol{n}) = (\cos\tau/2 + \mathrm{i}\sin\tau/2\cos\theta)^{2j}, \qquad \boldsymbol{nv} = \cos\theta. \tag{4.3.43}$$

Hence we get an integral representation for the character of the representation $T^j(g)$:

$$\chi_j(g)=\chi_j(\tau)=\frac{2j+1}{4\pi}\int(\cos\tau/2+\mathrm{i}\sin\tau/2\cos\theta)^{2j}\,d\boldsymbol{n}. \tag{4.3.44}$$

A similar expression arises also for the operator $\exp[-\beta(\boldsymbol{vJ})]$

$$Q(\boldsymbol{n})=(\cosh\beta/2+\sinh\beta/2\cos\theta)^{2j}. \tag{4.3.45}$$

11. When j is an integer, $j=l$, another realization of the representation space $\mathscr{H}^l$ may be used, namely, the space of the eigenfunctions of the Laplace operator on the sphere $S^2=\{\boldsymbol{v}:\boldsymbol{v}^2=1\}$:

$$-\Delta_{\theta\varphi}f(\boldsymbol{v})=l(l+1)f(\boldsymbol{v}),\qquad \boldsymbol{v}=(\sin\theta\cos\varphi,\sin\theta\sin\varphi,\cos\theta). \tag{4.3.46}$$

In this realization the coherent states are

$$\langle\boldsymbol{v}|\boldsymbol{n}\rangle=(\boldsymbol{e}_1-\mathrm{i}\boldsymbol{e}_2,\boldsymbol{v})^l. \tag{4.3.47}$$

Here $\boldsymbol{e}_1$ and $\boldsymbol{e}_2$ are unit vectors, mutually orthogonal and lying in the plane normal to the vector $\boldsymbol{n}$. Hence one easily gets an expression for the generating function for the spherical functions $Y_{lm}(\theta,\varphi)$:

$$K(z;\theta,\varphi)=\sum_{l,m}u_{lm}(z)\,Y_{lm}(\theta,\varphi),\quad u_{lm}(z)=\sqrt{\frac{(2l)!}{(l-m)!(l+m)!}}\,z^{l+m} \tag{4.3.48}$$

$$K(z;\theta,\varphi)=C_l(\sin\theta\,\mathrm{e}^{-\mathrm{i}\varphi}+2z\cos\theta+z^2\sin\theta\,\mathrm{e}^{\mathrm{i}\varphi})^l,$$

$$C_l=\frac{\sqrt{(2l+1)!}}{\sqrt{4\pi}}\,\frac{1}{2^l\,l!}. \tag{4.3.49}$$

12. The CS system may be used to determine the generating function for the representation matrix elements $T^j_{\mu\nu}=\langle\mu|T^j(g)|\nu\rangle$:

$$\langle\xi|T^j(g)|\eta\rangle=[(1+|\xi|^2)(1+|\eta|^2)]^{-j}K(\xi,\eta)$$

$$K(\xi,\eta)=\sum_{\mu,\nu}\bar{u}_{j\mu}(\xi)u_{j\nu}(\eta)\,T^j_{\mu\nu}(g),$$

$$u_{j\mu}(\xi)=\sqrt{\frac{(2j)!}{(j+\mu)!(j-\mu)!}}\,\xi^{j+\mu}. \tag{4.3.50}$$

Evidently,

$$K(\xi,\eta)=(\bar{\alpha}+\beta\xi-\bar{\beta}\eta+\alpha\xi\eta)^{2j}. \tag{4.3.51}$$

(Other generating functions for the matrix elements $T^j_{\mu\nu}$ are presented in [78].)

5. The Most Elementary Noncompact Non-Abelian Simple Lie Group: $SU(1,1)$

CS systems for the simplest noncompact non-Abelian Lie group $SU(1,1)$ $\sim SO(2,1)$ are introduced and studied in this chapter, following [15]. Unlike the case of the simplest compact non-Abelian Lie group $SU(2)$, there are several types of CS systems, since $SU(1,1)$ has several series of unitary irreducible representations. Some applications of the CS systems for $SU(1,1)$ are presented in Part III.

5.1 Group $SU(1,1)$ and Its Representations

5.1.1 Fundamental Properties of $SU(1,1)$

The group $SU(1,1)$ consists of all unimodular 2×2 matrices leaving invariant the Hermitian form $|z_1|^2-|z_2|^2$. Evidently, elements of $SU(1,1)$ are parametrized with a pair of complex numbers,

$$g=\begin{pmatrix}\alpha & \beta\\ \bar{\beta} & \bar{\alpha}\end{pmatrix}, \quad |\alpha|^2-|\beta|^2=1. \tag{5.1.1}$$

Group $SU(1,1)$ is locally isomorphic to $SO(2,1)$, the group of "rotations" of the three-dimensional pseudo-Euclidean space, also named the three-dimensional Lorentz group. To be more precise, $SO(2,1)=SU(1,1)/\mathbb{Z}_2$, where $\mathbb{Z}_2$ is the cyclic group containing two elements, $\mathbb{Z}_2=\{\mathrm{I},-\mathrm{I}\}$, I is the unit matrix. It is locally isomorphic also to the real symplectic group $\mathrm{Sp}(2,\mathbb{R})$, as well as to $SL(2,\mathbb{R})$, the group of real 2×2 matrices with the unit determinant. This group is considered in detail and its representations are described exhaustively in [83] and in a number of monographs [78, 79, 84]. Here I present only an overview of those properties necessary for our purpose.

Note the following substantial differences between the groups $SU(1,1)$ and $SU(2)$. First, $SU(1,1)$ is noncompact, while $SU(2)$ is compact. Second, $SU(2)$ is simply connected, while $SU(1,1)$ is not.

By definition, the group is simply connected if any closed contour in it may be continuously deformed into a point. Meanwhile, one may show that in $SU(1,1)$ a loop corresponding to the rotation by $2\pi n$, n is an integer, $0\leq t\leq 2\pi n$,

$$g(t)=\mathrm{diag}\,[\exp(\mathrm{i}t/2), \exp(-\mathrm{i}t/2)]$$

cannot be deformed continuously into a point. This fact is quite evident, because group G has the topological structure of the direct product $D\times S^1$, where D is the planar disk bounded with the unit circle S^1. Hence the fundamental group of G is not trivial: $\pi_1(G)=\mathbb{Z}$.

It is known that by combining a sufficient number of copies (sometimes called sheets) of a group G, which is not simply connected, and by splicing the copies appropriately, one may construct a simply connected group $\tilde{G}$, called the universal covering of G. Here $\tilde{G}$ contains an infinite number of sheets.

Our method is similar to that used for $SU(2)$. The group $G=SU(1,1)$ is embedded into $G^c=SL(2,\mathbb{C})$ and the Gaussian decomposition (4.1.6) is valid:

$$g=z_+hz_-,\qquad z_+\in Z_+,\qquad z_-\in Z_-,\qquad h\in H^c. \tag{5.1.2}$$

As in that case, the group G acts in Z_+ and in Z_-. For instance, its action in Z_- is given by

$$g=\begin{pmatrix}\alpha & \beta\\ \bar{\beta} & \bar{\alpha}\end{pmatrix}:\quad z\to z_g=\frac{\alpha z+\bar{\beta}}{\beta z+\bar{\alpha}}. \tag{5.1.3}$$

However, in contrast to the case of $SU(2)$, now the action of the group is not transitive; the complex z plane is foliated into three orbits:

$$\begin{aligned}&1.\ X_+=\{z:|z|<1\},\\ &2.\ X_-=\{z:|z|>1\}\\ &3.\ X_0=\{z:|z|=1\},\end{aligned} \tag{5.1.4}$$

i.e., the interiority of the unit circle, the circle itself, and the external region.

This statement is easily verified, since according to (5.1.3) the action of any element g leads to the transformation

$$(1-|z|^2)\to(1-|z_g|^2)=|\beta z+\bar{\alpha}|^{-2}(1-|z|^2).$$

In this construction the spaces X_+ and X_- may be considered as those obtained from the upper sheet of the two-sheet hyperboloid and from the one-sheet hyperboloid, respectively, embedded into the three-dimensional space with the pseudo-Euclidean metrics $s^2=x_0^2-x_1^2-x_2^2$, through the stereographical projection.

Let us consider, for example, the space X_+. It is not difficult to see that, as in Sect. 4.1, this space may be identified with the set of elements of the group G, written as

$$g=\begin{pmatrix}\alpha & \beta\\ \bar{\beta} & \alpha\end{pmatrix},\qquad \alpha^2-\beta_1^2-\beta_2^2=1,\qquad \bar{\alpha}=\alpha>0.$$

Setting

$$\alpha=\cosh\frac{\tau}{2},\qquad \beta=\sinh\frac{\tau}{2}\,\mathrm{e}^{-\mathrm{i}\varphi},\qquad \tau>0$$

we see that X_+ is isomorphic to the upper sheet of the hyperboloid $\{(\alpha,\beta_1,\beta_2):\alpha^2-\beta_1^2-\beta_2^2=1,\alpha>0\}$, i.e., to the set of unit (in the pseudo-Euclidean metrics) vectors, which are of the form

$$\boldsymbol{n}=(\cosh\tau,\sinh\tau\cos\varphi,\sinh\tau\sin\varphi),$$

$$\boldsymbol{n}^2=n_0^2-n_1^2-n_2^2=1,\qquad n_0>0.$$

An element of the space X_+ is written as

$$g_{\boldsymbol{n}}=\begin{pmatrix}\cosh\frac{\tau}{2} & \sinh\frac{\tau}{2}\,\mathrm{e}^{-\mathrm{i}\varphi}\\ \sinh\frac{\tau}{2}\,\mathrm{e}^{\mathrm{i}\varphi} & \cosh\frac{\tau}{2}\end{pmatrix}=\exp\left[\tau(m_1\sigma_1+m_2\sigma_2)/2\right],\tag{5.1.5}$$

where $\boldsymbol{m}=(0,\sin\varphi,-\cos\varphi)$, and $\sigma_1=\begin{pmatrix}0&1\\1&0\end{pmatrix}$, $\sigma_2=\begin{pmatrix}0&-\mathrm{i}\\ \mathrm{i}&0\end{pmatrix}$ are the Pauli matrices. Thus the matrix $g_{\boldsymbol{n}}$ describes a hyperbolic rotation around the vector $\boldsymbol{m}=(0,\sin\varphi,-\cos\varphi)$, the rotation "angle" being τ.

As in the case of $SU(2)$ (Sect. 4.1) the Gaussian decomposition leads to an isomorphism between these spaces. For example, the isomorphism between the unit circle $X_+=\{\zeta:|\zeta|<1\}$, and the upper sheet of the hyperboloid $\mathbb{H}^2=\{\boldsymbol{n}:\boldsymbol{n}^2=n_0^2-n_1^2-n_2^2=1,n_0>0\}$ is established through

$$\zeta=\tanh\frac{\tau}{2}\,\mathrm{e}^{-\mathrm{i}\varphi},\qquad \boldsymbol{n}=(\cosh\tau,\sinh\tau\cos\varphi,\sinh\tau\sin\varphi).\tag{5.1.6}$$

As in Sect. 4.1, the considered construction provides a complex structure on the upper sheet of the hyperboloid $\mathbb{H}^2$. Thus the upper sheet of the hyperboloid may be treated as a noncompact complex manifold.

The G-invariant metrics on $\mathbb{H}^2$ is

$$ds^2=d\boldsymbol{n}\,d\boldsymbol{n}=\frac{4\,d\zeta\cdot d\bar{\zeta}}{(1-|\zeta|^2)^2}\tag{5.1.7}$$

while the closed G-invariant 2-form is

$$\omega=2\mathrm{i}\,\frac{d\zeta\wedge d\bar{\zeta}}{(1-|\zeta|^2)^2}\,.\tag{5.1.8}$$

It is remarkable that $\mathbb{H}^2$ is a Kählerian manifold, as well as the sphere S^2.

Therefore (4.1.18′, 19′) are in analogy to

$$ds^2 = 4 \frac{\partial^2 F}{\partial \zeta \partial \bar{\zeta}} d\zeta \cdot d\bar{\zeta} \tag{5.1.9}$$

$$\omega = 2\mathrm{i} \frac{\partial^2 F}{\partial \zeta \partial \bar{\zeta}} d\zeta \wedge d\bar{\zeta}, \quad \text{where} \tag{5.1.10}$$

$$F = -\ln(1 - |\zeta|^2). \tag{5.1.11}$$

The group $SU(1,1)$ is noncompact, so unlike the case of $SU(2)$ all its unitary irreducible representations are infinite-dimensional. This group has a number of series of unitary irreducible representations: principal, discrete and supplementary. We shall consider here only representations of discrete and principal series.

5.1.2 Discrete Series

There are two discrete series of representations T^+ and T^-. It is sufficient to discuss anyone of them, since all the results can be transferred at once to the other.

The representations of the discrete series are infinite-dimensional, but in many aspects they are analogous to finite-dimensional representations of $SU(2)$. For instance, a basis vector $|m\rangle$ in representation space may be fixed by an integer m, running from zero to infinity.

The Lie algebra corresponding to the Lie group $SU(1,1)$ has three generators K_1, K_2 and K_0 as its basis elements. The commutation relations are

$$[K_1, K_2] = -\mathrm{i}K_0, \quad [K_2, K_0] = \mathrm{i}K_1, \quad [K_0, K_1] = \mathrm{i}K_2. \tag{5.1.12}$$

As for $SU(2)$, another basis is appropriate:

$$K_\pm = \pm\mathrm{i}(K_1 \pm \mathrm{i}K_2), \quad K_0, \tag{5.1.13}$$

and the corresponding commutation relations are

$$[K_0, K_\pm] = \pm K_\pm, \quad [K_-, K_+] = 2K_0. \tag{5.1.14}$$

It is not difficult to verify that the quadratic operator

$$\hat{C}_2 = K_0^2 - K_1^2 - K_2^2 = K_0^2 - \tfrac{1}{2}(K_+K_- + K_-K_+) \tag{5.1.15}$$

is invariant (the Casimir operator), i.e., it commutes with every K_j. By virtue of Schur's lemma, for any irreducible representation this operator is unity times a number,

$$\hat{C}_2 = k(k-1)\hat{1}. \tag{5.1.16}$$

Thus a representation of $SU(1,1)$ is determined by a single number k; for the discrete series this number acquires discrete values, $k=1, 3/2, 2, 5/2, \ldots$. (The corresponding representations of the universal covering group $\widetilde{SU}(1,1)$ are also given by a single number k, but it goes continuously on the positive axis from 0 to ∞.) Basis vectors $|k,\mu\rangle$ in the space where the representation $T^k(g)$ acts are marked by a number μ, which is the eigenvalue of the operator K_0:

$$K_0|k,\mu\rangle=\mu|k,\mu\rangle, \qquad \mu=k+m, \tag{5.1.17}$$

where m is an integer, $m\geq 0$.

As in Sect. 4.2, we consider a suitable realization of the representation $T^k(g)$ in the space of functions $\mathscr{F}^k=\{f(z)\}$ analytical inside the unit circle and which satisfy the condition

$$\|f\|_k<\infty,$$
$$\|f\|_k^2=\frac{2k-1}{\pi}\int_D |f(z)|^2(1-|z|^2)^{2k-2}d^2z, \qquad D=\{z:|z|<1\}. \tag{5.1.18}$$

The action of the operators $T^k(g)$ in the space $\mathscr{F}^k$ is given by

$$T^k(g)f(z)=(\beta z+\bar{\alpha})^{-2k}f(z_g), \qquad z_g=\frac{\alpha z+\bar{\beta}}{\beta z+\bar{\alpha}}. \tag{5.1.19}$$

In this realization the generators $K_\pm$ and K_0 act as first-order differential operators:

$$\tilde{K}_+=z^2\frac{d}{dz}+2kz \tag{5.1.20}$$

$$\tilde{K}_-=\frac{d}{dz} \tag{5.1.21}$$

$$\tilde{K}_0=z\frac{d}{dz}+k. \tag{5.1.22}$$

As for $SU(2)$, the representation $T^k(g)$ can be written in the semiclassical form

$$T^k(g)f(z)=\exp[\mathrm{i}S(g,z)]f(z_g). \tag{5.1.23}$$

The phase $S(g,z)$ here is an analog of the classical action, and it is given in (4.2.8), where the Kähler potential should be put in the form $F=-k\ln(1-|z|^2)$.

5.1.3 Principal (Continuous) Series

Besides the discrete series, two principal (or continuous) series of representations of $SU(1,1)$ and a supplementary series exist. We are concerned with the class–I representations, i.e., those containing vector $|\psi_0\rangle$ in the representation space, invariant under the action of the maximal compact subgroup K of group G. Here $K=U(1)$,

$$K=\{k: k=\operatorname{diag}[\exp(i\varphi/2), \exp(-i\varphi/2)]\}.$$

One of the principal series and the supplementary series have this property. Representations of the principal series are realized in the space of functions on the orbit of type X_0, i.e., the space of functions of the unit circle. Any class–I representation is specified by a nonnegative number λ; its realization is given by

$$T^\lambda(g)f(z)=|\beta z+\bar{\alpha}|^{-1+2i\lambda}f(z_g),$$
$$z_g=(\alpha z+\bar{\beta})(\beta z+\bar{\alpha})^{-1}, \qquad z=e^{i\theta}. \tag{5.1.24}$$

It can easily be verified that the operators $T^\lambda(g)$ do indeed form a representation of the group G, and that for the usual choice of the scalar product in the functional space

$$\langle f_1|f_2\rangle=\frac{1}{2\pi}\int_0^{2\pi} f_1(\theta)f_2(\theta)d\theta, \qquad f(z)\equiv f(\theta), \tag{5.1.25}$$

this representation is unitary. The Casimir operator $\hat{C}_2=K_0^2-K_1^2-K_2^2$ is a multiple of the unity operator $\hat{C}_2=-(\lambda^2+\frac{1}{4})\hat{1}$.

For representations of the supplementary series the action of the operator $T^\lambda(g)$ is given in the same manner, by (5.1.24), where λ should be a purely imaginary number: $\lambda=i\tau$, $-1/2<\tau<1/2$. Then the scalar product is given by a more complicated expression [78]

$$\langle f_1|f_2\rangle=N\int_0^{2\pi}\int_0^{2\pi}\left|\sin\frac{\theta_1-\theta_2}{2}\right|^{-2-2\sigma}\bar{f}_1(\theta_1)f_2(\theta_2)d\theta_1 d\theta_2$$
$$\sigma=-\tfrac{1}{2}+\tau, \qquad -1<\sigma<0, \qquad N=\frac{\Gamma(\sigma+1)\Gamma(-\sigma)}{2^{2\sigma+4}\pi^2\Gamma(-2\sigma-1)}. \tag{5.1.26}$$

The representation space is infinite-dimensional; in this space one can choose the basis consisting of eigenvectors of the operator K_0, which is the infinitesimal operator for the subgroup K,

$$K_0|\lambda,\mu\rangle=\mu|\lambda,\mu\rangle, \qquad \mu=0,\pm 1,\pm 2,\ldots. \tag{5.1.27}$$

The operators $K_\pm=\pm i(K_1\pm iK_2)$ act in this basis in a simple way,

$$K_+|\lambda,\mu\rangle=(\tfrac{1}{2}-i\lambda+\mu)|\lambda,\mu+1\rangle$$
$$K_-|\lambda,\mu\rangle=(-\tfrac{1}{2}+i\lambda+\mu)|\lambda,\mu-1\rangle. \tag{5.1.28}$$

In the above formulae $K_{1,2}$ are infinitesimal operators for the representation $T(g)$ corresponding to the elements $i\sigma_{1,2}/2$ of the Lie algebra $\mathcal{G}$. For the considered realization of $T^\lambda(g)$ the infinitesimal operators $K_\pm$, K_0 are written as the first-order differential operators,

$$K_0 = -\mathrm{i}d/d\theta, \qquad K_\pm = -\mathrm{i}e^{\pm \mathrm{i}\theta} d/d\theta \mp (-\tfrac{1}{2} + \mathrm{i}\lambda) e^{\pm \mathrm{i}\theta}. \tag{5.1.29}$$

5.2 Coherent States

5.2.1 Discrete Series

Applying the operators of $T^k(g)$ to a fixed vector $|\psi_0\rangle$ gives a CS system.

It is suitable to take the vector $|k, k\rangle$ as the fixed vector $|\psi_0\rangle$, since in this case the isotropy subalgebra $\mathscr{B}$ in $\mathscr{G}^c$ is maximal, so that such coherent states are closest to the classical states.

Further procedure is analogous to that described in Sect. 4.3. As is well known, any operator in the representation may be written as

$$T^k(g) = T^k(g_{\boldsymbol{n}}) T^k(h), \qquad h \in H.$$

Hence the coherent state is fixed by a unit pseudo-Euclidean vector

$$\boldsymbol{n} = (\cosh \tau, \sinh \tau \cos \varphi, \sinh \tau \sin \varphi),$$

$$|\boldsymbol{n}\rangle = \exp[\mathrm{i}\alpha(\boldsymbol{n})] T^k(g_{\boldsymbol{n}}) |\psi_0\rangle. \tag{5.2.1}$$

This parametrization of the CS system agrees with the general fact that a CS is determined by a point in the coset space G/H, which is the upper sheet of the two-sheet hyperboloid in our case, $\mathbb{H}^2 = \{\boldsymbol{n}; \boldsymbol{n}^2 = n_0^2 - n_1^2 - n_2^2 = 1, n_0 > 0\}$. The phase factor $\exp[\mathrm{i}\alpha(\boldsymbol{n})]$ may be taken equal to unity, so that

$$|n\rangle = D(n)|\psi_0\rangle, \qquad \text{where}$$

$$D(n) = T^k(g_n) = \exp[\mathrm{i}\tau(\boldsymbol{m}\boldsymbol{K})], \qquad \tau > 0 \tag{5.2.2}$$

and $\boldsymbol{m}$ is a unit vector orthogonal both to $\boldsymbol{n}$ and to $\boldsymbol{n}_0 = (1, 0, 0)$: $\boldsymbol{m} = (0, \sin \varphi, -\cos \varphi)$,

The operator $D(\boldsymbol{n})$ may also be written in another form, which is similar to that in (4.3.2),

$$D(\boldsymbol{n}) = D(\xi) = \exp(\xi K_+ - \bar{\xi} K_-), \qquad \xi = \mathrm{i}\frac{\tau}{2}(m_1 - \mathrm{i}m_2). \tag{5.2.3}$$

Note that the operators $D(\boldsymbol{n})$ do not form a group, but the multiplication law is

$$D(\boldsymbol{n}_1) D(\boldsymbol{n}_2) = D(\boldsymbol{n}_3) \exp(\mathrm{i}\Phi(\boldsymbol{n}_1 \boldsymbol{n}_2) K_0). \tag{5.2.4}$$

Direct calculation shows that the function $\Phi(\boldsymbol{n}_1, \boldsymbol{n}_2)$ is just the area $A(\boldsymbol{n}_0, \boldsymbol{n}_1, \boldsymbol{n}_2)$

of the geodesic triangle on the hyperboloid with vertices at the points $\boldsymbol{n}_0, \boldsymbol{n}_1, \boldsymbol{n}_2$:

$$\Phi(\boldsymbol{n}_1, \boldsymbol{n}_2) = A(\boldsymbol{n}_0, \boldsymbol{n}_1, \boldsymbol{n}_2). \tag{5.2.5}$$

This indicates the semiclassical structure of the CS system considered, Sect. 4.3.

The operator $D(\boldsymbol{n})$ may be written in the "normal form"

$$D(\xi) = \exp(\zeta K_+)\exp(\eta K_0)\exp(\zeta' K_-), \text{ where} \tag{5.2.6}$$

$$\zeta = \tanh|\xi|e^{i\varphi}, \quad \eta = 2\ln\cosh|\xi| = -\ln(1-|\zeta|^2), \quad \zeta' = -\bar{\zeta}. \tag{5.2.7}$$

The "antinormal" form of the operator is

$$D(\xi) = \exp(\zeta' K_-)\exp(-\eta K_0)\exp(\zeta K_+), \tag{5.2.8}$$

where the parameters ζ, η and ζ' are given in (5.2.7). Since the parameters are independent of k, it is sufficient to verify the above formulae for $K_0 = \sigma_3/2$, $K_{1,2} = i\sigma_{1,2}/2$, where σ_j are the Pauli matrices.

Applying the operator $D(\xi)$ to the state vector $|\psi_0\rangle$ and using the normal form (5.2.6), we get another representation (another parameter system) for the coherent states:

$$|\zeta\rangle = (1-|\zeta|^2)^k \exp(\zeta K_+)|0\rangle. \tag{5.2.9}$$

Expanding the exponential and using

$$|k, k+m\rangle = \left[\frac{\Gamma(2k)}{m!\,\Gamma(m+2k)}\right]^{1/2} (K_+)^m |k, k\rangle, \tag{5.2.10}$$

we obtain the decomposition of the CS over the orthonormal basis:

$$|\zeta\rangle = (1-|\zeta|^2)^k \sum_{m=0}^{\infty} \left(\frac{\Gamma(m+2k)}{m!\,\Gamma(2k)}\right)^{1/2} \zeta^m |k, k+m\rangle. \tag{5.2.11}$$

Another expression for the CS in the z representation is

$$\langle z|\zeta\rangle = \psi_\zeta(z) = (1-|\zeta|^2)^k (1-\zeta z)^{-2k}. \tag{5.2.12}$$

It is notable that the considered CS is an eigenvector for the operator

$$(\boldsymbol{n}\boldsymbol{K}) = n_0 K_0 - n_1 K_1 - n_2 K_2 \tag{5.2.13}$$

and, besides,

$$(K_- - 2\zeta K_0 + \zeta^2 K_+)|\zeta\rangle = 0. \tag{5.2.14}$$

The proof of these two relations arises from the definition $K_0|\boldsymbol{n}_0\rangle = k|\boldsymbol{n}_0\rangle$, $\boldsymbol{n}_0 = (1, 0, 0)$, and from the relation

$$D(\boldsymbol{n}) K_0 D^{-1}(\boldsymbol{n}) = (\boldsymbol{n}\boldsymbol{K}). \tag{5.2.15}$$

The equalities (5.2.13, 14) determine the coherent state up to an indefinite phase factor $\exp(\mathrm{i}\alpha)$.

The CS system obtained has all the properties of the spin CS system described in Chap. 4. Here we mention some of them, numbered as in Chap. 4.

3′) The CS are not mutually orthogonal

$$|\langle \boldsymbol{n}_1|\boldsymbol{n}_2\rangle|^2=\left(\frac{1+(\boldsymbol{n}_1\boldsymbol{n}_2)}{2}\right)^{-2k}. \tag{5.2.16}$$

5′) At $k>\frac{1}{2}$ the resolution of unity is

$$\int d\mu_k(\zeta)|\zeta\rangle\langle\zeta|=\hat{1}, \quad \text{where} \tag{5.2.17}$$

$$d\mu_k(\zeta)=\frac{2k-1}{\pi}\,\frac{d^2\zeta}{(1-|\zeta|^2)^2}. \tag{5.2.18}$$

10′) The generating function for the matrix elements of the operator $T^k(g)$ is

$$G(\xi,\eta;g)=\sum_{m,n=0}^{\infty} T^k_{k+m,k+n}(g)\bar{u}_m(\xi)u_n(\eta)$$

$$=(\alpha\bar{\xi}\eta+\beta\bar{\xi}+\bar{\beta}\eta+\bar{\alpha})^{-2k}, \quad \text{where} \tag{5.2.19}$$

$$u_m(\xi)=\left(\frac{\Gamma(m+2k)}{m!\,\Gamma(2k)}\right)^{1/2}\xi^m.$$

The last item to be considered in this section is a useful realization of the representations of $SU(1,1)$ by means of the operators bilinear in the boson creation and annihilation operators a^+ and a, satisfying the canonical commutation relations $[a,a^+]=1$. (This realization is possible because the groups $SU(1,1)$ and $\mathrm{Sp}(2,\mathbb{R})$ are isomorphic. An analogous realization for $\mathrm{Sp}(2N,\mathbb{R})$ is treated in Chap. 8.) A calculation shows that three operators

$$K_+=\tfrac{1}{2}(a^+)^2, \quad K_-=\tfrac{1}{2}a^2, \quad K_0=\tfrac{1}{4}(aa^+ +a^+a) \tag{5.2.20}$$

satisfy the commutation relations (5.1.14). For the realization (5.2.20), the Casimir operator (5.1.15) is

$$\hat{C}_2=-\tfrac{3}{16}\hat{1}=k(k-1)\hat{1}, \tag{5.2.21}$$

that corresponds to two solutions

$$k=\tfrac{1}{4} \quad \text{and} \quad k=\tfrac{3}{4}. \tag{5.2.22}$$

It is not difficult to see that the states $|n\rangle=(n!)^{-1/2}(a^+)^n|0\rangle$ for even n are a basis for the unitary representation space of group $SU(1,1)$ corresponding to $k=\frac{1}{4}$; respectively, the states with odd n provide with a basis for the case $k=\frac{3}{4}$.

The representation matrix elements for $SU(1,1)$ are expressed in terms of the hypergeometrical function [83]. In [85] a simpler expression was obtained for the representations considered, corresponding to $k=\frac{1}{4}$ and $k=\frac{3}{4}$:

$$T^k_{mn}(\tau)=\left[\frac{n_<!}{n_>!}\right]^{1/2}\left(\cosh\frac{\tau}{2}\right)^{-1/2}\mathrm{P}^{|m-n|/2}_{(m+n)/2}\left(\left(\cosh\frac{\tau}{2}\right)^{-1}\right), \tag{5.2.23}$$

where P^m_n is the associated Legendre function, $n_<=\min(m,n)$, $n_>=\max(m,n)$.

It is possible to introduce two sorts of bosonic operators, say, a_+ and a_-, satisfying the canonical commutation relations. The corresponding representations belonging to the discrete series for the group $SU(1,1)$ are constructed with the following bilinear operators

$$K_+=a_+^+a_-^+, \quad K_-=a_+a_-, \quad K_0=\tfrac{1}{2}(a_+^+a_+ +a_-^+a_- +1) \tag{5.2.24}$$

which satisfy the commutation relations (5.1.14), as may be verified directly. The corresponding Casimir operator (5.1.16) is

$$\hat{C}_2=-\tfrac{1}{4}+\tfrac{1}{4}(a_+^+a_+ -a_-^+a_-)^2. \tag{5.2.25}$$

Thus for the states

$$|m,n\rangle=(m!n!)^{-1/2}(a_+^+)^m(a_-^+)^n|0,0\rangle$$

with $m-n=n_0=\text{const}$, $\hat{C}_2=k(k-1)\hat{\mathrm{I}}$, $k=\frac{1}{2}(1+|n_0|)$. So the states $\{|n+n_0,n\rangle\}$ with a fixed n_0 form a basis for an irreducible representation T^k of $SU(1,1)$ belonging to the discrete series, and

$$k=\tfrac{1}{2}(1+|n_0|). \tag{5.2.26}$$

Any representation of the discrete series for $SU(1,1)$ can be realized in this way. The matrix elements are known; for the simplest case $k=\frac{1}{2}$ where the vacuum is the initial state, one has

$$|\langle n,n|T^{1/2}(g)|0,0\rangle|^2=(1-\varrho)\varrho^n, \quad \varrho=\frac{|\beta|^2}{|\alpha|^2}. \tag{5.2.27}$$

The purpose of this section was to describe the CS systems related to representations of $SU(1,1)$ belonging to the discrete series. The CS systems for representations of discrete series for other Lie groups are discussed in Sect. 12.1.

5.2.2 Principal (Continuous) Series

Let us consider a unitary irreducible class–I representation $T(g)$ of the group $G=SU(1,1)$, and let $|\psi_0\rangle$ be the vector in the representation space, invariant under the action of $K=U(1)$, the maximal compact subgroup. Applying the operators $T(g)$ to this vector gives a CS system parametrized by points of the coset space $X=G/K=SU(1,1)/U(1)$, which is the Lobachevsky plane (An analogous CS system can be constructed for representations $T^j(g)$ of $SU(2)$ in the case where $j=l$ is an integer. A CS system for the imaginary Lobachevsky plane was constructed by *Molchanov* [86].) It can be realized, say, as the interior of the unit disk,

$$X=D=\{\zeta:|\zeta|<1\}. \tag{5.2.28}$$

Selecting an element $g_\zeta \in G$ in the equivalence class $\zeta \in X=G/K$, we get the CS system

$$|\zeta\rangle=T(g_\zeta)|0\rangle, \quad |0\rangle=|\psi_0\rangle. \tag{5.2.29}$$

It is suitable to take the element

$$g_\zeta=\frac{1}{\sqrt{1-|\zeta|^2}}\begin{pmatrix}1 & -\zeta \\ -\bar{\zeta} & 1\end{pmatrix}, \quad |\zeta\rangle=T(g_\zeta)|0\rangle \tag{5.2.30}$$

as a representative of the class. Another equivalent definition of the CS system is

$$|\zeta\rangle=\exp(\xi K_+-\bar{\xi}K_-)|0\rangle, \quad \xi=|\xi|e^{i\varphi}, \quad \zeta=\tanh|\xi|e^{i\varphi}. \tag{5.2.31}$$

The CS system is overcomplete and not orthogonal. It has a number of remarkable properties; some of which are listed below.

1. States $|\zeta\rangle$ are normalized: $\langle\zeta|\zeta\rangle=1$.
2. Group G acts on the coherent state set transitively:

$$T^\lambda(g)|\zeta\rangle=T^\lambda(g)T^\lambda(g_\zeta)|0\rangle=T^\lambda(gg_\zeta)|0\rangle=|\zeta_g\rangle, \tag{5.2.32}$$

where

$$\zeta_g=\zeta'=\frac{\alpha\zeta-\beta}{-\bar{\beta}\zeta+\bar{\alpha}}; \quad gg_\zeta=g_{\zeta'}k, \quad k\in K. \tag{5.2.33}$$

3. States $|\zeta_1\rangle$ and $|\zeta_2\rangle$ are, in general, nonorthogonal to each other. Their scalar product is determined by the function $\Phi_\lambda(\tau)$, the so-called zonal spherical function

$$\langle\zeta_1|\zeta_2\rangle=\langle 0|T^\lambda(g_{\zeta_1}^{-1}g_{\zeta_2})|0\rangle=\Phi_\lambda(\tau). \tag{5.2.34}$$

Here τ is the distance between points ζ_1 and ζ_2 in the standard Lobachevsky metric. It is determined by

$$\cosh\frac{\tau}{2}=\frac{1}{\sqrt{1-|\zeta_1|^2}}\,\frac{1}{\sqrt{1-|\zeta_2|^2}}\,|1-\bar{\zeta}_1\zeta_2|. \tag{5.2.35}$$

Function $\Phi_\lambda(\tau)$, in turn, is given by

$$\Phi_\lambda(\tau)=\langle 0|T^\lambda(a(\tau))|0\rangle, \tag{5.2.36}$$

$$a(\tau)=\begin{pmatrix}\alpha & \beta\\ \beta & \alpha\end{pmatrix}, \quad \alpha=\cosh\frac{\tau}{2}, \quad \beta=-\sinh\frac{\tau}{2}, \quad \tau>0. \tag{5.2.37}$$

4. Let $\{|\mu\rangle\equiv|k,\mu\rangle\}$ be the orthonormal basis in the representation space $\mathscr{H}^\lambda$ given by the eigenvectors of the operator K_0:

$$K_0|\mu\rangle=\mu|\mu\rangle. \tag{5.2.38}$$

Expanding coherent states in this basis gives

$$|\zeta\rangle=\sum_{n=-\infty}^{\infty} u_n(\zeta)|n\rangle, \quad \text{where} \tag{5.2.39}$$

$$u_n(\zeta)=\langle n|\zeta\rangle=\langle n|T^\lambda(g_\zeta)|0\rangle=u_n(\tau,\varphi), \quad \zeta=-\tanh\frac{\tau}{2}\,e^{i\varphi}. \tag{5.2.40}$$

The state $|\zeta\rangle$ is normalized, therefore

$$\sum_{n=-\infty}^{\infty} |u_n^\lambda(\zeta)|^2=1. \tag{5.2.41}$$

Note also that

$$u_n^\lambda(0,\varphi)=\delta_{n0} \tag{5.2.42}$$

$$u_n^\lambda(\tau,\varphi)=e^{-in\varphi}R_n^\lambda(\tau). \tag{5.2.43}$$

5. Since functions $u_n^\lambda(\tau,\varphi)$ are matrix elements of the operator $T^\lambda(g)$, they are eigenfunctions of the Laplace-Beltrami operator for the Lobachevsky plane

$$-\tilde{\Delta}u_n(\tau,\varphi)=\Lambda u_n(\tau,\varphi), \quad \text{where} \tag{5.2.44}$$

$$\tilde{\Delta}=\frac{\partial^2}{\partial\tau^2}+\cosh\tau\,\frac{\partial}{\partial\tau}+\frac{1}{\sinh^2\tau}\,\frac{\partial^2}{\partial\varphi^2}. \tag{5.2.45}$$

For representations of the principal series

$$\Lambda=\tfrac{1}{4}+\lambda^2 \tag{5.2.46}$$

and for representations of supplementary series

$$\Lambda=\tfrac{1}{4}-\sigma^2, \quad -\tfrac{1}{2}<\sigma<\tfrac{1}{2}. \tag{5.2.47}$$

6. The zonal spherical function $u_0^\lambda(\tau, \varphi)$, as we have seen, is independent of $\varphi : u_0^\lambda(\tau, \varphi) = \Phi_\lambda(\tau)$ is the eigenfunction of the radial part of the Laplace-Beltrami operator

$$-\left(\frac{d^2}{d\tau^2} + \coth\tau \frac{d}{d\tau}\right) \Phi_\lambda(\tau) = \left(\lambda^2 + \frac{1}{4}\right) \Phi_\lambda(\tau). \tag{5.2.48}$$

It satisfies the normalization condition

$$\Phi_\lambda(0) = 1. \tag{5.2.49}$$

7. Functions $u_n^\lambda(\tau, \varphi)$ are eigenfunctions of commuting self-adjoint operators $\tilde{\Delta}$ and $p_\varphi = -\mathrm{i}\partial/\partial\varphi$ and form an orthogonal system of functions on the Lobachevsky plane. For the class–I principal series

$$\frac{1}{2\pi} \int_0^\infty \int_0^{2\pi} \bar{u}_n^\lambda(\tau, \varphi) u_{n'}^{\lambda'}(\tau, \varphi) \sinh\tau \, d\tau \, d\varphi = N_n(\lambda) \delta_{nn'} \delta(\lambda - \lambda'). \tag{5.2.50}$$

8. Coefficient $N_n(\lambda)$ is determined by the asymptotic behavior of the function $R_n^\lambda(\tau)$ at $\tau \to \infty$

$$R_n^\lambda(\tau) \sim [c_n(\lambda) \mathrm{e}^{\mathrm{i}\lambda\tau} + c_n(-\lambda) \mathrm{e}^{-\mathrm{i}\lambda\tau}] \mathrm{e}^{-\tau/2} \tag{5.2.51}$$

and, as easily seen,

$$N_n(\lambda) = N_n(-\lambda) = \pi |c_n(\lambda)|^2. \tag{5.2.52}$$

9. It is known also that coefficient $N_0(\lambda)$ determines the Plancherel measure for representation of the principal series [78]. Namely, an arbitrary function $f(\tau)$ can be expanded in the integral over $\Phi_\lambda(\tau)$

$$f(\tau) = \int_0^\infty \hat{f}(\lambda) \Phi_\lambda(\tau) d\mu(\lambda), \qquad \text{where} \tag{5.2.53}$$

$$d\mu(\lambda) = \frac{d\lambda}{N_0(\lambda)} = \lambda \tanh \pi\lambda \, d\lambda, \tag{5.2.54}$$

and $\hat{f}(\lambda)$ can be found from

$$\hat{f}(\lambda) = \int_0^\infty \Phi_\lambda(\tau) f(\tau) \sinh\tau \, d\tau. \tag{5.2.55}$$

10. Using (5.2.36) one can easily obtain an expression for the generating function of the matrix elements $T_{mn}^\lambda(g) = \langle m | T^\lambda(g) | n \rangle$. To this end, we consider a matrix element $\langle \xi | T^\lambda(g) | \eta \rangle$. It can be easily seen that

$$\langle \xi | T^\lambda(g) | \eta \rangle = \Phi_\lambda(\tau), \tag{5.2.56}$$

where τ is the distance between points ξ_g and η

$$\cosh\frac{\tau}{2} = \frac{1}{\sqrt{1-|\xi|^2}}\,\frac{1}{\sqrt{1-|\eta|^2}}\,|\bar{\alpha}+\beta\bar{\xi}-\bar{\beta}\eta-\alpha\bar{\xi}\eta|. \tag{5.2.57}$$

Let us now consider a realization of the representation $T^\lambda(g)$. It acts in the space of square integrable functions on the unit circle $\{z: z=e^{i\theta}\}$:

$$T^\lambda(g)f(z) = |\beta z+\bar{\alpha}|^{-1+2i\lambda} f(z_g), \qquad z_g = \frac{\alpha z+\bar{\beta}}{\beta z+\bar{\alpha}}. \tag{5.2.58}$$

It can be readily seen that the Hilbert space vector $f_0(z)\equiv 1$ is invariant with respect to the maximal compact subgroup. The CS system for the principal series is obtained with the action of the operator $T^\lambda(g)$ applied to f_0,

$$\begin{aligned} T^\lambda(g_\zeta)f_0 = f_\zeta(z) &= |\beta z+\bar{\alpha}|^{-1+2i\lambda} \\ &= (1-|\zeta|^2)^{1/2-i\lambda}|1-\zeta z|^{-1+2i\lambda}. \end{aligned} \tag{5.2.59}$$

Going over with the help of stereographic projection from the plane ζ to the upper sheet of hyperboloid $\{\boldsymbol{n}: \boldsymbol{n}^2=1, n_0>0\}$ we get

$$f_\zeta(z) \to f_{\tau,\varphi}(\theta) = [\cosh\tau - \sinh\tau\cos(\varphi-\theta)]^{-1/2+i\lambda}. \tag{5.2.59'}$$

For representations of the supplementary series

$$f_\zeta(z) = (1-|\zeta|^2)^{-\sigma}|1-\bar{\zeta}z|^{2\sigma}, \qquad -1<\sigma<0, \qquad \text{where} \tag{5.2.60}$$

$$\zeta = -\beta/\bar{\alpha}, \qquad |\zeta|<1, \qquad z=e^{i\theta}, \qquad |z|=1. \tag{5.2.61}$$

Thus, in accord with the general theory, any CS is determined by a point of the coset space $X=G/K$, which is the Lobachevsky plane $\{\zeta: |\zeta|<1\}$. Note some properties of functions $f_\zeta(z)$.

11. At fixed $z=e^{i\theta}$ function $\Phi_\theta(\zeta)=f_\zeta(\theta)$ is constant on horocycles of the Lobachevsky plane which in this case are circles tangent to the circle $\{\zeta: |\zeta|=1\}$ at the point $\zeta=z=e^{i\theta}$ [84, 87]. The horocycle equation is of the form

$$\frac{1+|\zeta|^2-2|\zeta|\cos(\theta-\varphi)}{1-|\zeta|^2} = [\cosh\tau-\sinh\tau\cos(\varphi-\theta)] = \text{const}. \tag{5.2.62}$$

Thus the kernel $K(\zeta,z)=f_\zeta(z)$ can be called the horospherical kernel and, consequently, when realizing representation $T^\lambda(g)$, the CS system is given by the horospherical kernels.

This statement remains valid also for representations of the principal series of an arbitrary Lie group (Part II).

12. We have

$$|f_\zeta(z)|^2 = \frac{1-|\zeta|^2}{1+|\zeta|^2-2|\zeta|\cos(\theta-\varphi)} = P(\zeta, z), \tag{5.2.63}$$

where $P(\zeta, z)$ is the Poisson kernel for the disk $\{\zeta : |\zeta| < 1\}$. Thus

$$f_\zeta(z) = [P(\zeta, z)]^{1/2-i\lambda} = \Phi_z(\zeta)$$

$$\Phi_z^\lambda(\zeta) \equiv \Phi_\theta^\lambda(\tau, \varphi) = [\cosh\tau - \sinh\tau\cos(\theta-\varphi)]^{-1/2+i\lambda}. \tag{5.2.64}$$

13. At fixed z the function $\Phi_z^\lambda(\zeta) = f_\zeta^\lambda(z)$ is an eigenfunction for the Laplace–Beltrami operator $\tilde{\Delta} = (1-|\zeta|^2)^2\Delta$ (Δ is the usual Laplace operator) on the Lobachevsky plane,

$$-\tilde{\Delta}\Phi_z^\lambda(\zeta) = (\lambda^2 + \tfrac{1}{4})\Phi_z^\lambda(\zeta). \tag{5.2.65}$$

As such functions are constant on horocycles, which are analogs of straight lines in the Euclidean plane, the CS are in a sense generalizations of the standard plane waves, $\Phi_{\boldsymbol{k}}(\boldsymbol{r}) = \exp(i\boldsymbol{kr}) = \exp[ikr\cos(\theta-\varphi)]$.

14. The orthonormalized basis state vectors $\{|n\rangle\}$, which are eigenvectors for the operators $T(k)$, $k \in K$ [$K = U(1)$ is the maximal compact subgroup of $SU(1,1)$] correspond to the functions

$$f_n(z) = z^n = e^{in\theta}. \tag{5.2.66}$$

15. Decomposing the coherent states over this basis gives an integral representation

$$u_n^\lambda(\zeta) = \frac{1}{2\pi}\int_0^{2\pi} e^{-in\theta}(1-|\zeta|^2)^{1/2-i\lambda}|1-\zeta z|^{-1+2i\lambda}d\theta$$

$$= \frac{1}{2\pi}\int_0^{2\pi} e^{-in\theta}\Phi_\theta^\lambda(\zeta)d\theta. \tag{5.2.67}$$

Applying once more the stereographical projection $\zeta = \tanh\frac{\tau}{2}e^{i\varphi}$, one gets

$$u_n^\lambda(\tau, \varphi) = \frac{1}{2\pi}\int_0^{2\pi}(\cosh\tau - \sinh\tau\cos(\theta-\varphi))^{-1/2+i\lambda}e^{-in\theta}d\theta. \tag{5.2.68}$$

Thus we have reconstructed the familiar integral representation for the matrix elements $T_{n0}(g)$ [78].

Note that for all values of λ and n functions $u_n^\lambda(\tau, \varphi)$ are a complete system of functions on the upper sheet of the hyperboloid $\mathbb{H}^2 = \{\boldsymbol{\xi} : \boldsymbol{\xi}^2 = \xi_0^2 - \xi_1^2 - \xi_2^2 = 1, \xi_0 > 0\}$.

16. For zonal spherical functions ($n=0$) this representation is somewhat simplified

$$\Phi_\lambda(\tau)=\frac{1}{\pi}\int_0^\pi (\cosh\tau-\sinh\tau\cos\varphi)^{-1/2+i\lambda}d\varphi. \tag{5.2.69}$$

17. The CS system satisfies the orthogonality relations

$$\frac{1}{(2\pi)^2}\int_0^\infty\int_0^{2\pi}\Phi_\theta^\lambda(\tau,\varphi)\Phi_{\theta'}^{\lambda'}(\tau,\varphi)\sinh\tau\,d\tau\,d\varphi=N_0(\lambda)\delta(\lambda-\lambda')\delta(\theta-\theta') \tag{5.2.70}$$

$$\frac{1}{(2\pi)^2}\int_0^\infty\int_0^{2\pi}\Phi_\theta^\lambda(\tau,\varphi)\Phi_\theta^\lambda(\tau',\varphi')d\mu(\lambda)d\theta=(\sinh\tau)^{-1}\delta(\tau-\tau')\delta(\varphi-\varphi'),$$

$$d\mu(\lambda)=[N_0(\lambda)]^{-1}d\lambda=\lambda'\tanh\pi\lambda\,d\lambda. \tag{5.2.71}$$

18. Making use of these relations we can expand an arbitrary function on the Lobatschevsky plane over the basis $\Phi_\theta^\lambda(\tau,\varphi)$

$$f(\tau,\varphi)=\frac{1}{2\pi}\int\hat{f}(\lambda,\theta)\Phi_\theta^\lambda(\tau,\varphi)d\theta\,d\mu(\lambda), \quad \text{where} \tag{5.2.72}$$

$$\hat{f}(\lambda,\theta)=\frac{1}{2\pi}\int\bar{\Phi}_\theta^\lambda(\tau,\varphi)\,\mathrm{f}\,(\tau,\varphi)\sinh\tau\,d\tau\,d\varphi. \tag{5.2.73}$$

19. Functions $\Phi_\theta^\lambda(\tau,\varphi)$ are closely related to a scattering problem. To see this, let us rewrite the Laplace–Beltrami operator as

$$\begin{aligned}\tilde{\Delta}&=\frac{\partial^2}{\partial\tau^2}+c\tanh\tau\frac{\partial}{\partial\tau}+\frac{1}{\sinh^2\tau}\frac{\partial^2}{\partial\varphi^2}\\&=\frac{1}{\sqrt{\sinh\tau}}\frac{\partial^2}{\partial\tau^2}\sqrt{\sinh\tau}+\frac{1}{4\sinh^2\tau}-\frac{1}{4}+\frac{1}{\sinh^2\tau}\frac{\partial^2}{\partial\varphi^2}.\end{aligned} \tag{5.2.74}$$

Hence it follows that the function $\psi_n^\lambda(\tau)=\sqrt{\sinh\tau}\,u_n^\lambda(\tau)$ satisfies the equation

$$\left(-\frac{d^2}{d\tau^2}+\frac{n^2-1/4}{\sinh^2\tau}\right)\psi_n^\lambda(\tau)=\lambda^2\psi_n^\lambda(\tau) \tag{5.2.75}$$

which is the Schrödinger equation describing the scattering problem with the potential $V(\tau)=(n^2-\frac{1}{4})\sinh^{-2}\tau$. The asymptotic behavior of functions $\psi_n^\lambda(\tau)$ at $\tau\to\infty$ is written as

$$\psi_n^\lambda(\tau)\sim c_n(\lambda)e^{i\lambda\tau}+c_n(-\lambda)e^{-i\lambda\tau}. \tag{5.2.76}$$

The S matrix of this problem is

$$S_n(\lambda) = -\frac{c_n(\lambda)}{c_n(-\lambda)}. \tag{5.2.77}$$

20. A simple method to find $c_n(\lambda)$ and $S_n(\lambda)$ is to go over to the limit $\tau \to \infty$ under the integral. For instance,

$$u_0^\lambda(\tau) = \Phi_\lambda(\tau) = \frac{1}{\pi}\int_0^\pi (\cosh\tau - \sinh\tau\cos\varphi)^{-1/2+\mathrm{i}\lambda}\,d\varphi$$

$$\sim [c_0(\lambda)\mathrm{e}^{\mathrm{i}\lambda\tau} + c_0(-\lambda)\mathrm{e}^{-\mathrm{i}\lambda\tau}]\mathrm{e}^{-\tau/2}, \qquad \tau \to \infty$$

$$c_0(\lambda) = \frac{1}{\pi}\int_0^\pi \left(\frac{1-\cos\varphi}{2}\right)^{-1/2+\mathrm{i}\lambda} d\varphi = \frac{1}{\sqrt{\pi}}\,\frac{\Gamma(\mathrm{i}\lambda)}{\Gamma(\mathrm{i}\lambda+\frac{1}{2})}. \tag{5.2.78}$$

21. Thus, the S matrix for the corresponding scattering problem is

$$S_0(\lambda) = -\frac{\Gamma(\mathrm{i}\lambda)\Gamma(\frac{1}{2}-\mathrm{i}\lambda)}{\Gamma(-\mathrm{i}\lambda)\Gamma(\frac{1}{2}+\mathrm{i}\lambda)}. \tag{5.2.79}$$

It has poles at the points $\lambda = \mathrm{i}n$, $n = 1, 2, 3, \ldots$ and $\lambda = -\mathrm{i}(n-\frac{1}{2})$, $n = 1, 2, \ldots$.

22. Note, finally, that from the integral representation (5.2.69) one can easily get an explicit expression for $\Phi_\lambda(\tau)$:

$$\Phi_\lambda(\tau) = F(a, b; c; -\sinh^2\tau),$$

$$2a = \tfrac{1}{2} + \mathrm{i}\lambda, \qquad 2b = \tfrac{1}{2} - \mathrm{i}\lambda, \qquad c = 1. \tag{5.2.80}$$

By means of a known transformation of the hypergeometric function one gets an expression appropriate for calculating the asymptotic behavior,

$$\Phi_\lambda(\tau) = (2\cosh\tau)^{-1/2}\left[(2\cosh\tau)^{\mathrm{i}\lambda}\frac{\Gamma(\mathrm{i}\lambda)}{\sqrt{\pi}\,\Gamma(\frac{1}{2}+\mathrm{i}\lambda)}\right.$$

$$\left.\cdot F\left(-\frac{\sigma}{2}, -\frac{\sigma-1}{2}, -\sigma+\frac{1}{2}; \cosh^{-2}\tau\right) + \text{c.c.}\right]$$

$$\sigma = -\tfrac{1}{2} + \mathrm{i}\lambda. \tag{5.2.81}$$

Hence one can easily get expressions for the S matrix and for the Plancherel measure.

6. The Lorentz Group: $SO(3,1)$

In this chapter we study the CS system for $SO(3,1)$, the Lorentz group. The states in this system are parametrized with points in the upper sheet of the two-sheeted hyperboloid in the Minkowski space. From another point of view, this CS system has been considered in [89]. It was found to be suitable for a number of problems in relativistic physics and representation theory.

6.1 Representations of the Lorentz Group

The Lorentz group and its representations have been described thoroughly in a number of books [77, 78, 84, 88], so here I present only the information necessary for the problem in view.

The Lorentz group is the group of linear transformations of four-dimensional space, leaving invariant the pseudo-Euclidean form $x^2=x_0^2-x_1^2-x_2^2-x_3^2$. It is isomorphic locally to the group $G=SL(2,\mathbb{C})=\{g\}$, i.e., the group of complex unimodular 2×2 matrices. The isomorphism is evident, if one compares the four-dimensional vector $x=(x_0,x_1,x_2,x_3)$ and the 2×2 matrix

$$\hat{x}=x_0\mathrm{I}+x_1\sigma_1+x_2\sigma_2+x_3\sigma_3, \tag{6.1.1}$$

where I is the unit 2×2 matrix, and σ_j are the Pauli matrices,

$$\sigma_1=\begin{pmatrix}0 & 1\\ 1 & 0\end{pmatrix},\quad \sigma_2=\begin{pmatrix}0 & -\mathrm{i}\\ \mathrm{i} & 0\end{pmatrix},\quad \sigma_3=\begin{pmatrix}1 & 0\\ 0 & -1\end{pmatrix}. \tag{6.1.2}$$

The matrix transformation $\hat{x}\to\hat{x}'=g\hat{x}g^+$ corresponds to the transformation in the vector space, leaving x^2 invariant. Two matrix transformations, g and $-g$, are mapped to the same vector transformation in $SO(3,1)$:

$$SO(3,1)=SL(2,\mathbb{C})/\mathbb{Z}_2, \tag{6.1.3}$$

where the cyclic group $\mathbb{Z}_2=\{\mathrm{I},-\mathrm{I}\}$is the center of $SL(2,\mathbb{C})$. Therefore one can consider $SL(2,\mathbb{C})$ instead of the Lorentz group.

It is known [77, 88] that two series of unitary irreducible representations of the Lorentz group exist, the principal series and the supplementary series. A representation belonging to the principal series is specified by a nonnegative halfinteger number j_0 and a real number λ: $T(g)=T^{j_0\lambda}(g)$. For $j_0=0$, the representations $T^{0\lambda}$ and $T^{0,-\lambda}$ are equivalent. When the representation $T^{j_0\lambda}$ is restricted to the maximal compact subgroup $K=SU(2)$, it becomes reducible, and the $SU(2)$-group representations $T^{j_0}, T^{j_0+1}, T^{j_0+2},\dots$ are present in its decomposition. For $j_0=0$ the representation space contains a vector $|\psi_0\rangle$ invariant under transformations belonging to the subgroup K, so it is a class–I representation, denoted by $T^{\lambda}(g)$.

Representations of the supplementary series correspond to $j_0=0$ and are specified by a purely imaginary parameter λ; $\lambda=\mathrm{i}\sigma$, $-1<\sigma<0$. All of them are class–I representations. Two standard realizations of the unitary irreducible representations are well known [84, 88].

First, they can be realized in the space $\mathscr{F}_\chi$, $\chi=(n_1, n_2)$, where n_1 and n_2 are complex numbers, such that (n_1-n_2) is an integer. The space $\mathscr{F}_\chi$ is the space of functions $f(z,\bar{z})$ satisfying two conditions.

1. Function $f(z,\bar{z})$ is infinite-differentiable at all values of z

$$f(z,\bar{z})\in C^{\infty}. \tag{6.1.4}$$

2. $$f(z,\bar{z})=z^{n_1-1}\bar{z}^{n_2-1}f\left(-\frac{1}{z}, -\frac{1}{\bar{z}}\right)\in C^{\infty}. \tag{6.1.5}$$

Representation $T^{\chi}(g)$ is determined by

$$T^{\chi}(g)f(z)=(\beta z+\delta)^{n_1-1}(\bar{\beta}\bar{z}+\bar{\delta})^{n_2-1}f(z_g)$$

$$z_g=(\alpha z+\gamma)(\beta z+\delta)^{-1}. \tag{6.1.6}$$

For representations of the principal series one has

$$n_1=(j_0+\mathrm{i}\lambda), \qquad n_2=(-j_0+\mathrm{i}\lambda) \tag{6.1.7}$$

and the scalar product is of the form

$$\langle f_1|f_2\rangle=\frac{1}{\pi}\int \bar{f}_1 f_2\, dx\, dy, \qquad z=x+\mathrm{i}y. \tag{6.1.8}$$

For representations of the supplementary series

$$n_1=n_2=\sigma, \qquad -1<\sigma<0, \tag{6.1.9}$$

and the scalar product is of a more complicated form,

$$\langle f_1|f_2\rangle=\frac{1}{\pi^2}\iint d^2z_1 d^2z_2 |z_1-z_2|^{-2-2\sigma}\bar{f}_1(z)f_2(z). \tag{6.1.10}$$

Thus for the class-I representations $n_1 = n_2 = n$, $n = i\lambda$, or $n = \sigma$, so

$$T^{(n)}(g) f(z) = |\beta z + \delta|^{2(-1+n)} f(z_g). \tag{6.1.6$'$}$$

It can be easily verified also that a normalized vector invariant under the maximal compact subgroup is of the form

$$f_0(z) = (1 + |z|^2)^{-1+n}. \tag{6.1.11}$$

To obtain the second realization of the representation, let us map the z plane on the sphere $\{\boldsymbol{n}: \boldsymbol{n}^2 = 1\}$, using the transformation inverse to stereographic projection

$$|z| = \tan\frac{\theta}{2}, \qquad \cos\theta = \frac{1-|z|^2}{1+|z|^2}, \qquad \sin\theta\, e^{i\varphi} = \frac{2z}{1+|z|^2}. \tag{6.1.12}$$

We obtain a realization of representations in the space of functions $f(\boldsymbol{n})$ on the unit sphere $S^2 = \{\boldsymbol{n}: \boldsymbol{n}^2 = 1\}$. Here

$$\|\psi\|^2 = \frac{1}{4\pi} \int |\psi(\boldsymbol{n})|^2 d\boldsymbol{n}, \tag{6.1.13}$$

where function $\psi(\boldsymbol{n})$ is related to function $f(z)$ by

$$f(z) = f_0(z)\psi(\boldsymbol{n}) = (1 + |z|^2)^{-1+i\lambda} \psi(\boldsymbol{n}). \tag{6.1.14}$$

Hence we get the transformation law of functions $\psi(\boldsymbol{n})$:

$$T^\lambda(g)\psi(\boldsymbol{n}) = (\xi_0 - \boldsymbol{\xi n})^{-1+i\lambda} \psi(\boldsymbol{n}_g), \tag{6.1.15}$$

where the four-dimensional vector $(\xi_0, \boldsymbol{\xi})$, $\xi_0^2 - \boldsymbol{\xi}^2 = 1$, $\xi_0 > 0$ is determined by

$$\begin{aligned}
\xi_0 &= \tfrac{1}{2}(|\alpha|^2 + |\beta|^2 + |\gamma|^2 + |\delta|^2) = \tfrac{1}{2}\operatorname{tr}\{g^+ g\}\\
\xi_1 &= \tfrac{1}{2}(\alpha\bar{\gamma} + \beta\bar{\delta} + \bar{\alpha}\gamma + \bar{\beta}\delta) = \tfrac{1}{2}\operatorname{tr}\{g^+ \sigma_1 g\}\\
\xi_2 &= +\tfrac{i}{2}(\alpha\bar{\gamma} - \beta\bar{\delta} - \bar{\alpha}\gamma - \bar{\beta}\delta) = \tfrac{1}{2}\operatorname{tr}\{g^+ \sigma_2 g\}\\
\xi_3 &= \tfrac{1}{2}(|\alpha|^2 + |\beta|^2 - |\gamma|^2 - |\delta|^2) = \tfrac{1}{2}\operatorname{tr}\{g^+ \sigma_3 g\}.
\end{aligned} \tag{6.1.16}$$

6.2 Coherent States

With the second realization of the representation, the vector invariant under the maximal compact subgroup is $\psi_0(\boldsymbol{n}) \equiv 1$, and acting on it by operators $T^\lambda(g)$ we obtain the CS system

$$|\xi\rangle \to \psi_\xi(\boldsymbol{n}) = (\xi_0 - \boldsymbol{\xi n})^{-1+i\lambda}. \tag{6.2.1}$$

Thus, coherent states are parametrized by points of the upper sheet of the two-sheet hyperboloid. This system of functions was considered originally (from another point of view) in [89] and turned out to be very convenient for considering a number of problems of the Lorentz group representation theory and of theoretical physics. *Gelfand* and *Graev* [84] related this approach to the horospherical transformation. The so-called horospherical method for the general case was formulated and complex semisimple Lie groups were considered in detail.

The CS system has all the properties listed in Chap. 5, with slight modifications. Mention some of them.

2′. Group G acts on the CS space transitively; and this action is given by

$$T^{\lambda}(g)|\xi\rangle=|\xi_g\rangle, \qquad \xi_g=g\xi, \tag{6.2.2}$$

where g is a real 4×4 matrix satisfying the condition

$$g'sg=s, \qquad s=\mathrm{diag}\,[1,-1,-1,-1]. \tag{6.2.3}$$

3′. The scalar product for two coherent states is given by

$$\langle\xi|\eta\rangle=\Phi_{\lambda}(\tau)=\frac{\sin\lambda\tau}{\lambda\sinh\tau}, \qquad \text{where} \tag{6.2.4}$$

$$\cosh\tau=\xi\eta=\xi_0\eta_0-\boldsymbol{\xi}\boldsymbol{\eta}. \tag{6.2.5}$$

4′. In the representation space $\mathscr{H}^{\lambda}$ there exists a basis $\{|l,m\rangle\}$; its elements are eigenvectors of the operators

$J^2=J_1^2+J_2^2+J_3^2$ and J_3:

$$J^2|l,m\rangle=l(l+1)|l,m\rangle, \qquad J_3|l,m\rangle=m|l,m\rangle \tag{6.2.6}$$

$$l=0,1,2,\dots, \qquad -l\leq m\leq l,$$

Here J_k is the operator of the infinitesimal rotation around the kth axis, $k=1,2,3$. The expansion of the CS over this basis is

$$|\xi\rangle=\sum_{l,m}u_{lm}^{\lambda}(\xi)|l,m\rangle, \qquad \text{where} \tag{6.2.7}$$

$$u_{lm}^{\lambda}(\xi)=\langle l,m|T^{\lambda}(g_{\xi})|0\rangle. \tag{6.2.8}$$

The state $|\xi\rangle$ is normalized, so

$$\sum_{l,m}|u_{lm}^{\lambda}(\xi)|^2=1. \tag{6.2.9}$$

Note also that

$$u_{lm}^{\lambda}(\xi)=R_l^{\lambda}(\tau)\,\bar{Y}_{lm}(\mathbf{v}). \tag{6.2.10}$$

Here $\mathbf{v}$ is a unit vector: $\boldsymbol{\xi}=\sinh\tau\mathbf{v}$, $Y_{lm}(\mathbf{v})$ are the standard normalized spherical functions $\int \bar{Y}_{lm}(\mathbf{v})\, Y_{l'm'}(\mathbf{v})\, d\mu(\mathbf{v})=\delta_{ll'}$, $d\mu(\mathbf{v})=\frac{1}{4\pi}\, d\mathbf{v}$, $Y_{00}\equiv 1$.

5′. Functions $u^{\lambda}_{lm}(\tau,\mathbf{v})$ are matrix elements of operator $T^{\lambda}(g)$, so they are eigenfunctions of the Laplace–Beltrami operator for the Lobatschevsky space

$$-\tilde{\Delta} u^{\lambda}_{lm}(\tau,\mathbf{v})=(\lambda^2+1)\, u^{\lambda}_{lm}(\tau,\mathbf{v}), \tag{6.2.11}$$

where

$$\tilde{\Delta}=\frac{\partial^2}{\partial\tau^2}+2\coth\tau\,\frac{\partial}{\partial\tau}+\frac{1}{\sinh^2\tau}\,\Delta_{\mathbf{v}}$$

$$\Delta_{\mathbf{v}}=\frac{\partial^2}{\partial\theta^2}+\cot\theta\,\frac{\partial}{\partial\theta}+\frac{1}{\sin^2\theta}\,\frac{\partial^2}{\partial\varphi^2}\,. \tag{6.2.12}$$

For representations of the supplementary series

$$\lambda=\mathrm{i}\sigma, \qquad -1<\sigma<1.$$

6′. The zonal spherical function $\Phi_\lambda(\tau)$ is the eigenfunction of the radial part of the Laplace–Beltrami operator

$$-\left(\frac{d^2}{d\tau^2}+2\coth\tau\,\frac{d}{d\tau}\right)\Phi_\lambda(\tau)=(\lambda^2+1)\,\Phi_\lambda(\tau). \tag{6.2.13}$$

7′. Functions $u^{\lambda}_{lm}(\tau,\mathbf{v})$ are eigenfunctions of commuting self-adjoint operators $\tilde{\Delta}$, $\mathbf{J}^2$ and J_3 and form an orthogonal system of functions in the Lobatschevsky space. For class–I representations of the principal series

$$\iint \bar{u}^{\lambda}_{lm}(\tau,\mathbf{v})\, u^{\lambda'}_{l'm'}(\tau,\mathbf{v})\sinh^2\tau\, d\tau\, d\mu(\mathbf{v})=N_l(\lambda)\,\delta_{ll'}\delta_{mm'}\delta(\lambda-\lambda'),$$

$$d\mu(\mathbf{v})=\frac{1}{4\pi}\, d\mathbf{v}. \tag{6.2.14}$$

8′. Coefficient $N_l(\lambda)$ is determined by the asymptotic behavior of function $R^{\lambda}_l(\tau)$ at $\tau\to\infty$

$$R^{\lambda}_l(\tau)\sim[c_l(\lambda)\mathrm{e}^{\mathrm{i}\lambda\tau}+c_l(-\lambda)\mathrm{e}^{-\mathrm{i}\lambda\tau}]\mathrm{e}^{-\tau}. \tag{6.2.15}$$

It can be easily proved that

$$N_l(\lambda)=N_l(-\lambda)=\frac{\pi}{2}|c_l(\lambda)|^2. \tag{6.2.16}$$

9′. Coefficient $N_0(\lambda)$ defines the Plancherel measure for representations of the principal series, namely,

$$f(\tau)=\sqrt{\frac{2}{\pi}}\int \hat{f}(\lambda)\Phi_\lambda(\tau)d\mu(\lambda), \qquad d\mu(\lambda)=\lambda^2 d\lambda, \tag{6.2.17}$$

where $\hat{f}(\lambda)$ is found from

$$\hat{f}(\lambda)=\sqrt{\frac{2}{\pi}}\int \bar{\Phi}_\lambda(\tau)f(\tau)\sinh^2\tau\, d\tau. \tag{6.2.18}$$

10′. Using (6.2.4) one can easily obtain an expression for generating function for the matrix elements $\langle lm|T^\lambda(g)|l'm'\rangle$. It is easy to see that

$$\langle\xi|T^\lambda(g)|\eta\rangle=\Phi_\lambda(\tau), \qquad \text{where} \tag{6.2.19}$$

$$\cosh\tau=(\xi,\eta_g)=(\xi,g\eta) \tag{6.2.20}$$

τ is the distance between points ξ and η_g.

Consider now a particular realization of the representation $T^\lambda(g)$. It acts in the space of the square integrable functions on the unit sphere $S^2=\{\boldsymbol{n}:\boldsymbol{n}^2=1\}$

$$T^\lambda(g)f(\boldsymbol{n})=(\xi_0-\boldsymbol{\xi n})^{-1+i\lambda}f(\boldsymbol{n}_g). \tag{6.2.21}$$

With such a realization of the representation, the vector invariant under the maximal compact subgroup $K=SO(3)$ is $f_0(\boldsymbol{n})\equiv 1$. Acting on it by operator $T^\lambda(g)$ we get the CS system

$$f_\xi^\lambda(\boldsymbol{n})=(\xi_0-\boldsymbol{\xi n})^{-1+i\lambda}=\Phi_{\boldsymbol{n}}^\lambda(\xi)=\Phi_{\boldsymbol{n}}^\lambda(\tau,\boldsymbol{v}). \tag{6.2.22}$$

Note a number of properties of the functions $\Phi_{\boldsymbol{n}}^\lambda(\xi)=\Phi_{\boldsymbol{n}}^\lambda(\zeta)$, $\xi_0=\dfrac{1}{\sqrt{1-|\zeta|^2}}$, $\boldsymbol{\xi}=\boldsymbol{\zeta}/\sqrt{1-|\zeta|^2}$.

11′. At fixed $\boldsymbol{n}$ the function $\Phi_{\boldsymbol{n}}^\lambda(\zeta)$ is constant on horospheres of the Lobatschevsky space, which in this case are spheres tangent to the sphere $\{\zeta:|\zeta|=1\}$ at the point $\boldsymbol{\zeta}=\boldsymbol{n}$ [84]. The horosphere equation is

$$\frac{1+|\zeta|^2-2(\boldsymbol{\zeta n})}{1-|\zeta|^2}=(\cosh\tau-\sinh\tau\,\boldsymbol{nv})=\text{const}. \tag{6.2.23}$$

Therefore the kernel $K(\zeta,\boldsymbol{n})=\Phi_n(\zeta)$ can be called a horospherical kernel and, consequently, at the realization of representation $T^\lambda(g)$ used, the CS system is given by horospherical kernels.

12′. Note that

$$|\Phi_{\boldsymbol{n}}(\zeta)|^2=\left|\frac{1-|\zeta|^2}{1+|\zeta|^2-2\boldsymbol{\zeta n}}\right|^2=P(\zeta,\boldsymbol{n}), \tag{6.2.24}$$

where $P(\zeta, \boldsymbol{n})$ is the Poisson kernel for the ball

$$\{\zeta : |\zeta| < 1\} : \Delta_\zeta P(\zeta, \boldsymbol{n}) = 0, \quad |\zeta| < 1, \quad |\boldsymbol{n}| = 1.$$

Thus

$$\Phi_{\boldsymbol{n}}(\zeta) = P(\zeta, \boldsymbol{n})^{(1 - i\lambda)/2}. \tag{6.2.25}$$

13′. At fixed $\boldsymbol{n}$ function $\Phi_n(\zeta)$ is the eigenfunction of the Laplace-Beltrami operator $\tilde{\Delta} = (1 - |\zeta|^2)^2 \Delta$ (Δ is the usual Laplace operator for three-dimensional space) on the Lobatschevsky space

$$-\tilde{\Delta}\Phi_{\boldsymbol{n}}(\zeta) = (\lambda^2 + 1)\Phi_{\boldsymbol{n}}(\zeta). \tag{6.2.26}$$

These functions are a natural generalization of plane waves for three-dimensional Euclidean space.

14′. The orthonormalized basis $\{|l, m\rangle\}$ which is proper for operators $\boldsymbol{J}^2$ and J_3, where J_1, J_2 and J_3 are infinitesimal operators of the maximal compact subgroup $G = SO(3)$, has the form

$$\{Y_{lm}(\boldsymbol{n})\}, \quad \int \bar{Y}_{lm}(\boldsymbol{n}) Y_{l'm'}(\boldsymbol{n}) d\mu(\boldsymbol{n}) = \delta_{ll'}\delta_{mm'}, \tag{6.2.27}$$

15′. Expanding coherent states in this basis $|\xi\rangle = \sum_{l,m} u_{lm}(\xi)|l, m\rangle$ and noting that $u_{lm}(\xi) = \langle l, m|\xi\rangle$, we get an integral representation for the function $u^\lambda_{lm}(\xi)$:

$$u^\lambda_{lm}(\xi) = \int d\mu(\boldsymbol{n}) \bar{Y}_{lm}(\boldsymbol{n}) (\xi_0 - \boldsymbol{\xi}\boldsymbol{n})^{-1 + i\lambda}. \tag{6.2.28}$$

16′. For zonal spherical function ($l = 0, m = 0$) this representation is simplified to

$$\Phi_\lambda(\tau) = \frac{1}{2}\int_0^\pi (\cosh\tau - \sinh\tau\cos\theta)^{-1 + i\lambda} \sin\theta \, d\theta. \tag{6.2.29}$$

17′. The system of functions describing coherent states satisfies the orthogonality conditions

$$\frac{1}{(2\pi)^3} \int \bar{\Phi}^\lambda_{\boldsymbol{n}}(\tau, \boldsymbol{v}) \Phi^{\lambda'}_{\boldsymbol{n}'}(\tau, \boldsymbol{v}) \sinh^2\tau \, d\tau \, d\boldsymbol{v} = N(\lambda)\delta(\lambda - \lambda')\delta(\boldsymbol{n}, \boldsymbol{n}') \tag{6.2.30}$$

$$\frac{1}{(2\pi)^3} \int \bar{\Phi}^\lambda_{\boldsymbol{n}}(\tau, \boldsymbol{v}) \Phi^\lambda_{\boldsymbol{n}}(\tau', \boldsymbol{v}') d\mu(\lambda) d\boldsymbol{n} = (\sinh\tau)^{-2}\delta(\tau - \tau')\delta(\boldsymbol{v}, \boldsymbol{v}') \tag{6.2.31}$$

$$d\mu(\lambda) = N^{-1}(\lambda) d\lambda = \lambda^2 d\lambda.$$

18′. Using these relations we can expand the arbitrary function in the Lobatschevsky space in functions $\Phi^\lambda_{\boldsymbol{n}}(\tau, \boldsymbol{v})$

$$f(\tau, \mathbf{v}) = \frac{1}{(2\pi)^{3/2}} \int \hat{f}(\lambda, \mathbf{n}) \Phi_{\mathbf{n}}^{\lambda}(\tau, \mathbf{v}) d\mu(\lambda) d\mathbf{n}, \qquad d\mu(\lambda) = \lambda^2 d\lambda, \tag{6.2.32}$$

where

$$\hat{f}(\lambda, \mathbf{n}) = \frac{1}{(2\pi)^{3/2}} \int f(\tau, \mathbf{v}) \bar{\Phi}_{\mathbf{n}}^{\lambda}(\tau, \mathbf{v}) \sinh^2 \tau \, d\mathbf{v} \, d\tau. \tag{6.2.33}$$

19′. Functions are closely related to a scattering problem. To see this, let us rewrite the Laplace-Beltrami operator as

$$\begin{aligned} \tilde{\Delta} &= \frac{d^2}{d\tau^2} + 2 \coth \tau \frac{d}{d\tau} + \frac{1}{\sinh^2 \tau} \Delta_{\theta\varphi} \\ &= (\sinh \tau)^{-1} \frac{d^2}{d\tau^2} (\sinh \tau) - 1 + \sinh^{-2} \tau \, \Delta_{\theta\varphi}. \end{aligned} \tag{6.2.34}$$

Hence it follows that functions $\psi_l^{\lambda}(\tau) = \sinh \tau \, R_l^{\lambda}(\tau)$ satisfy the differential equation

$$\left(-\frac{d^2}{d\tau^2} + \frac{l(l+1)}{\sinh^2 \tau} \right) \psi_l^{\lambda}(\tau) = \lambda^2 \psi_l^{\lambda}(\tau) \tag{6.2.35}$$

which is just the Schrödinger equation with the potential $V_l(\tau) = l(l+1) \sinh^{-2} \tau$. Hence it is also seen that the asymptotic of the function $\psi_l^{\lambda}(\tau)$ at $\tau \to \infty$ is

$$\psi_l^{\lambda}(\tau) \sim c_l(\lambda) \mathrm{e}^{\mathrm{i}\lambda\tau} + c_l(-\lambda) \mathrm{e}^{-\mathrm{i}\lambda\tau} \tag{6.2.36}$$

and the S matrix for this problem is

$$S_l(\lambda) = -c_l(\lambda)/c_l(-\lambda). \tag{6.2.37}$$

Note that the problem considered is intimately related to the scattering problem of a charged particle in the Coulomb field [90, 91].

20′. To find the coefficients $c_l(\lambda)$, it is appropriate to consider the $\tau \to \infty$ limit for the decomposition of $\Phi_n^{\lambda}(\tau, \mathbf{v})$ over the spherical harmonics $Y_{lm}(n)$. The result is

$$\frac{1}{4\pi} \int \left(\frac{1 - \mathbf{n}\mathbf{v}}{2} \right)^{-1 + \mathrm{i}\lambda} \bar{Y}_{ll}(\mathbf{n}) d\mathbf{n} = c_l(\lambda) \, \bar{Y}_{ll}(\mathbf{v}). \tag{6.2.38}$$

Calculating this integral explicitly, one gets

$$c_l(\lambda) = \frac{2^{2l} \Gamma(l + \frac{3}{2})}{\Gamma(\frac{3}{2})} \frac{\Gamma(1 + \mathrm{i}\lambda)}{\mathrm{i}\lambda \Gamma(l + 1 + \mathrm{i}\lambda)}. \tag{6.2.39}$$

21. Thus the S matrix for the scattering problem considered here is

$$S_l(\lambda)=\frac{\Gamma(1+\mathrm{i}\lambda)\,\Gamma(l+1-\mathrm{i}\lambda)}{\Gamma(1-\mathrm{i}\lambda)\,\Gamma(l+1+\mathrm{i}\lambda)}. \tag{6.2.40}$$

It has simple poles at the points $\lambda=\mathrm{i}m,\ -\mathrm{i}(l+k);\ m,k=1,2,\ldots$.

22. Note, in conclusion, that it is easy to get an explicit expression for the function $R_l^\lambda(\tau)$, using the integral representation (6.2.28),

$$R_l^\lambda(\tau)=B_l(\sinh\tau)^l F(a,b,c;\ -\sinh^2\tau)$$

$$a=\tfrac{1}{2}(l+1+\mathrm{i}\lambda),\qquad b=\tfrac{1}{2}(l+1-\mathrm{i}\lambda),\qquad c=l+\tfrac{3}{2}. \tag{6.2.41}$$

7. Coherent States for the $SO(n, 1)$ Group: Class-I Representations of the Principal Series

This chapter extends the results of Chaps. 5 and 6 to the case of the group $SO(n, 1)$ for arbitrary n.

7.1 Class-I Representations of $SO(n, 1)$

The group $G = SO(n, 1)$ consists of linear homogeneous transformations of the $(n+1)$-dimensional space under which the quadratic form

$$x^2 = x_0^2 - \boldsymbol{x}^2, \quad x = (x_0, \boldsymbol{x}), \quad \boldsymbol{x} = (x_1, \dots, x_n) \tag{7.1.1}$$

is invariant. The invariance of this form is written as

$$g' s g = s \quad \text{or} \quad g^{-1} = s g' s. \tag{7.1.2}$$

Here g and s are real $(n+1) \times (n+1)$ matrices, g' is transposed g, and $s = \mathrm{diag}(1, -1, -1, \dots, -1)$.

The group $SO(n, 1)$ is noncompact; its maximal compact subgroup is $SO(n)$. The coset space $X = G/K$ is the familiar Lobachevsky space of n dimensions. It can be realized in various manners; realization using the interior of the unit sphere is the most appropriate for our purpose, i.e., $\{\boldsymbol{x} : |\boldsymbol{x}| < 1\}$, as well as the realization in the upper sheet of the two-sheet hyperboloid:

$$\{\xi = (\xi_0, \boldsymbol{\xi}); \quad \xi_0^2 - \boldsymbol{\xi}^2 = 1, \quad \xi_0 > 0\}.$$

Space X must also be embedded into the group G. The most suitable way is

$$\xi = (\xi_0, \boldsymbol{\xi}), \quad \rightarrow g_\xi = \begin{pmatrix} \xi_0, \boldsymbol{\xi} \\ \boldsymbol{\xi}', \mathrm{I} + (\xi_0 + 1)^{-1} \boldsymbol{\xi} \otimes \boldsymbol{\xi} \end{pmatrix}. \tag{7.1.3}$$

It is easily seen that (7.1.2) holds, so $g_\xi \in G$.

All the unitary irreducible class-I representations belong to either the principal or the supplementary series. Let us consider their structure.

Let us start from the realization in the unit ball $D = \{\boldsymbol{x} : |\boldsymbol{x}| < 1\}$. The boundary is the $(n-1)$-dimensional sphere $S^{n-1} = \{\boldsymbol{n} : \boldsymbol{n}^2 = 1\}$, and it is easily seen that this space is homogeneous with respect to the action of the group G. Class-I representations of the principal and supplementary series can be realized in the space of the square-integrable functions $\{f(\boldsymbol{n})\}$ on S^{n-1}. Such a representation is characterized by a real number λ and is given by

$$T^{\lambda}(g) f(\boldsymbol{n}) = (\xi_0^g - \boldsymbol{\xi}^g \boldsymbol{n})^{\sigma} f(\boldsymbol{n}_g), \quad \text{where} \tag{7.1.4}$$

$$\sigma = -\varrho + \mathrm{i}\lambda, \quad \varrho = \frac{n-1}{2}, \quad \xi_0^g = g_{00}, \quad \xi_j^g = g_{j0}. \tag{7.1.5}$$

For representation of the principal series $\lambda \geq 0$, and the scalar product is given by

$$\langle f_1 | f_2 \rangle = \int_{S^{n-1}} \bar{f}_1(\boldsymbol{n}) f_2(\boldsymbol{n}) d\mu(\boldsymbol{n}), \quad \int d\mu(\boldsymbol{n}) = 1. \tag{7.1.6}$$

For representations of the supplementary series λ is purely imaginary: $\lambda = \mathrm{i}\nu$, $-\varrho < \nu < \varrho$; consequently, σ is a real number

$$-2\varrho < \sigma < 0, \quad -(n-1) < \sigma < 0. \tag{7.1.7}$$

The scalar product is given here by a more complicated formula [78]

$$\langle f_1 | f_2 \rangle = C_{\sigma} \int_{S^{n-1}} \int_{S^{n-1}} (1 - \boldsymbol{n}\boldsymbol{n}')^{\sigma+1-n} \bar{f}_1(\boldsymbol{n}) f_2(\boldsymbol{n}) d\mu(\boldsymbol{n}) d\mu(\boldsymbol{n}') \tag{7.1.8}$$

$$C_{\sigma} = \frac{\sqrt{\pi}\,\Gamma(\sigma)}{2^{\sigma-1}\Gamma\left(\frac{n}{2}\right)\Gamma\left(\sigma - \frac{n-1}{2}\right)}. \tag{7.1.9}$$

7.2 Coherent States

We define the CS system for representations of class-I by

$$|\xi\rangle = T^{\lambda}(g_{\xi}) |\psi_0\rangle, \tag{7.2.1}$$

where $|\psi_0\rangle$ is a vector in $\mathscr{H}^{\lambda}$, invariant under the action of $T^{\lambda}(k)$, $k \in K$, and g_{ξ} is given by (7.1.3).

It can be readily seen that in the $\boldsymbol{n}$ representation the function $f_0(\boldsymbol{n}) \equiv 1$ plays the role of $|\psi_0\rangle$. Acting on it by operators we obtain the CS system

$$f_{\xi}^{\lambda}(n) = (\xi_0 - \boldsymbol{\xi}\boldsymbol{n})^{-\varrho + \mathrm{i}\lambda} = \Phi_{\boldsymbol{n}}^{\lambda}(\xi), \quad \xi_0^2 - \boldsymbol{\xi}^2 = 1, \quad \boldsymbol{n}^2 = 1. \tag{7.2.2}$$

The properties of this CS system are analogous to those of the CS system for the Lorentz group (Chap. 6). We present here only some of them.

1. The scalar product of two coherent states is given by

$$\langle\xi|\eta\rangle=\langle 0|T^{\lambda}(a(\tau))|0\rangle=\Phi_{\lambda}(\tau), \tag{7.2.3}$$

where τ is the distance between the points ξ and η, $\cosh\tau=(\xi\eta)=\xi_0\eta_0-\boldsymbol{\xi\eta}$, $a(\tau)$ is hyperbolic "rotation" on "angle" τ in plane $(0,1)$.

2. If the representation T^{λ} is restricted to the maximal compact subgroup $K=SO(n)$, it is decomposed over irreducible components. This decomposition contains only those representations of the group K which are characterized by a single integer number l:

$$\mathcal{H}^{\lambda}=\mathcal{H}_0^{\lambda}\oplus\mathcal{H}_1^{\lambda}\oplus\ldots\oplus\mathcal{H}_l^{\lambda}\oplus\ldots. \tag{7.2.4}$$

Therefore in the space of representation one can choose the basis $\{|l,\mu\rangle\}$, where μ are numbers characterizing the basis vector in $\mathcal{H}_l^{\lambda}$. These vectors are eigenvectors of the operator $\boldsymbol{J}^2$ for the group K $\left(\boldsymbol{J}^2=\sum_{k<l}\hat{J}_{kl}\hat{J}_{kl}\right)$,

$$\boldsymbol{J}^2|l,\mu\rangle=l(l+n-2)|l,\mu\rangle. \tag{7.2.5}$$

The number of vectors in $\mathcal{H}_l^{\lambda}$ equals the dimension of representation T^l of group $SO(n)$ and is given by

$$d_l=\frac{(2l+n-2)(l+n-3)!}{l!(n-2)!}. \tag{7.2.6}$$

3. Expanding coherent states on basis $|l,\mu\rangle$ yields

$$|\xi\rangle=\sum u_{l\mu}^{\lambda}(\xi)|l,\mu\rangle, \quad \text{where} \tag{7.2.7}$$

$$u_{l\mu}^{\lambda}(\xi)=\langle l,\mu|\xi\rangle=\langle l,\mu|T^{\lambda}(g_{\xi})|0\rangle. \tag{7.2.8}$$

4. The state $|\xi\rangle$ is normalized, therefore

$$\sum_{l,\mu}|u_{l\mu}^{\lambda}(\xi)|^2=1. \tag{7.2.9}$$

Note also that

$$u_{l\mu}^{\lambda}(\tau,\boldsymbol{v})=u_{l\mu}^{\lambda}(\xi)=R_l^{\lambda}(\tau)\,\bar{Y}_{l\mu}(\boldsymbol{v}), \quad \xi_0=\cosh\tau, \quad \boldsymbol{\xi}=\boldsymbol{v}\sinh\tau. \tag{7.2.10}$$

Here $\boldsymbol{v}$ is the unit vector; $\boldsymbol{v}^2=1$, $Y_{l\mu}(\boldsymbol{v})$ are standard normalized spherical functions:

$$\int\bar{Y}_{l\mu}(\boldsymbol{v})\,Y_{l'\mu'}(\boldsymbol{v})\,d\mu(\boldsymbol{v})=\delta_{ll'}\delta_{\mu\mu'}, \quad \int d\mu(\boldsymbol{v})=1.$$

5. Functions $u_{l\mu}^{\lambda}(\tau,\boldsymbol{v})$ are matrix elements of operator $T^{\lambda}(g)$, so they are eigenfunctions of the Laplace-Beltrami operator for the Lobachevsky space

$$-\tilde{\Delta} u_{l\mu}^{\lambda}(\tau, \mathbf{v}) = (\lambda^2 + \varrho^2) u_{l\mu}^{\lambda}(\tau, \mathbf{v}), \qquad \text{where} \tag{7.2.11}$$

$$\tilde{\Delta} = \frac{\partial^2}{\partial \tau^2} + 2\varrho \coth\tau \frac{\partial}{\partial \tau} + \frac{1}{\sinh^2\tau} \Delta_{\mathbf{v}}, \tag{7.2.12}$$

Here $\Delta_{\mathbf{v}}$ is the Laplace-Beltrami operator on the $(n-1)$-dimensional sphere

$$S^{n-1} = \{\mathbf{v} : \mathbf{v}^2 = 1\}.$$

Hence it follows that the function $R_l^{\lambda}(\tau)$ satisfies the equation

$$\left\{ -\left[\frac{d^2}{d\tau^2} + (n-1)\coth\tau \frac{d}{d\tau} \right] + \frac{l(l+n-2)}{\sinh^2\tau} \right\} R_l^{\lambda}(\tau) = \left[\lambda^2 + \left(\frac{n-1}{2} \right)^2 \right] R_l^{\lambda}. \tag{7.2.13}$$

6. Functions $R_l^{\lambda}(\tau)$ are related to the problem of particle scattering in potential $V(\tau) = g^2 \sinh^{-2}\tau$. To elucidate this point, let us consider a new function

$$\psi_l^{\lambda}(\tau) = (\sinh\tau)^{(n-1)/2} R_l^{\lambda}(\tau). \tag{7.2.14}$$

It can be easily seen that it satisfies the Schrödinger equation

$$\left[-\frac{d^2}{d\tau^2} + V_l(\tau) \right] \psi_l^{\lambda}(\tau) = \lambda^2 \psi_l^{\lambda}(\tau), \qquad \text{where} \tag{7.2.15}$$

$$V_l(\tau) = \left[\left(\frac{n}{2} - 1 + l \right)^2 - \frac{1}{4} \right] \sinh^{-2}(\tau). \tag{7.2.16}$$

7. Hence it is seen that $\psi_l^{\lambda}(n;\tau) = C\psi_0^{\lambda}(n+2l;\tau)$ and, correspondingly,

$$R_l^{\lambda}(n;\tau) = B_{nl}(\sinh\tau)^l R_0^{\lambda}(n+2l;\tau) = B_{nl}(\sinh\tau)^l \Phi_{\lambda}(n+2l;\tau). \tag{7.2.17}$$

8. Solving (7.2.13), we get an explicit expression for the function $R_l^{\lambda}(\tau)$:

$$R_l^{\lambda}(\tau) = B_{nl}(\sinh\tau)^l F(a,b,c; -\sinh^2\tau), \qquad a = \tfrac{1}{2}(\varrho + l + \mathrm{i}\lambda),$$

$$b = \tfrac{1}{2}(\varrho + l - \mathrm{i}\lambda), \qquad c = \varrho + l + \tfrac{1}{2} \tag{7.2.17'}$$

where $F(a,b,c;x)$ is the hypergeometric function.

9. The asymptotic behavior of function $R_l^{\lambda}(\tau)$ is of the form

$$R_l^{\lambda}(\tau) \sim [C_l(\lambda)\mathrm{e}^{\mathrm{i}\lambda\tau} + C_l(-\lambda)\mathrm{e}^{-\mathrm{i}\lambda\tau}]\mathrm{e}^{-\varrho\tau}. \tag{7.2.18}$$

10. From (7.2.17) we find the following expression for the coefficient $C_l(\lambda)$

$$C_l(\lambda) = B_{nl} \frac{2^{2l}\Gamma(\varrho + l + \frac{1}{2})\Gamma(\varrho + \mathrm{i}\lambda)}{\Gamma(\varrho + \frac{1}{2})\Gamma(\varrho + l + \mathrm{i}\lambda)} C_0(\lambda). \tag{7.2.19}$$

11. Functions $u^{\lambda}_{l\mu}(\tau, \mathbf{v})$ form an orthogonal system of functions in the Lobachevsky space

$$\iint \bar{u}^{\lambda}_{l\mu}(\tau, \mathbf{v}) u^{\lambda'}_{l'\mu'}(\tau, \mathbf{v}) (\sinh \tau)^{n-1} d\tau \, d\mu(\mathbf{v}) = N_l(\lambda) \delta_{ll'} \delta_{\mu\mu'} \delta(\lambda - \lambda'). \tag{7.2.20}$$

Here

$$d\mu(\mathbf{v}) = A^{-1} \cdot d\mathbf{v}, \qquad \int d\mu(\mathbf{v}) = 1. \tag{7.2.21}$$

12. Coefficient $N_l(\lambda)$ is taken from the asymptotics of function $R^{\lambda}_l(\tau)$ at $\tau \to \infty$:

$$N_l(\lambda) = \frac{\pi}{2^{n-2}} |C_l(\lambda)|^2. \tag{7.2.22}$$

13. Function $\Phi_\lambda(\tau)$, the zonal spherical function, is the eigenfunction of the radial part of the Laplace-Beltrami operator

$$-\left[\frac{d^2}{d\tau^2} + (n-1)\coth \tau \frac{d}{d\tau}\right] \Phi_\lambda(\tau) = (\lambda^2 + \varrho^2) \Phi_\lambda(\tau), \qquad \varrho = \frac{n-1}{2}. \tag{7.2.23}$$

It is normalized by the condition

$$\Phi_\lambda(0) = 1. \tag{7.2.24}$$

14. From the explicit expression

$$\Phi_\lambda(\tau) = F(a, b, c; -\sinh^2 \tau), \quad a = \frac{\varrho + i\lambda}{2}, \quad b = \frac{\varrho - i\lambda}{2}, \quad c = \varrho + \frac{1}{2} \tag{7.2.25}$$

we get

$$C_0(\lambda) = \frac{2^{2\varrho - 1} \Gamma(\varrho + \frac{1}{2})}{\sqrt{\pi}} \frac{\Gamma(i\lambda)}{\Gamma(\varrho + i\lambda)} \tag{7.2.26}$$

$$N_0(\lambda) = 2^{n-2} |\Gamma(\varrho + \tfrac{1}{2})|^2 \frac{|\Gamma(i\lambda)|^2}{|\Gamma(\varrho + i\lambda)|^2}. \tag{7.2.27}$$

15. Coefficient $N_0(\lambda)$ determines the Plancherel measure for representations of the principal series, namely,

$$f(\tau) = a_n \int \hat{f}(\lambda) \Phi_\lambda(\tau) d\mu(\lambda), \tag{7.2.28}$$

$$d\mu(\lambda) = \frac{d\lambda}{N(\lambda)}, \qquad N(\lambda) = |\Gamma(i\lambda)|^2 / |\Gamma(\varrho + i\lambda)|^2, \quad a_n = \frac{1}{2^{(n-2)/2} \Gamma\left(\dfrac{n}{2}\right)} \tag{7.2.29}$$

$$\hat{f}(\lambda) = a_n \int f(\tau) \Phi_\lambda(\tau) (\sinh \tau)^{n-1} \, d\tau \tag{7.2.30}$$

$$\int |f(\tau)|^2 (\sinh \tau)^{n-1} \, d\tau = \int |\hat{f}(\lambda)|^2 \, d\mu(\lambda). \tag{7.2.31}$$

Consider now the properties of the CS system in the $\boldsymbol{n}$ representation

$$\Phi_{\boldsymbol{n}}^\lambda(\xi) = \langle \xi | \lambda, \boldsymbol{n} \rangle = (\xi_0 - \boldsymbol{\xi n})^{-\varrho + i\lambda}, \qquad \varrho = \frac{n-1}{2}. \tag{7.2.32}$$

16. The system of functions $\Phi_{\boldsymbol{n}}^\lambda(\xi)$ is complete and orthogonal

$$\frac{1}{(2\pi)^n} \int \bar{\Phi}_{\boldsymbol{n}}^\lambda(\xi) \Phi_{\boldsymbol{n}}^\lambda(\xi') \, d\boldsymbol{n} \, d\mu(\lambda) = \delta(\xi, \xi') \tag{7.2.33}$$

$$\frac{1}{(2\pi)^n} \int \bar{\Phi}_{\boldsymbol{n}}^\lambda(\xi) \Phi_{\boldsymbol{n}'}^{\lambda'}(\xi) \, d\mu(\xi) = N(\lambda) \delta(\lambda - \lambda') \delta(\boldsymbol{n}, \boldsymbol{n}'), \quad d\mu(\xi) = \frac{d\boldsymbol{\xi}}{\xi_0}. \tag{7.2.34}$$

17. Using these relations one can expand any function $f(\xi)$ on a hyperboloid in functions $\Phi_{\boldsymbol{n}}^\lambda(\xi)$:

$$f(\xi) = \frac{1}{(2\pi)^{n/2}} \int \hat{f}(\lambda, \boldsymbol{n}) \Phi_{\boldsymbol{n}}^\lambda(\xi) \, d\mu(\lambda) \, d\boldsymbol{n}, \tag{7.2.35}$$

where

$$\hat{f}(\lambda, \boldsymbol{n}) = \frac{1}{(2\pi)^{n/2}} \int \bar{\Phi}_{\boldsymbol{n}}^\lambda(\xi) f(\xi) \, d\mu(\xi). \tag{7.2.36}$$

Moreover

$$\int |\hat{f}(\lambda, \boldsymbol{n})|^2 \, d\mu(\lambda) \, d\boldsymbol{n} = \int |f(\xi)|^2 \, d\mu(\xi). \tag{7.2.37}$$

18. Averaging the coherent state over the sphere $\{\boldsymbol{n} : \boldsymbol{n}^2 = 1\}$ yields an integral representation for the zonal spherical function

$$\Phi_\lambda(\tau) = \int_0^\pi (\cosh \tau - \sinh \tau \cos \theta)^\sigma \, d\mu(\theta);$$

$$d\mu(\theta) = \frac{\Gamma\left(\frac{n}{2}\right)}{\sqrt{\pi}\, \Gamma\left(\frac{n-1}{2}\right)} (\sin \theta)^{n-2} \, d\theta, \qquad \sigma = -\varrho + i\lambda, \quad \varrho = \frac{n-1}{2}. \tag{7.2.38}$$

19. Hence one gets at once an expression in terms of the hypergeometric function,

$$\Phi_\lambda(\tau)=(\cosh\tau)^{-\varrho+i\lambda}F(a,b,c;\tanh^2\tau);$$

$$a=\tfrac{1}{2}(\varrho-i\lambda),\qquad b=\tfrac{1}{2}(\varrho-i\lambda+1),\qquad c=\varrho+\tfrac{1}{2}. \tag{7.2.39}$$

Clearly this expression is equivalent to that in (7.2.25).

20. To get the $\tau\to\infty$ asymptotic behavior, we use a transformation for the hypergeometric function; the result is

$$\Phi_\lambda(\tau)=2^{\varrho-1}\pi^{-1/2}\Gamma\left(\frac{n}{2}\right)(\cosh\tau)^{-\varrho}$$

$$\cdot\left[\frac{\Gamma(i\lambda)}{\Gamma(i\lambda+\varrho)}(2\cosh\tau)^{i\lambda}F(a,b,c;\cosh^{-2}\tau)+\text{c.c.}\right],$$

$$a=\tfrac{1}{2}(\varrho-i\lambda),\qquad b=\tfrac{1}{2}(\varrho+1-i\lambda),\qquad c=1-i\lambda. \tag{7.2.40}$$

21. Hence one gets immediately the desired asymptotics, at $\tau\to\infty$

$$\Phi_\lambda(\tau)\sim 2^{2\varrho-1}\pi^{-1/2}\Gamma\left(\frac{n}{2}\right)\left|\frac{\Gamma(i\lambda)}{\Gamma(i\lambda+\varrho)}\right|e^{-\varrho\tau}(e^{i(\lambda\tau+\delta_0)}+\text{c.c.}) \tag{7.2.41}$$

where the phase shift is given by

$$\delta_0(\lambda)=\arg\Gamma(i\lambda)-\arg\Gamma(\varrho+i\lambda). \tag{7.2.42}$$

This is just the scattering phase shift for the potential

$$V_0(\tau)=\left[\left(\frac{n}{2}-1\right)^2-\frac{1}{4}\right]\sinh^{-2}\tau.$$

22. An expression for the coefficient $C(\lambda)$ can be obtained more simply; namely, setting the limit $\tau\to\infty$ in the integrand in (7.2.38). The result is

$$C(\lambda)=\int_0^\pi\left(\frac{1-\cos\theta}{2}\right)^\sigma d\mu(\theta)=\frac{2^{2\varrho-1}\Gamma(\varrho+\frac{1}{2})}{\sqrt{\pi}}\,\frac{\Gamma(i\lambda)}{\Gamma(\varrho+i\lambda)}. \tag{7.2.43}$$

23. The expansion of the zonal function $\Phi_n^\lambda(\xi)$, $\xi=(\cosh\tau,\ \boldsymbol{v}\sinh\tau)$, is obtained from (7.2.7, 10),

$$\Phi_{\boldsymbol{n}}^\lambda(\xi)=[\cosh\tau-\sinh\tau(\boldsymbol{vn})]^{-\varrho+i\lambda}=\sum_l R_l^\lambda(\tau)\sum_\mu \bar{Y}_{l\mu}(\boldsymbol{v})\,Y_{l\mu}(\boldsymbol{n}). \tag{7.2.44}$$

It can be easily seen that

$$\sum_\mu \bar{Y}_{l\mu}(\boldsymbol{v})\,Y_{l\mu}(\boldsymbol{n})=A_l C_l^\alpha(\boldsymbol{vn}),\qquad \alpha=\frac{n-2}{2}=\varrho-\frac{1}{2}, \tag{7.2.45}$$

where $C_l^\alpha(x)$ is the Gegenbauer polynomial of degree l. Setting $\boldsymbol{v}=\boldsymbol{n}$ and integrating both parts of (7.2.41) over $d\mu(\boldsymbol{n})$, then

$$\sum_\mu \bar{Y}_{l\mu}(\boldsymbol{v})\, Y_{l\mu}(\boldsymbol{n}) = d_l \frac{C_l^\alpha(\boldsymbol{vn})}{C_l^\alpha(1)}, \tag{7.2.46}$$

where $Y_{00}\equiv 1$, $\int d\mu(n)=1$, and d_l is given by (7.2.6).

Finally we get an expansion for the horospherical wave

$$[\cosh\tau - \sinh\tau(\boldsymbol{vn})]^{-\varrho+\mathrm{i}\lambda} = \sum_l R_l^\lambda(\tau)\frac{(l+\varrho-\frac{1}{2})}{(\varrho-\frac{1}{2})} C_l^\alpha(\boldsymbol{vn}). \tag{7.2.47}$$

24. Let us now take into account the fact that the Gegenbauer polynomials form the orthonormalized system $\{\hat{C}_l^\alpha(\cos\theta)\}$ on the interval $(0,\pi)$, with the weight $(\sin\theta)^{2\alpha}$,

$$\hat{C}_l^\alpha(\cos\theta) = \left[\frac{(\alpha+l)\,l!\,\Gamma(\alpha)\Gamma(2\alpha)}{\sqrt{\pi}\,\Gamma(2\alpha+l)\,\Gamma(\alpha+\frac{1}{2})}\right]^{1/2} C_l^\alpha(\cos\theta). \tag{7.2.48}$$

From (7.2.47) we obtain an expression for the zonal spherical functions

$$R_l^\lambda(\tau) = \frac{\Gamma(\alpha)\Gamma(2\alpha)\,l!}{2\sqrt{\pi}\,\Gamma(\alpha+\frac{1}{2})\Gamma(2\alpha+l)} \int_0^\pi (\cosh\tau - \sinh\tau\cos\theta)^{-\varrho+\mathrm{i}\lambda} \cdot C_l^\alpha(\cos\theta)(\sin\theta)^{2\alpha}\, d\theta. \tag{7.2.49}$$

25. Now one can easily write an expression for the $\tau\to\infty$ asymptotics of the adjoint spherical function,

$$R_l^\lambda(\tau) \sim B_l^\lambda \mathrm{e}^{-\varrho\tau}\mathrm{e}^{\mathrm{i}(\lambda\tau+\delta_l)} + \text{c.c.}), \tag{7.2.50}$$

where

$$\delta_l = \arg\Gamma(\mathrm{i}\lambda) - \arg\Gamma(\varrho+l+\mathrm{i}\lambda). \tag{7.2.51}$$

26. The expansion of the CS at $\tau\to\infty$ now takes the form

$$\left(\frac{1-\cos\theta}{2}\right)^{-\varrho+\mathrm{i}\lambda} = C_0(\lambda)\sum_{l=0}^\infty \frac{(l+\varrho-\frac{1}{2})}{(\varrho-\frac{1}{2})}\,\frac{\Gamma(\varrho+\mathrm{i}\lambda)\Gamma(\varrho+l-\mathrm{i}\lambda)}{\Gamma(\varrho-\mathrm{i}\lambda)\Gamma(\varrho+l+\mathrm{i}\lambda)}\, C_l^\alpha(\cos\theta),$$

$$C_0(\lambda) = [2^{2\varrho-1}\Gamma(\varrho+\tfrac{1}{2})\Gamma(\mathrm{i}\lambda)]/[\sqrt{\pi}\,\Gamma(\varrho+\mathrm{i}\lambda)]. \tag{7.2.52}$$

27. Note that at $\lambda\to\infty$, $\tau\to 0$, $\lambda\tau=\text{const}$ there is a transition to the flat space case and our formulae are replaced by those for expansion of usual plane waves. In particular, $\exp[\mathrm{i}\delta_l(\lambda)] \sim (-1)^l \exp[\mathrm{i}\delta_0(\lambda)]$.

8. Coherent States for a Bosonic System with a Finite Number of Degrees of Freedom

In this chapter the CS system is constructed for a bosonic system with a finite number of degrees of freedom. The system state is specified with a complex symmetrical matrix satisfying an additional condition. The exposition here follows mainly works [35, 94].

8.1 Canonical Transformations

Recall that the basic operators used in describing a bosonic system with N degrees of freedom are the creation operators a_j^+ and annihilation operators a_k ($j, k=1, \ldots, N$). These operators act in the standard Hilbert space $\mathscr{H}_B$, the so-called Fock space, and satisfy the commutation relations

$$[a_j, a_k^+]=\delta_{jk}, \quad [a_j, a_k]=[a_j^+, a_k^+]=0. \tag{8.1.1}$$

We shall consider the transformations preserving the commutation relations, i.e., the canonical transformations.

The simplest canonical transformations are the displacements $a_j \to a_j+\alpha_j$, where α_j are arbitrary complex numbers, Chaps. 1, 3. Now let us treat the most general linear homogeneous transformation in the operator basis $\{a_j, a_k^+\}$:

$$a_i \to \tilde{a}_i = u_{ij} a_j + v_{ik} a_k^+, \quad a_i^+ \to \tilde{a}_i^+ = \bar{v}_{ij} a_j + \bar{u}_{ik} a_k^+, \tag{8.1.2}$$

or in compact notation,

$$a \to \tilde{a} = Ua + Va^+, \quad a^+ \to \tilde{a}^+ = \bar{V}a + \bar{U}a^+.$$

The conditions under which the transformation is canonical are

$$UU^+ - VV^+ = \mathrm{I}, \quad UV' = VU', \quad U=(u_{ij}), \quad V=(v_{ij}). \tag{8.1.3}$$

Here and in the following $^+$ means the Hermitian conjugation of operators, $^-$ stands for the complex conjugation, $'$ means transposition of matrices. The sum over repeated indices is implied everywhere. These conditions are equivalent to a condition for the $2N \times 2N$ matrix

$$M=\begin{pmatrix} U & V \\ \bar{V} & \bar{U} \end{pmatrix}, \quad MKM^+ = K, \quad K=\begin{pmatrix} \mathrm{I} & 0 \\ 0 & -\mathrm{I} \end{pmatrix}, \tag{8.1.4}$$

where 0 and I stand for zero and unit $N \times N$ matrices, respectively.

Hence one gets at once an expression for the inverse matrix

$$M^{-1} = KM^{+}K = \begin{pmatrix} U^{+} & -V' \\ -V^{+} & U' \end{pmatrix} \tag{8.1.5}$$

and the conditions may be rewritten as

$$M^{+}KM = K, \qquad U^{+}U - V'\bar{V} = \mathrm{I}, \qquad U^{+}V = V'\bar{U}. \tag{8.1.6}$$

Through (8.1.3, 6), a matrix inverse to U exists, so that

$$U^{-1}V = V'(U^{-1})', \qquad \bar{V}U^{-1} = (U^{-1})'V^{+}. \tag{8.1.7}$$

The last relation is equivalent to the symmetry of the matrices

$$Z = U^{-1}V \quad \text{and} \quad W = \bar{V}U^{-1}. \tag{8.1.8}$$

It follows also from (8.1.3, 6) that

$$(1 - ZZ^{+}) = (U^{+}U)^{-1}, \qquad (1 - W^{+}W) = (UU^{+})^{-1}. \tag{8.1.9}$$

So the matrices $1 - ZZ^{+}$ and $1 - W^{+}W$ are positively definite.

The matrices M defined by (8.1.4, 6) form a group G with respect to the standard matrix multiplication. This group is called the real symplectic group; the conventional notation is $\mathrm{Sp}(2N, \mathbb{R})$.

Note some properties of this group [6, 35, 36].

1. The group G is noncompact (its invariant volume is infinite) so all its unitary irreducible representations are infinite-dimensional.

2. The group G is not simply connected. In other words, there are closed paths in the group space which cannot be continuously deformed into a point. An example is the path corresponding to the transformation $a_j \to \exp(2\pi i n t) a_j$, $0 \le t < 1$, it is closed for any integer nonzero n, but cannot be contracted into a point.

Splicing together appropriately an infinite number of copies of G, one gets a simply connected group $\tilde{G}$, the so-called universal covering of G [36].

As shown in the following, also important is a group $\bar{G}$ which covers the group G twice and is called the methaplectic group, the notation is $\mathrm{Mp}(2n, \mathbb{R})$. *Weil* [92] generalized this group to the case when the real numbers are replaced by elements of an arbitrary locally compact group.

3. The maximal compact subgroup K of $\mathrm{Sp}(2N, \mathbb{R})$ is the group of matrices of the form $k = \begin{pmatrix} U & 0 \\ 0 & \bar{U} \end{pmatrix}$, where U is a unitary matrix, so the subgroup is isomorphic to the group $U(N)$.

4. The matrices Z (respectively, W) in (8.1.8), as is readily verifiable, are representatives of the left (respectively, right) cosets of the group G by the

subgroup K. The quotient space G/K is a symmetric space of the noncompact type [81]. As follows from (8.1.9), it may be realized as a complex homogeneous bounded domain (the so-called Siegel unit disk [81, 93])

$$\{Z : \mathrm{I} - ZZ^{+} > 0\}, \tag{8.1.10}$$

where Z is a complex symmetrical $N \times N$ matrix, the symbol >0 meaning that the matrix is positive definite. The group $G = \mathrm{Sp}(2N, \mathbb{R})$ acts in the matrix space as the group of linear-fractional transformations,

$$g : Z \to \tilde{Z} = (U + Z\bar{V})^{-1}(V + Z\bar{U}). \tag{8.1.11}$$

Turning back to the canonical transformation (8.1.2), note that according to the Stone-von Neumann theorem [25, 26], the operators $\tilde{a}_j$ and a_j are unitary equivalent. In other words, a unitary operator $T = T(g)$ exists such that

$$\tilde{a}_j = T a_j T^{+}. \tag{8.1.12}$$

It is easily seen that the operators $T(g)$ realize a representation of the group $G = \mathrm{Sp}(2N, \mathbb{R})$. It is not difficult to describe also the infinitesimal operators, i.e., the corresponding representation of the Lie algebra. Suppose $T \approx \mathrm{I} + \mathrm{i}\varepsilon H$, where H is an Hermitian infinitesimal operator. Then the commutator $[H, a_j]$ must be linear in a_j and a_k^{+}. This is the case if H is quadratic in a_j and a_k^{+}. Thus we are led to the algebra of the operators

$$X_{ij} = a_i a_j, \qquad X^{ij} = a_j^{+} a_i^{+}, \qquad X_l^k = \tfrac{1}{2}(a_k^{+} a_l + a_l a_k^{+}) \tag{8.1.13}$$

which is isomorphic to the Lie algebra of the symplectic group. The commutation relations are

$$[X_{ij}, X_{kl}] = [X^{ij}, X^{kl}] = 0$$

$$[X_{ij}, X^{kl}] = X_i^k \delta_j^l + X_i^l \delta_j^k + X_j^k \delta_i^l + X_j^l \delta_i^k$$

$$[X_{ij}, X_l^k] = X_{il} \delta_j^k + X_{jl} \delta_i^k$$

$$[X^{ij}, X_l^k] = -X^{ik} \delta_l^j - X^{jk} \delta_l^i$$

$$[X_i^j, X_l^k] = X_l^j \delta_i^k - X_i^k \delta_l^j. \tag{8.1.14}$$

Consider now the infinitesimal symplectic transformations. The matrix given in (8.1.4) is rewritten as $M = \mathrm{I} + \varepsilon\tilde{M}$, and ε is an infinitesimal parameter, $\varepsilon \to 0$. The generator of the transformation satisfies the condition

$$\tilde{M}^{+} K + K\tilde{M} = 0 \tag{8.1.15}$$

or, equivalently,

$$\tilde{M}^{+} = -K\tilde{M}K. \tag{8.1.16}$$

Hence

$$\tilde{M}=\begin{pmatrix} C & A \\ \bar{A} & \bar{C} \end{pmatrix}, \qquad A'=A, \qquad C^{+}=-C, \tag{8.1.17}$$

where A and C are $N\times N$ matrices.

The considered representation $T(g)$ is unitary; it acts in the standard Hilbert space $\mathscr{H}_B$, the Fock space (Chaps. 1, 3). A basis in this space consists of the vectors

$$|[n]\rangle=|n_1,\ldots,n_N\rangle=[n_1!n_2!\ldots n_N!]^{-1/2}(a_1^+)^{n_1}\ldots(a_N^+)^{n_N}|0\rangle,$$

where the state vector $|0\rangle$ is the vacuum: $|0\rangle=|0,\ldots,0\rangle$, $a_j|0\rangle=0$.

It is not difficult to see that the Hilbert space is reducible with respect to the action of the operators given in (8.1.3). Namely, the states with even (odd) numbers $n=\sum_{j=1}^{N} n_j$ form an irreducible subspace for the representation T^+ (the space $\mathscr{H}_B^{(+)}$) and, respectively, T^- (the space $\mathscr{H}_B^{(-)}$).

One should have in mind, however, that the above representations are two-valued representations of the group $\mathrm{Sp}(2N, \mathbb{R})$. To be more precise, they are (single-valued) representations of the methaplectic group $\mathrm{Mp}(2N, \mathbb{R}) = \overline{\mathrm{Sp}(2N, \mathbb{R})}$.

8.2 Coherent States

Now we are in position to construct the CS systems for the representations T^+ and T^-. Note, first of all, that for any $g\in\mathrm{Sp}(2N, \mathbb{R})$ one has the decomposition

$$T(g)=\mathscr{N}\exp(-\tfrac{1}{2}\xi_{ij}X^{ij})\exp(\alpha_k^l X_l^k)\exp(-\tfrac{1}{2}\eta^{ij}X_{ij}), \tag{8.2.1}$$

where $\mathscr{N}$ is a normalization factor. In both subspaces $\mathscr{H}^{(+)}$ and $\mathscr{H}^{(-)}$ there is a vector $|\varphi_0\rangle$ annihilated by any operator X_{ij},

$$X_{ij}|\varphi_0\rangle=0. \tag{8.2.2}$$

Namely,

$$|\varphi_0\rangle=|0\rangle\equiv|0,\ldots,0\rangle \quad \text{for} \quad T^+ \tag{8.2.3}$$

and

$$|\varphi_0\rangle=|1,0,\ldots,0\rangle \qquad \text{for} \quad T^-; \tag{8.2.4}$$

in the latter case such a vector is not unique.

The constructions of the CS systems in both cases are similar; we are concerned here only with the case of T^+. In this case the isotropy subgroup H of the vector $|\varphi_0\rangle$ is $U(N)$:

$$H=\left\{g:g=\begin{pmatrix}U & 0\\ 0 & \bar{U}\end{pmatrix}\right\}, \qquad U\in U(N). \tag{8.2.5}$$

Applying various operators $T(g)$ to the state vector $|\varphi_0\rangle$ one gets, as usual, a CS system $\{T, |\varphi_0\rangle\}$ related to the bosonic operators.

As follows from (8.2.1), a CS vector $|\xi\rangle$ is defined by

$$|\xi\rangle=\mathcal{N}\exp\left(-\tfrac{1}{2}\xi_{ij}X^{ij}\right)|0\rangle, \tag{8.2.6}$$

where ξ_{ij} are elements of a complex symmetrical matrix ξ, and $\mathcal{N}$ is a normalization factor. Moreover, it follows from (8.2.1) that the CS vector satisfies

$$\tilde{a}_j|\xi\rangle=T(g)a_jT^{-1}(g)|\xi\rangle=0 \tag{8.2.7}$$

or, equivalently,

$$(Ua+Va^+)|\xi\rangle=0, \tag{8.2.8}$$

where a and a^+ are columns composed of the operators a_j, a_k^+, respectively. Hence

$$(a+\xi a^+)|\xi\rangle=0, \qquad \text{where} \tag{8.2.9}$$

$$\xi=U^{-1}V \tag{8.2.10}$$

is a symmetrical matrix, and (8.2.6) is valid. Recalling the condition (8.1.3), $UU^+-VV^+=\mathrm{I}$, it is clear that the Hermitian matrix $(\mathrm{I}-\xi\xi^+)$ must be positive definite. This is just the condition necessary for normalization of the state vector $|\xi\rangle$ given in (8.2.6). The set of complex symmetrical matrices satisfying this condition will be denoted by S_N.

Thus in the case in view a CS is determined by a point in the set S_N. The normalization factor $\mathcal{N}$ can be directly calculated:

$$\mathcal{N}=[\det(\mathrm{I}-\xi\xi^+)]^{1/4}. \tag{8.2.11}$$

The CS system has properties similar to those of the standard CS sytsem considered in Chap. 1. A few of them are presented below.

The states in the system are not orthogonal to one another; the scalar product is

$$\langle\xi|\eta\rangle=[\det(\mathrm{I}-\xi\xi^+)\det(\mathrm{I}-\eta\eta^+)]^{1/4}\,[\det(\mathrm{I}-\xi^+\eta)]^{-1/2}. \tag{8.2.12}$$

Expanding the CS over the standard basis gives

$$|\xi\rangle = \mathcal{N} \sum_{[n]} U_{[n]}(\xi)|[n]\rangle, \tag{8.2.13}$$

where the sum is over all $[n]=(n_1, \ldots, n_N)$, such that $|n|=\sum n_j$ is an even number. Hence

$$\sum_{[n]} \bar{U}_{[n]}(\xi)\, U_{[n]}(\eta) = [\det(\mathrm{I}-\xi^+\eta)]^{-1/2}. \tag{8.2.14}$$

Note an integral representation for the coefficient function resulting from (8.2.13):

$$U_{[m]}(\xi) = \int d\mu(Z)\, \frac{\bar{Z}^{[m]}}{\sqrt{[m]!}} \exp\left(-\tfrac{1}{2}\, \xi_{jk} z_j z_k\right), \tag{8.2.15}$$

where the measure is

$$d\mu(Z) = \frac{1}{\pi^N} \exp\left[-\sum_1^N |z_j|^2\right] \prod_1^N dx_j\, dy_j, \qquad z_j = x_j + \mathrm{i} y_j$$

and $|m| = \sum m_j$ is even.

The function $U_{[m]}(\xi)$ is a homogeneous polynomial of elements of the matrix ξ, its degree is $|m|/2$. As shown in [35], the polynomial satisfies the following set of differential equations

$$(K_{ij}K_{lm} - K_{im}K_{jl})\, U_{[n]}(\xi) = 0, \tag{8.2.16}$$

where $1 \le i, j, l, m \le N$ and

$$K_{ij} = \frac{1}{2}\, \frac{\partial}{\partial \xi_{ij}}, \qquad i \neq j; \qquad K_{ii} = \frac{\partial}{\partial \xi_{ii}}. \tag{8.2.17}$$

The relevant class of functions of interest for our purpose, $\mathscr{F}_N^+$, is the set of functions analytical in the generalized Siegel disk $\mathscr{S}_N$ satisfying the set of equations (8.2.16) and having a finite norm,

$$f(\xi) = \sum_{|m|=\text{even}} C_{[m]} U_{[m]}(\xi), \qquad \|f\|^2 = \sum_{[m]} |C_{[m]}|^2 < \infty \Lambda \tag{8.2.18}$$

where $C_{[m]}$ are coefficients in the expansion $f(\xi) = \sum_{|m|=\text{even}} C_{[m]} U_{[m]}(\xi)$ Because of (8.2.18), the functions are bound by the inequality

$$|f(\xi)| \le \|f\|\, [\det(\mathrm{I}-\xi\xi^+)]^{-1/4}. \tag{8.2.19}$$

The following lemma is useful.

Lemma. The entire function exp $[(z\xi z)/2]$ belongs to the space $\mathscr{F}_N^+$, if and only if the matrix

$$\mathrm{I}-\xi\xi^+=\mathrm{I}-\xi\bar{\xi} \tag{8.2.20}$$

is positive definite.

Finally, let me mention a formula showing the action of the representation $T^{(+)}(g)$ in the space $\mathscr{F}_N^+$,

$$T^{(+)}(g)f(\xi)=[\det(U+\xi\bar{V})]^{-1/2}f((U+\xi\bar{V})^{-1}(V+\xi\bar{U})). \tag{8.2.21}$$

The CS system for the odd representation $T^{(-)}(g)$ is constructed similarly; all the relevant formulae of this section are easily extended to the odd case.

8.3 Operators in the Space $\mathscr{H}_B^{(+)}$

Let $\hat{A}$ be an arbitrary operator acting in $\mathscr{H}_B^{(+)}$. Following the general pattern, we associate a covariant symbol with it

$$A(\bar{\xi},\xi)=\langle\xi|\hat{A}|\xi\rangle,\qquad \langle\xi|\xi\rangle=1. \tag{8.3.1}$$

Consider first the bilinear operators

$$X_{jk}=a_ja_k,\qquad X^{jk}=a_j^+a_k^+,\qquad X_k^j=\tfrac{1}{2}(a_j^+a_k+a_ka_j^+). \tag{8.3.2}$$

The semiclassical limit is pertinent here, so we renormalize the operators, and rewrite the commutation relations (8.1.1),

$$[a_j,a_k^+]=\hbar\delta_{jk},\quad [a_j,a_k]=[a_j^+,a_k^+]=0,\quad |\xi\rangle=\mathscr{N}\exp\left(-\frac{1}{2\hbar}\xi_{ij}X^{ij}\right)|0\rangle. \tag{8.1.1$'$}$$

A direct calculation shows [94] that the corresponding covariant symbols are elements of the matrices

$$A=-\hbar\xi(\mathrm{I}-\xi^+\xi)^{-1},\qquad A^+,\qquad B=\frac{\hbar}{2}\mathrm{I}+\hbar\xi(\mathrm{I}-\xi^+\xi)^{-1}\xi^+. \tag{8.3.3}$$

Note that the matrices A and B are not independent; the following relations hold:

$$B^2-A\bar{A}=\frac{\hbar^2}{4}\mathrm{I},\qquad BA=A\bar{B}, \tag{8.3.4}$$

$$B^+=B,\qquad A'=A. \tag{8.3.5}$$

The inverse statement is also valid. If two matrices A and B satisfy (8.3.4, 5), and A^{-1} exists, then such a symmetrical matrix ξ exists so that A and B are given by (8.3.3) [94].

Further, it is not difficult to see that an arbitrary element of the Lie algebra of the group $\mathrm{Sp}(2N, \mathbb{R})$ is given by (8.1.17), where A is complex, $A = A'$, and C is anti-Hermitian. Substituting there $C = \mathrm{i}B$, and introducing the matrix ξ by means of (8.3.3), the manifold $S_N = \{\xi\}$ embeds into the Lie algebra $\mathscr{G}$. Evidently, the image of S_N in $\mathscr{G}$ depends on Planck's constant, and we use the explicit notation $S_N(\hbar)$. Suppose $\mathrm{i}x \in S_N(h)$; then it is easily seen that

$$g x(\xi, \bar{\xi}) g^{-1} = x(g\xi, \overline{g\xi}), \tag{8.3.6}$$

where $g \in G$. Thus the manifolds $S_N(\hbar)$ are orbits of the adjoint representation for the group $G = \mathrm{Sp}(2N, \mathbb{R})$. Now (8.3.4) may be rewritten as

$$[x(\xi, \bar{\xi})]^2 = -\left(\frac{\hbar}{2}\right)^2 \mathrm{I}, \tag{8.3.7}$$

where I is the unit matrix.

Note that as $\hbar \to 0$, one gets the space of complex unitary symmetrical matrices,

$$\xi^+ \xi = \mathrm{I}, \tag{8.3.8}$$

so the manifold $S_N(0)$ belongs to the boundary of the Siegel disk $\{Z : \mathrm{I} - \bar{Z}Z > 0\}$.

The manifold S_N is Kählerian. This statement means that it is a complex manifold possessing Riemannian metrics, which may be written in local coordinates as

$$ds^2 = \sum_{a,b} \frac{\partial^2 F}{\partial z^a \partial \bar{z}^b} dz^a d\bar{z}^b, \tag{8.3.9}$$

where the function $F(z, \bar{z})$ is called the Kähler potential.

The metrics (8.3.9) is invariant under a transformation group G, if the potential F satisfies the condition

$$F(gz, \overline{gz}) = F(z, \bar{z}) + \alpha(g, z) + \overline{\alpha(g, z)}, \tag{8.3.10}$$

where $\alpha(g, z)$ is an analytical function of z (i.e., independent of $\bar{z}$).

Consider the manifold S_N, and let Z be its element. In view of (8.1.10) the function

$$f(Z, \bar{Z}) = \det(\mathrm{I} - Z\bar{Z}) \tag{8.3.11}$$

does exist, and because of (8.1.11) its transformation is

$$f(gZ, \overline{gZ}) = f(Z, \bar{Z}) \alpha(g, Z) \overline{\alpha(g, Z)}, \quad \text{where} \tag{8.3.12}$$

$$\alpha(g, z) = [\det(U + Z\bar{V})]^{-1}. \tag{8.3.13}$$

Therefore

$$F(Z,\bar{Z})=\tfrac{1}{2}\ln f(Z,\bar{Z}) \tag{8.3.14}$$

is the invariant Kähler potential. Hence

$$ds^2=\tfrac{1}{2}\operatorname{tr}\{dZ(\mathrm{I}-\bar{Z}Z)^{-1}d\bar{Z}(\mathrm{I}-Z\bar{Z})^{-1}\}. \tag{8.3.15}$$

Defining the scalar product $(x, y)=\operatorname{tr}\{xy^+\}$ for the $N\times N$ matrices, the interval can be rewritten as

$$ds^2=(dZ, H\,dZ), \qquad H\,dZ=\tfrac{1}{2}K\,dZ\,\bar{K}, \qquad K=(\mathrm{I}-Z\bar{Z})^{-1}. \tag{8.3.16}$$

The explicit form of the Laplace-Beltrami operator on the manifold S_N looks like

$$\Delta=2\operatorname{tr}\left\{\frac{\partial}{\partial\bar{Z}}(\mathrm{I}-\bar{Z}Z)\frac{\partial}{\partial Z}(\mathrm{I}-Z\bar{Z})\right\}. \tag{8.3.17}$$

where

$$\frac{\partial}{\partial Z}=\begin{pmatrix} \dfrac{\partial}{\partial z_{11}} & \cdots\cdots & \dfrac{1}{2}\dfrac{\partial}{\partial z_{1N}} \\ \cdots & \cdots & \cdots \\ \dfrac{1}{2}\dfrac{\partial}{\partial z_{1N}} & \cdots\cdots & \dfrac{\partial}{\partial z_{NN}} \end{pmatrix}. \tag{8.3.18}$$

Consider the Hilbert space $\mathscr{F}_k(S_N)$ of analytical functions on S_N with the scalar product

$$(f_1, f_2)=C_N(k)\int\bar{f}_1(Z)f_2(Z)\,[\det(\mathrm{I}-Z\bar{Z})]^{k/2}\,d\mu_N(Z,\bar{Z}). \tag{8.3.19}$$

Here k is a parameter, the invariant measure in S_N is

$$d\mu_N(z,\bar{z})=[\det(\mathrm{I}-Z\bar{Z})]^{-(N+1)}\prod_{j\le k}\frac{dx_{jk}dy_{jk}}{\pi}, \quad z_{jk}=x_{jk}+\mathrm{i}y_{jk} \tag{8.3.20}$$

and the normalization constant $C_N(k)$ is determined by the condition $(f_0, f_0)=1$, where $f_0(z)\equiv 1$. Using the results from [93], one gets an expression for the normalization constant

$$C_N(k)=2^{-N}\frac{\Gamma(k-1)\Gamma(k-2)\ldots\Gamma(k-N)}{\Gamma(k-2)\Gamma(k-4)\ldots\Gamma(k-2N)}. \tag{8.3.21}$$

The integral in (8.3.19) is convergent for $k>2N$, and for $k\le 2N$ it is obtained through the analytical continuation. The resulting expression defines a nonnegative scalar product on the ray $k > 2N$, as well as on certain points to the left of it,

$$k = 0, 1, 2, \dots (N-1). \tag{8.3.22}$$

Only these values of k will be considered.

The set of functions

$$f_{\bar{\zeta}}(Z) = C(\zeta, \bar{\zeta})\,[\det(\mathrm{I} - Z\bar{\zeta})]^{-k/2}, \qquad Z, \quad \zeta \in S_N \tag{8.3.23}$$

on the manifold S_N is just the generalized CS system, written in the z representation.

9. Coherent States for a Fermionic System with a Finite Number of Degrees of Freedom

This chapter studies the CS system for a fermionic system with a finite number of degrees of freedom. The system state is specified with a complex skew-symmetrical matrix. The exposition follows [94].

9.1 Canonical Transformations

We shall consider the Fock space $\mathscr{H}_F$ for a fermionic system with N degrees of freedom. The system contains identical particles subject to Fermi-Dirac statistics, and its states are described by completely antisymmetrical wave functions. (A rigorous mathematical approach was given in [6].) The fermion creation-annihilation operators c_j^+, c_k satisfy the anticommutation relations

$$\{c_k, c_j^+\} \equiv c_k c_j^+ + c_j^+ c_k = \delta_{jk}, \qquad \{c_j, c_k\} = \{c_j^+, c_k^+\} = 0. \tag{9.1.1}$$

Consider linear canonical transformations, i.e., the linear homogeneous transformations of the operators c_j, c_k^+, that do not change the commutation relations (9.1.1),

$$c_i \to \tilde{c}_i = u_{ij} c_j + v_{ik} c_k^+, \qquad c_k^+ \to \tilde{c}_k^+ = \bar{v}_{kj} c_j + \bar{u}_{kl} c_l^+. \tag{9.1.2}$$

It is easily seen that the following relations are necessary for the transformation (9.1.2) to be canonical

$$UV' + VU' = 0, \qquad UU^+ + VV^+ = \mathrm{I}. \tag{9.1.3}$$

Evidently, these conditions are equivalent to the unitarity of the matrix

$$M = \begin{pmatrix} U & V \\ \bar{V} & \bar{U} \end{pmatrix}, \qquad MM^+ = \mathrm{I}. \tag{9.1.4}$$

Hence the inverse matrix is

$$M^{-1} = \begin{pmatrix} U^+ & V' \\ V^+ & U' \end{pmatrix}. \tag{9.1.5}$$

Furthermore, we suppose that U^{-1} does exist. Then

$$U^{-1}V = -V'(U^{-1})', \qquad \bar{V}U^{-1} = -(U^{-1})'V^{+} \tag{9.1.6}$$

is equivalent to the skew symmetry of the matrices

$$Z = U^{-1}V \quad \text{and} \quad W = \bar{V}U^{-1}. \tag{9.1.7}$$

It follows from (9.1.3) that the matrices $(\mathrm{I} + ZZ^{+})$ and $(\mathrm{I} + W^{+}W)$ are positive definite.

Evidently, the matrices M in (9.1.4) form a group with respect to the conventional matrix multiplication. This group is isomorphic to $SO(2N, \mathbb{R})$, the group of real orthogonal $2N \times 2N$ matrices. A few properties of this group are mentioned.

1. The group $SO(2N, \mathbb{R})$ is compact (its invariant volume is finite), so all its unitary irreducible representations are finite-dimensional.

2. The group $SO(2N, \mathbb{R})$ is doubly connected: a closed path corresponding to a rotation by the angle 2π cannot be continuously deformed into a point, while the rotation by 4π can be contracted. So taking two copies of the group and *splicing* them appropriately gives a simply connected group, the universal covering $\tilde{G}$, called the spinor group Spin$(2N)$.

3. The group G contains an important subgroup K, isomorphic to the unitary group $U(N)$; K contains matrices of the form

$$K = \begin{pmatrix} U & 0 \\ 0 & \bar{U} \end{pmatrix}, \qquad U \in U(N).$$

4. The matrices Z (respectively, W) in (9.1.7) are representatives of the left (respectively, right) cosets with respect to the subgroup K. The quotient space G/K is a symmetrical space of the compact type. It is a complex manifold, and the matrices Z (respectively, W), $Z' = -Z$, determine a local coordinate system in this manifold.

The group G acts in the matrix space as a group of linear-fractional transformations,

$$g\colon Z \to \tilde{Z} = (U + Z\bar{V})^{-1}(V + Z\bar{U}). \tag{9.1.8}$$

Note also that the operators $\tilde{c}_j$ and c_j are unitary equivalent

$$\tilde{c}_j = T^{(+)}(g)c_jT(g), \qquad g = \begin{pmatrix} U & V \\ \bar{V} & \bar{U} \end{pmatrix} \in G, \tag{9.1.9}$$

and the operators $T(g)$ form a representation of the group $SO(2N, \mathbb{R})$.

As in the preceding chapter, the infinitesimal operators of the representation are quadratic in the operators c_j, c_k^{+}. Namely, the operators

$$X_{ij}=c_i c_j, \qquad X^{ij}=c_j^+ c_i^+, \qquad X_i^j=\tfrac{1}{2}(c_i c_j^+ - c_j^+ c_i) \tag{9.1.10}$$

form the Lie algebra isomorphic to that for the group $SO(2N, \mathbb{R})$. The commutation relations are

$$\begin{aligned}
&[X_{ij}, X_{kl}]=[X^{ij}, X^{kl}]=0\\
&[X_{ij}, X^{kl}]=X_i^k\delta_j^l - X_i^l\delta_j^k - X_j^k\delta_i^l + X_j^l\delta_i^k\\
&[X_{ij}, X_l^k]=X_{il}\delta_j^k - X_{jl}\delta_i^k, \qquad [X^{ij}, X_l^k]= -X^{ik}\delta_l^j + X^{ik}\delta_l^i\\
&[X_i^j, X_l^k]=X_l^j\delta_i^k - X_i^k\delta_l^j.
\end{aligned} \tag{9.1.11}$$

Consider now the infinitesimal orthogonal transformations, the corresponding matrices are $M=\mathrm{I}+\mathrm{i}\varepsilon\tilde{M}$, $\varepsilon\to 0$. The infinitesimal rotation matrix $\tilde{M}$ is Hermitian, $\tilde{M}^+=M$. Hence

$$\tilde{M}=\begin{pmatrix} C & A \\ \bar{A} & \bar{C}\end{pmatrix}, \qquad A'=-A, \qquad C^+=C. \tag{9.1.12}$$

The considered representation $T(g)$ is unitary and acts in the standard finite-dimensional (of the dimensionality 2^N) space. It is the Fock space with the basis $|n_1,\dots,n_N\rangle=|[n]\rangle$, where the numbers n_j acquire only two values, 0 and 1,

$$|[n]\rangle=|n_1,\dots,n_N\rangle=(c_1^+)^{n_1}\dots(c_N^+)^{n_N}|0,0,\dots,0\rangle$$

and $|0\rangle=|0,\dots,0\rangle$ is the vacuum state vector, satisfying the conditions $c_j|0\rangle=0$.

It is easy to see that the space $\mathscr{H}_F$ is reducible with respect to the action of operators of the representation $T(g)$. Namely, there are two irreducible subspaces $\mathscr{H}_F^{(+)}$ and $\mathscr{H}_F^{(-)}$ with the basis vectors corresponding to even and odd $|n|=\sum n_j$, respectively, where the irreducible representations $T^{(+)}$ and $T^{(-)}$ act.

The representations in view are the so-called spinor representations of the group $SO(2N, \mathbb{R})$. To be more precise, they are two-valued representations of the doubly connected group $SO(2N, \mathbb{R})$, and realize the one-valued representations of the universal covering group Spin$(2N)$.

9.2 Coherent States

Constructing the CS systems for the representations $T^{(+)}(g)$ and $T^{(-)}(g)$ is analoguous to that presented in Sect. 8.2. So here I give only some formulae for the CS system related to the representation $T^{(+)}(g)$. The coherent state in view is defined by

$$|\xi\rangle=\mathscr{N}\exp\left(-\tfrac{1}{2}\xi_{ij}X^{ij}\right)|0\rangle, \tag{9.2.1}$$

where ξ is a complex skew-symmetrical matrix, and $\mathcal{N}$ is a normalization factor. This state $|\xi\rangle$ satisfies

$$(c_j+\xi_{jk}c_k^+)|\xi\rangle=0, \tag{9.2.2}$$

and the normalization factor is

$$\mathcal{N}=[\det(\mathrm{I}+\xi\xi^+)]^{-1/4}=[\det(\mathrm{I}-\xi\bar{\xi})]^{-1/4}. \tag{9.2.3}$$

The coherent states are not orthogonal to one another; the scalar product is

$$\langle\xi|\eta\rangle=[\det(\mathrm{I}+\xi\xi^+)\det(\mathrm{I}+\eta\eta^+)]^{-1/4}[\det(\mathrm{I}+\xi^+\eta)]^{1/2}. \tag{9.2.4}$$

Expanding CS over the canonical basis gives

$$|\xi\rangle=\mathcal{N}\sum_{[n]}u_{[n]}(\xi)|[n]\rangle, \tag{9.2.5}$$

where $[n]=(n_1,\ldots,n_N)$, $n_j=0, 1$, and the sum is over all n with even $|n|=\sum n_j$. Hence one gets

$$\sum_{[n]}\bar{u}_{[n]}(\xi)u_{[n]}(\eta)=[\det(\mathrm{I}+\xi^+\eta)]^{1/2}. \tag{9.2.6}$$

In an analogy to Sect. 8.2, one has to consider the space $\mathcal{F}_N$, the space of polynomials of elements of the matrix ξ, which may be expanded over the basis polynomials

$$u(\xi)=\sum_{|m|=\text{even}}C_{[m]}u_{[m]}(\xi).$$

Finally, I give the formula describing the action of the representation $T^{(+)}(g)$ in the space $\mathcal{F}_N$,

$$T^{(+)}(g)f(\xi)=[\det(U+\xi\bar{V})]^{1/2}f((U+\xi\bar{V})^{-1}(V+\xi\bar{U})). \tag{9.2.7}$$

9.3 Operators in the Space $\mathcal{H}_F^{(+)}$

The procedure is quite similar to that of Sect. 8.3. For an arbitrary operator $\hat{A}$, acting in $\mathcal{H}_F^{(+)}$, there is a covariant symbol

$$A(\bar{\xi},\xi)=\langle\xi|\hat{A}|\xi\rangle,\qquad \langle\xi|\xi\rangle=1. \tag{9.3.1}$$

Consider the bilinear operators,

$$X_{jk}=c_jc_k,\qquad X^{jk}=c_k^+c_j^+,\qquad X_j^k=\tfrac{1}{2}(c_k^+c_j-c_jc_k^+). \tag{9.3.2}$$

It is suitable to renormalize the operators, so that the commutation relations contain Planck's constant $\hbar$ explicitly

$$\{c_j, c_k^+\} = \hbar\delta_{jk}, \quad \{c_j, c_k\} = 0, \quad \{c_j^+, c_k^+\} = 0. \tag{9.1.1'}$$

It is not difficult to show [94] that their covariant symbols are elements of the matrices

$$A = -\hbar\xi(\mathrm{I} + \xi^+\xi)^{-1}, \quad A^+,$$
$$B = \hbar\xi(\mathrm{I} + \xi^+\xi)^{-1}\xi^+ - \frac{\hbar}{2}\,\mathrm{I}, \tag{9.3.3}$$

and the matrices A and B satisfy the conditions

$$B^2 - A\bar{A} = \frac{\hbar^2}{4}\,\mathrm{I}, \quad BA = A\bar{B}, \quad B = B^+, \quad A' = -A. \tag{9.3.4}$$

The inverse statement is also true: if the matrices A, B satisfy (9.3.4), a matrix ξ exists such that A and B are expressed by (9.3.3).

An arbitrary element of the Lie algebra for the group $SO(2N, \mathbb{R})$ is written as

$$\mathrm{i}\begin{pmatrix} C & A \\ -\bar{A} & -\bar{C} \end{pmatrix}, \quad C^+ = C, \quad A' = -A, \tag{9.3.5}$$

where A is a complex skew-symmetrical matrix, and C is an Hermitian matrix. As in Sect. 8.3, it is suitable to substitute $C = B$, and to express A and B in terms of ξ and $\bar{\xi}$; the result is an embedding of M_N into the Lie algebra $\mathscr{G}$. The embedding depends on $\hbar$, and the image of M_N in $\mathscr{G}$, evidently, is the orbit of the adjoint representation of the group $SO(2N, \mathbb{R})$. In the limit $\hbar \to 0$ the manifold $M_N(\hbar)$ is reduced to a single point. The reason is that no nontrivial classical limit of the quantum theory exists in the fermionic case.

It is remarkable that the manifold M_N is Kählerian as in the bosonic case. The metric in local coordinates is

$$ds^2 = \sum_{a,b} \frac{\partial^2 F}{\partial z^a \partial \bar{z}^b}\, dz^a\, d\bar{z}^b, \tag{9.3.6}$$

where the Kähler potential is

$$F = \ln\det(\mathrm{I} + ZZ^+). \tag{9.3.7}$$

The potential F (so also the metric) is invariant under the action of the group $G = SO(2N, \mathbb{R})$. The corresponding expression for the metric is

$$ds^2 = \tfrac{1}{2}\operatorname{tr}\{dZ(\mathrm{I} + Z^+Z)^{-1}dZ^+(\mathrm{I} + ZZ^+)^{-1}\}. \tag{9.3.8}$$

Introducing, as usual, the scalar product $(x, y)=\operatorname{tr}\{xy^+\}$ in the matrix space, the metric is rewritten as

$$ds^2=(dZ, H\,dZ), \quad H\,dZ=\tfrac{1}{2}K\,dZ\,K^+, \quad K=(\mathrm{I}+ZZ^+)^{-1}. \tag{9.3.9}$$

The Laplace-Beltrami operator in the space M_N looks like

$$\Delta=-2\operatorname{tr}\left\{\frac{\partial}{\partial Z}(1+Z^+Z)\frac{\partial}{\partial \bar{Z}}(\mathrm{I}+ZZ^+)\right\}, \tag{9.3.10}$$

where

$$\frac{\partial}{\partial Z}=\frac{1}{2}\begin{pmatrix} 0 & \dfrac{\partial}{\partial z_{12}} & \cdots\cdots & \dfrac{\partial}{\partial z_{1N}} \\ -\dfrac{\partial}{\partial z_{12}} & 0 & \cdots\cdots\cdots & \\ -\dfrac{\partial}{\partial z_{1N}} & \cdots\cdots\cdots & & 0 \end{pmatrix}. \tag{9.3.11}$$

Consider the Hilbert space $\mathscr{F}_k(M_N)$ of analytical functions on M_N with the scalar product

$$(f_1, f_2)=C_N(k)\int \overline{f_1(Z)}\,f_2(Z)\,[\det(\mathrm{I}+Z^+Z)]^{-k/2}\,d\mu_N(Z,\bar{Z}). \tag{9.3.12}$$

Here the invariant measure is

$$d\mu_N(Z,\bar{Z})=[\det(\mathrm{I}+ZZ^+)]^{-(N-1)}\prod_{i<j}\frac{dx_{ij}dy_{ij}}{\pi},$$

$$z_{jk}=x_{jk}+\mathrm{i}y_{jk} \tag{9.3.13}$$

and the normalization constant $C_N(k)$ is determined by the condition $(f_0, f_0)=1$ with $f_0(z)=1$. Using some results from [93] one gets

$$C_N(k)=\frac{\Gamma(k+N+1)\Gamma(k+N+2)\ldots\Gamma(k+2N)}{\Gamma(k+1)\Gamma(k+3)\ldots\Gamma(k+2N-1)}. \tag{9.3.14}$$

Note that from the point of view of the quantization theory (Part II) the spaces $\mathscr{F}_k(M_N)$ are of interest for integer $k=0, 1, 2, \ldots$.

Part II

General Case

10. Coherent States for Nilpotent Lie Groups

This chapter considers square-integrable CS systems related to connected simply connected nilpotent Lie groups. These states are related to the simplest orbits of coadjoint group representations, namely, to orbits which are linear spaces. In this case, as for the Heisenberg-Weyl group $G=W_N$, any coherent state is parametrized by a point of the orbit.

10.1 Structure of Nilpotent Lie Groups

Let G be a Lie group. Let us consider the sequence of subgroups of this group

$$G_0=G\supset G_1\supset\ldots\supset G_k\supset\ldots, \tag{10.1.1}$$

where $G_k=[G_{k-1},G]$, i.e., G_k is the closed subgroup in G_{k-1} generated by commutators of the type $aba^{-1}b^{-1}$, $a\in G$, $b\in G_{k-1}$. It is easy to see that subgroup G_k is an invariant subgroup not only for group G_{k-1} but also for the whole group G.

Definition 1. Group G is called nilpotent if sequence (10.1.1) is terminated, $G_l=\{e\}$ for all $l\geq k$, i.e., higher members of the sequence are trivial subgroups containing a single element (the unity) only.

For Lie algebra $\mathscr{G}$ we consider the sequence of subalgebras

$$\mathscr{G}_0=\mathscr{G}\supset\mathscr{G}_1\supset\ldots\supset\mathscr{G}_k\supset\ldots, \tag{10.1.2}$$

where $\mathscr{G}_k=[\mathscr{G}_{k-1},\mathscr{G}]$, i.e., $\mathscr{G}_k$ is the subalgebra in $\mathscr{G}_{k-1}$ generated by commutators of the type $[\alpha,\beta]=\alpha\beta-\beta\alpha$, $\alpha\in\mathscr{G}$, $\beta\in\mathscr{G}_{k-1}$. Analogously to the case of the Lie group, the Lie algebra $\mathscr{G}_k$ is an invariant subalgebra not only for Lie algebra $\mathscr{G}_{k-1}$ but also for the whole Lie algebra $\mathscr{G}$.

Definition 2. Lie algebra $\mathscr{G}$ is called nilpotent if sequence (10.1.2) is terminated, $\mathscr{G}_l=\{0\}$ for all $l\geq k$, i.e., subalgebras $\mathscr{G}_l$ contain the zero element only. Note that a connected Lie group is nilpotent if the corresponding Lie algebra is nilpotent.

In the following we shall consider only connected simply connected linear Lie groups. Such groups are determined completely by the corresponding Lie algebras. A useful criterion of nilpotence is given by the Engel theorem.

Engel Theorem. Let X be a linear Lie algebra and let each element of X satisfy the condition $x^m = 0$. Then there is a basis in which all matrices $x \in X$ are strictly triangular, i.e., they have zeros on the main diagonal and below it.

Hence, a linear Lie algebra satisfying the condition of the Engel theorem is nilpotent. Note also that any nilpotent Lie algebra has a nontrivial center and that the center is necessarily connected.

All unitary irreducible representations of nilpotent Lie groups may be found by the so-called "orbit method" developed by *Kirillov* [19, 95]. The essence of this method is a one-to-one correspondence between unitary irreducible representations of the group and the orbits of this group in the space $\mathscr{G}^*$, dual to Lie algebra $\mathscr{G}$ of this group.

10.2 Orbits of Coadjoint Representation

We present here only formulations of those statements necessary for our purpose. A more detailed consideration and proofs of these statements may be found in [19, 95]. Let $G = \{g\}$ be an arbitrary Lie group, $\mathscr{G}$ be the corresponding Lie algebra and $\mathscr{G}^*$ be a space dual to $\mathscr{G}$, i.e., the space of linear functionals on $\mathscr{G}$. Group G acts in space $\mathscr{G}$ by means of the adjoint representation $Ad(g)$ and in space $\mathscr{G}^*$ by means of the coadjoint representation $Ad^*(g)$.

Space $\mathscr{G}^*$ is foliated to orbits under the action of $Ad(g)$. Let us denote the set of orbits of coadjoint representation of group G by $O(G) = \{O\}$. As shown in [95], every such orbit has a standard closed nondegenerate G-invariant 2-form B^O, the so-called Kirillov form.

The definition of this form is as follows.

Let ξ_x be a vector field on O corresponding to an element x of Lie algebra $\mathscr{G}$. Then for any point $f \in O$

$$B_f^O(\xi_x(f), \xi_y(f)) = \langle f, [x, y] \rangle. \tag{10.2.1}$$

Here $\langle f, z \rangle$ is the value of functional f on element $z \in \mathscr{G}$. Hence it follows, in particular, that all the orbits O are even-dimensional manifolds.

Recall that the real manifold of even dimensions with a closed nondegenerate 2-form is called a symplectic manifold. Therefore any orbit of coadjoint representation is a homogeneous (with respect to the action of G) symplectic manifold.

Hence the orbit can be considered as a phase space for a Hamiltonian system for which the given group G is the symmetry group. (A number of such systems is considered in [19].)

Actually, the orbits essentially exhaust all homogeneous symplectic manifolds, since the following statement holds.

Theorem [19]. Any homogeneous symplectic manifold for which the motion group is a Lie group G is locally isomorphic to an orbit of coadjoint representation of group G or of the central extension of this group by means of the additive group of real numbers.

10.3 Orbits of Nilpotent Lie Groups

One-to-one correspondence between the orbits of the coadjoint representation and nonequivalent unitary irreducible representations of the group considered has been established for nilpotent Lie groups in [95].

Let us consider, for example, the simplest non-Abelian nilpotent Lie group, the so-called Heisenberg-Weyl group W_1. This group can be considered as a matrix group $\{g\}$ with elements of the form

$$g=\begin{pmatrix}1 & a & c\\ 0 & 1 & b\\ 0 & 0 & 1\end{pmatrix}. \tag{10.3.1}$$

The Lie algebra $\mathscr{G}$ for this group consists of matrices like

$$\begin{pmatrix}0 & \alpha & \gamma\\ 0 & 0 & \beta\\ 0 & 0 & 0\end{pmatrix} \tag{10.3.2}$$

and space $\mathscr{G}^*$, dual to $\mathscr{G}$, may be realized as the space of matrices of the form

$$\begin{pmatrix}0 & 0 & 0\\ x & 0 & 0\\ z & y & 0\end{pmatrix}. \tag{10.3.3}$$

The coadjoint representation transforms coordinates x, y, z as follows

$$g:\ \begin{aligned} &x\to x+bz\\ &y\to y-az\\ &z\to z. \end{aligned} \tag{10.3.4}$$

So it is seen that orbits of the coadjoint representation are of two types, one composed of planes, the other of points.

1) The planes are defined by

$$z=\lambda\neq 0. \tag{10.3.5}$$

These orbits correspond to infinite-dimensional unitary irreducible representations $T^{\lambda}(g)$ (Sect. 1.1).

2) The points

$$x=\mu, \quad y=\nu, \quad z=0 \tag{10.3.6}$$

correspond to a one-dimensional representation $T^{\mu,\nu}(g)$.

Let us consider now an arbitrary simply connected nilpotent Lie group G. It is known [95] that any orbit of coadjoint representation is an algebraic manifold in space $\mathscr{G}^*$, i.e., it is given by the set of polynomial equations in the space $\mathscr{G}^*$, and any orbit is simply connected. If the dimension of the typical orbit is $2m$, then the set of such orbits is parametrized by r real numbers where the dimension of the group G is $n=2m+r$.

Let us note that the orbit to which an element $f \in \mathscr{G}^*$ belongs lies in the affine hyperplane $f+\mathscr{Z}^{\perp}$, where $\mathscr{Z}^{\perp}$ is the set of linear functionals vanishing at the center Z of algebra $\mathscr{G}$. This hyperplane depends only on restriction f_0 of the functional f to the center of Lie algebra $\mathscr{G}: f_0 \in \mathscr{Z}^*$.

It is known as well that a linear subspace of a dimensionality not less than m must belong to the 2-m-dimensional orbit O. Moreover, sometimes the orbit O is a linear space. This case is considered below, where $O=f+\mathscr{Z}^{\perp}$.

10.4 Representations of Nilpotent Lie Groups

The following theorem proved by *Kirillov* [95] is valid for connected nilpotent Lie groups.

Theorem. Any unitary irreducible representation of such a group G is monomial, i.e., this representation is induced by a one-dimensional representation of some closed subgroup H.

The monomial representation is determined by an element $f \in \mathscr{G}^*$, i.e., by a linear functional on the Lie algebra $\mathscr{G}$ of group G.

Representations corresponding to functionals f_1 and f_2 are equivalent if and only if f_1 and f_2 belong to the same orbit of the coadjoint representation.

Therefore, a one-to-one correspondence arises between the orbits of coadjoint representation of the group G and unitary irreducible representations of this group.

This construction is described in brief below.

Let f be a point of orbit O, i.e., an element of space $\mathscr{G}^*$. Subalgebra $\mathscr{G}_0$ is called subordinate to function f if

$$\langle f, [x, y]\rangle = 0 \tag{10.4.1}$$

for every pair $x, y \in \mathscr{G}_0$. Furthermore, we assume that the subalgebra $\mathscr{G}_0$ is maximal, and the exponential mapping generates the simply connected group G_0. Recall that for connected simply connected nilpotent Lie groups this mapping is one-to-one.

We can construct unitary representation of group G as a representation induced by a one-dimensional representation U^f of subgroup G_0:

$$U^f: g_0 \to \lambda(g_0) = \exp\,(\mathrm{i}\,\langle f, \ln g_0 \rangle). \tag{10.4.2}$$

Kirillov [95] showed that such representation is irreducible and that the representations corresponding to two different points f_1 and f_2 of the orbit are equivalent. Moreover, it was found that this construction gives all unitary irreducible representations of the connected simply connected nilpotent Lie group.

The representations in view can be realized in the space of functions of m real variables, where $m = \dim O/2$. For the generic orbits this representation depends on $(\dim G - 2m)$ real parameters.

We are interested in an important class of representations of Lie groups introduced in [96], the class of the so-called square-integrable representations defined as follows.

Let Z be the center of the Lie group G, i.e., the set of elements which commute with all elements of G. Let us call the unitary irreducible representation $T(g)$ of group G in the Hilbert space $\mathscr{H}$ square-integrable if nonzero vectors $|\psi_1\rangle$ and $|\psi_2\rangle$ exist such that

$$\int_{X=G/Z} |\langle \psi_1 | T(g) | \psi_2 \rangle|^2 \, d\mu(x) < \infty. \tag{10.4.3}$$

Here $d\mu(x)$ is an invariant measure on the homogeneous space $X = G/Z$ and it should be taken into account that $|\langle \psi_1 | T(g) | \psi_2 \rangle|$ depends not on g but only on $x(g)$, i.e., on projection of element g onto space X.

This definition is a natural generalization of the standard definition of square-integrable representations.

One should bear in mind that if condition (10.4.3) is satisfied for some vectors $|\psi_1\rangle$ and $|\psi_2\rangle$, it will, as known, also be fulfilled for every pair $|\psi_1\rangle$ and $|\psi_2\rangle$.

Let us formulate now an important statement proved in [96].

Theorem 1. Let f be a linear functional on $\mathscr{G}$, O be the orbit corresponding to it and $T^O(g)$ be the corresponding representation.

The following statements are equivalent:

i) Representation $T^O(g)$ is square-integrable;

ii) the orbit O is a linear space, $O = f + \mathscr{Z}^{\perp}$;

iii) the 2-form $B^f = \langle f, [x, y] \rangle$ is nondegenerate on the space $\mathscr{G}/\mathscr{Z}$.

Two more criteria for the existence of square-integrable representations are also known [96].

1) Let us define a polynomial $P(f)$ on $\mathscr{G}^*$ according to

$$P(f) = Pf(B^f), \tag{10.4.4}$$

where $Pf(B)=\sqrt{\det B}$ is the so-called Pfaffian of the skew-symmetric matrix B of an even order which is a polynomial in elements of B, as is well known. It can be shown [97] that the function $P(f)$ depends only on the restriction f_0 of the element f to the center $\mathscr{Z}$ of $\mathscr{G}$. It is a homogeneous polynomial function.

The condition that the representation is square-integrable is equivalent to condition

$$P(f) \neq 0. \tag{10.4.5}$$

Hence it follows that if a single square-integrable representation exists, then almost all irreducible representations (given in terms of the Plancherel measure) are also square-integrable.

2) Let $\mathscr{Z}$ be the center of the Lie algebra $\mathscr{G}$, $\mathscr{Z}(\mathscr{G})$ be the center of the universal enveloping algebra of the Lie algebra $\mathscr{G}$, $\mathscr{S}(\mathscr{Z})$ be the universal enveloping algebra for $\mathscr{Z}$.

Proposition [96]. Group G has square-integrable representations if and only if $\mathscr{Z}(\mathscr{G})=\mathscr{S}(\mathscr{Z})$.

Remark [96]. The definition of square-integrable representations can be replaced by an equivalent definition: representation $T(g)$ is square-integrable in the sense used in [96], if this representation determines square-integrable representations (in the usual sense) of the quotient group G/G_1, where G_1 is the kernel of representation $T(g)$.

10.5 Coherent States

In our case unitary irreducible representations are related to orbits of the coadjoint representation. Hence the coherent states are also related to orbits. Thus, the orbit O of the coadjoint representation corresponds to a unitary irreducible representation $T^O(g)$ of group G, and so it is related to a CS system $\{|x\rangle\}$ whose elements are parametrized by points of orbit $O=X=\{x\}$.

For the simplest case of the Heisenberg-Weyl group the typical orbit is the $2m$-dimensional Euclidean space. The corresponding CS system was also parametrized by points of the $2m$-dimensional Euclidean space, i.e., in this case $X=O$.

The problem in view is to describe all systems possessing this property. An exhaustive answer was given in [98].

Proposition. Suppose that orbit O of the coadjoint representation of connected simply connected Lie group G is a linear space itself. Then representation $T^O(g)$ admits a CS system which is parametrized by points of O.

Proof. As noted above, the representation $T^O(g)$ is square-integrable. Hence, from [96] it follows that:

i) $O \approx G/Z$, where Z is the center of G;
ii) representation $T(g)$ on Z is $T(z)=\lambda(z)\hat{I}$, where $\lambda(z)$ is the character of Z;
iii) $\int_X |\langle\psi|T^O(g)|\varphi\rangle|^2 d\mu(x) < \infty, \; X=G/Z$

for arbitrary vectors $|\psi\rangle, |\varphi\rangle$ of the Hilbert space $\mathscr{H}$. It is evident now that the action of operators $T(g)$ on the normalized vector $|\psi_0\rangle$ gives the system of coherent states parametrized by the points of the orbit.

The inverse statement is also valid.

Theorem [98]. Let $T(g)$ be an irreducible unitary representation of connected simply connected nilpotent Lie group G and let O be the corresponding orbit of the coadjoint representation. Suppose that representation T^O admits the CS system parametrized by the points of O. Then O is a linear manifold in $\mathscr{G}^*$.

The proof of this theorem is not too simple [98]. It is based on a lemma which will be given here also without a proof.

Lemma. Let H be a closed connected subgroup of connected simply connected nilpotent Lie group G and let $\lambda(h)$ be a unitary character of $H=\{h\}$.

Suppose that a unitary irreducible representation $T(g)$ of group G is contained in the representation induced by representation of group H with the character $\lambda(h)$. Let $T(g)$ contain a representation of H with the character $\lambda(h)$ under restriction of G to H. Then the following statements are valid.

i) Orbit O corresponding to $T(g)$ is a linear manifold in the space $\mathscr{G}^*$ dual to Lie algebra $\mathscr{G}$ of group G.
ii) Subgroup H is isomorphic to the isotropy subgroup G^f of arbitrary point f of orbit O.
iii) The representation of group G induced by the character of subgroup H is a multiple of representation $T(g)$.

Concluding this section, we would like to restate [98] the theorem formulated above. In the alternative formulation, the theorem is valid for a wider class of Lie groups, possibly for exponentially solvable Lie groups.

Theorem. The unitary irreducible representation $T^O(g)$ of simply connected nilpotent Lie group G admits a CS system parametrized by the points of the same orbit O of the coadjoint representation if and only if it is a square-integrable representation of the quotient group G/G_1, where G_1 is the kernel of representation $T(g)$.

Remark. A similar theorem is valid also for compact Lie groups.

11. Coherent States for Compact Semisimple Lie Groups

Coherent-state systems related to compact semisimple Lie groups have properties analogous to those for the rotation group of three-dimensional space (Chap. 4). The arguments, however, use a more sophisticated technique [19, 70, 79].

11.1 Elements of the Theory of Compact Semisimple Lie Groups

Note first of all that any compact semisimple Lie group is a direct product of compact simple Lie groups [19, 70, 79]. Therefore it is sufficient to consider the case of a compact simple Lie group.

So let G be a compact simple Lie group, i.e., a compact Lie group with no closed connected invariant subgroup. Let $\mathscr{G}$ be the Lie algebra of group G, $\{X_a\}$ a basis in $\mathscr{G}$, $\mathscr{G}^c = \mathscr{G} + i\mathscr{G}$ a complex extension of $\mathscr{G}$, i.e., the set of linear combinations of its elements with complex coefficients, H the Cartan subgroup of G, i.e., the maximal commutative semisimple subgroup in G, $\mathscr{H}$ the Cartan subalgebra in $\mathscr{G}$, i.e., the Lie algebra for group H.

As is well known [70, 79] it is possible to choose a canonical basis $\{E_\alpha, H_j\}$ in Lie algebra, the so-called Cartan basis, for which the commutation relations take the form

$$\begin{aligned}
&[H_j, E_\alpha] = \alpha_j E_\alpha, \qquad [H_j, H_k] = 0 \\
&[E_\alpha, E_\beta] = N_{\alpha\beta} E_{\alpha+\beta}, \quad \text{if} \quad \alpha+\beta \in R \\
&[E_\alpha, E_\beta] = 0, \quad \text{if} \quad \alpha+\beta \bar{\in} R \quad \text{and} \quad \alpha+\beta \neq 0 \\
&[E_\alpha, E_{-\alpha}] = \alpha_j H_j .
\end{aligned} \tag{11.1.1}$$

Here

$$\mathscr{G}^c = \mathscr{H}^c \bigoplus_{\alpha \in R} \mathscr{G}_\alpha$$

is the so-called Cartan decomposition of the algebra $\mathscr{G}^c$, $E_\alpha \in \mathscr{G}_\alpha$, $H_j \in \mathscr{H}$ ($j = 1, \ldots, r$; r is the rank of $\mathscr{G}$); $R = \{\alpha\}$ is a set of vectors $\{\alpha = (\alpha_1, \ldots \alpha_r)\}$ in the r-dimensional space, the so-called root system for the Lie algebra $\mathscr{G}^c$.

Compact simple Lie algebras can be completely classified: there are four series A_r, B_r, C_r, D_r (r is the rank of the algebra), and five exceptional algebras G_2, F_4, E_6, E_7, and E_8 [70, 79].

Any Lie algebra corresponds to a connected simply connected Lie group G; all connected (but not necessarily simply connected) Lie groups can be obtained from Lie group G by factorizing G by its center Z, which has a finite number of elements for compact simple groups.

Any compact simple Lie group G can be embedded into a connected complex Lie group G^c, which is obtained by exponential mapping from the Lie algebra $\mathscr{G}^c$. The group G^c has the following important subgroups.

Subgroups $B_\pm$ are obtained by exponential mapping of the subalgebras $\mathscr{B}_\pm$ spanned on elements E_α, H_j ($\alpha \in R_\pm$, $H_j \in \mathscr{H}^c$; $R_+(R_-)$ is a subsystem of positive (negative) roots). Subgroups $B_\pm$ are maximal connected solvable subgroups, the so-called Borel subgroups. Other important subgroups are subgroups $Z_\pm$, obtained by exponential mapping of subalgebras $\mathscr{Z}_\pm$ spanned on elements E_α ($\alpha \in R_\pm$, respectively). These subgroups are maximal connected nilpotent subgroups in G^c. Another subgroup of interest in G^c is the Cartan subgroup H^c.

As for $SU(2)$, Gaussian decomposition is possible for the complex group G^c (but not for the real group G).

In space G^c an everywhere dense subspace G_0^c exists, any element of which has a decomposition

$$g = \zeta h z = b_+ z = \zeta b_-$$

$$\zeta \in Z_+, \quad z \in Z_-, \quad h \in H^c, \quad b_+ \in B_+, \quad b_- \in B_-. \tag{11.1.2}$$

This decomposition is unique and in the matrix realization of group G^c, elements ζ, h and z are expressed in terms of the elements of the matrix g rationally.

The quotient spaces $X_+ = G^c/B_-$ and $X_- = B_+ \backslash G^c$ are compact complex homogeneous manifolds and the Gaussian decomposition determines a complex homogeneous structure in these spaces. The action of the Lie group G^c on these spaces is given by

i) for X_+

$$g: \zeta \to g\zeta = \zeta_1, \quad \text{where} \quad g\zeta = \zeta_1 h_1 z_1; \tag{11.1.3}$$

ii) for X_-

$$g: z \to zg = z_2, \quad \text{where} \quad zg = \zeta_2 h_2 z_2. \tag{11.1.4}$$

Both these spaces are isomorphic to the coset space $X = G/H$, the so-called flag manifold. (Here H is the Cartan subgroup of group G.) This isomorphism is determined as follows. Let us represent the Lie algebra $\mathscr{G}$ as an orthogonal sum $\mathscr{G} = \mathscr{K} + \mathscr{H}$ with respect to the Killing-Cartan metric. Using the exponential mapping we can embed space X into group G; $X = \exp \mathscr{K}$. The correspondence between an element $x = \exp \kappa$, $\kappa \in \mathscr{K}$, and the element ζ in the Gaussian decomposition is

$$x = \exp \kappa = \zeta h z. \tag{11.1.5}$$

We have thus obtained the mapping $X \to X_+$, which is a generalization of the stereographic projection.

It is known [99] that all spaces $X_\pm$ are orbits of maximal dimensionality for adjoint representation, all degenerate orbits of lower dimensionalities being not only complex but also Kählerian homogeneous manifolds. This means that these spaces admit a special Hermitian G-invariant metric,

$$ds_l^2 = h_{j\bar{k}}^l d\zeta_j \, d\bar{\zeta}_k, \qquad h_{j\bar{k}}^l = \frac{\partial^2 F^l(\zeta, \bar{\zeta})}{\partial \zeta_j \, \partial \bar{\zeta}_k}, \tag{11.1.6}$$

where function $F^l(\zeta, \bar{\zeta})$ is called the potential of Kählerian metrics and may be found from the Gaussian decomposition

$$\zeta^+ \zeta = \zeta_3 h_3 z_3 = \zeta_3 \exp(F^j H_j) z_3. \tag{11.1.7}$$

In the matrix realization, h_3 is diagonal and has elements $\delta_1, \ldots, \delta_r$, while

$$F^l = \ln \delta_l. \tag{11.1.8}$$

Respectively, there are r closed G-invariant 2-forms ω^l:

$$\omega^l = \frac{\mathrm{i}}{2} \, h_{j\bar{k}}^l d\zeta_j \wedge d\bar{\zeta}_k, \qquad l = 1, 2, \ldots, r. \tag{11.1.9}$$

11.2 Representations of Compact Simple Lie Groups

As is well known any unitary irreducible representation of simple compact Lie group G of rank r is characterized by an r-dimensional vector $\lambda = (\lambda_1, \ldots, \lambda_r)$, the so-called highest weight: $T(g) = T^\lambda(g)$, where $\lambda = \sum \lambda_l w_l$, w_l are simple weights, λ_l – integers.

Correspondingly, in the representation space $\mathscr{H}^\lambda$ the highest weight vector $|\lambda\rangle$ exists, i.e., a vector satisfying conditions

$$\hat{E}_\alpha |\lambda\rangle = 0, \qquad \alpha \in R_+, \qquad \hat{H}_j |\lambda\rangle = \lambda_j |\lambda\rangle, \tag{11.2.1}$$

where $\hat{E}_\alpha$ and $\hat{H}_j$ are operators in $\mathscr{H}^\lambda$, the representation operators for Lie algebra $\mathscr{G}$.

In space $\mathscr{H}^\lambda$ there exists a basis $\{|\mu\rangle\}$, where $|\mu\rangle$ is the weight vector, i.e., an eigenvector of all operators H_j:

$$H_j|\mu\rangle = \mu_j|\mu\rangle. \tag{11.2.2}$$

A general representation $T^\lambda(g)$ characterized by the highest weight $\lambda = (\lambda_1, \ldots, \lambda_r)$ corresponds to a fiber bundle over X, with the circle as a fiber, with the connection form

$$\theta = \left(\frac{\partial F}{\partial \zeta_j}\, d\zeta_j - \frac{\partial F}{\partial \bar{\zeta}_j}\, d\bar{\zeta}_j\right) \tag{11.2.3}$$

and the curvature form

$$\omega = \frac{\partial^2 F}{\partial \zeta_j \partial \bar{\zeta}_k}\, d\zeta_j \wedge d\bar{\zeta}_k = d\theta, \quad \text{where} \tag{11.2.4}$$

$$F = \sum_l \lambda_l F^l, \quad l = 1, 2, \ldots, r. \tag{11.2.5}$$

Representation $T^\lambda(g)$ with the highest weight λ may be realized in the space of polynomials $\mathscr{F}^\lambda$ over X_-, or, what is the same, over the group Z. Namely,

$$T^\lambda(g) f(z) = \alpha(z, g) f(z_g), \tag{11.2.6}$$

where quantities $\alpha(z, g)$ and z_g are given by the Gaussian decomposition

$$zg = \zeta_1 h_1 z_1, \tag{11.2.7}$$

$$z_g = z_1, \quad \alpha(z, g) = \alpha(\lambda) = \delta_1^{\lambda_1} \ldots \delta_r^{\lambda_r}. \tag{11.2.8}$$

The invariant scalar product in $\mathscr{F}^\lambda$ is introduced by

$$(f_1, f_2) = N d_\lambda \int \bar{f}_1(z) f_2(z)\, d\mu_\lambda(z), \tag{11.2.9}$$

where d_λ is the dimensionality of representation T^λ. In this case representation $T^\lambda(g)$ has a simple "semiclassical" meaning and may be rewritten as [80]

$$T^\lambda(g) f(z) = \exp\left[\mathrm{i}S(z, g)\right] f(z_g) \tag{11.2.10}$$

$$S(z, g) = \int_0^z (\theta - g_* \cdot \theta) + S(0, g), \tag{11.2.11}$$

$$\theta = \left(\frac{\partial F}{\partial z_j}\, dz_j - \frac{\partial F}{\partial z_j}\, d\bar{z}_j\right),$$

$$F = \sum \lambda_l F^l(z, \bar{z}) = -\ln \langle \lambda | T(zz^+) | \lambda \rangle. \tag{11.2.12}$$

Here θ is the connection form in the fiber bundle with base X and a circle as a fiber, and this fiber bundle is related to the representation $T^\lambda(g)$.

A similar construction works also for degenerate representations for which the highest weight λ is singular, i.e., $(\lambda, \alpha)=0$ for one or several roots α. Then the *isotropy* subgroup $\tilde{B}$ of vector $|\psi_0\rangle$ is one of the so-called parabolic subgroups; this means that $\tilde{B}$ contains the Borel subgroup B, i.e., the maximum solvable subgroup. The coset space $X=G^c/\tilde{B}$ is the degenerate orbit of the coadjoint representation, but this space is still a homogeneous Kählerian manifold [99]. Hence the construction considered above is valid completely also in this case.

11.3 Coherent States

Let us construct the CS system for an arbitrary compact Lie group following [15]. The case of the $SU(n)$ series was also considered by *Gilmore* [100].

To construct a CS system one has to take an initial vector $|0\rangle$ in space $\mathscr{H}^\lambda$. Note first of all that the isotropy subgroup H_μ for any state $|\mu\rangle$ corresponding to a weight vector μ contains the Cartan subgroup $H=u(1)\times \ldots \times u(1)=T^r$ [group $U(1)$ enters here r times, where r is the rank of group G] and for general weight vectors subgroup H_μ coincides with H.

The isotropy subgroup for a linear combination of weight vectors is, in general, a subgroup of the Cartan subgroup. Therefore, it is convenient to choose a weight vector $|\mu\rangle$ as an initial element of the CS system. And in the general case the isotropy subgroup H_μ is isomorphic to the Cartan subgroup H, and CS is characterized by the point of $X=G/H$.

For degenerate representation where the highest weight λ is orthogonal to some root α: $(\lambda, \alpha)=0$, the isotropy subgroup H_μ may be larger than T^r for some state vectors $|\mu\rangle$. Then the coherent state $|x\rangle$ is characterized by a point of degenerate orbit of adjoint representation. Indeed, in all cases

$$H_j'|x\rangle=[T(g)\,H_j T^{-1}(g)]|x\rangle=\mu_j|x\rangle, \qquad |x\rangle=T(g)|\mu\rangle. \tag{11.3.1}$$

Therefore, if we take a state vector $|\mu\rangle$ as the initial vector $|0\rangle$, then the coherent state $|x\rangle$ is characterized by a point of an orbit of adjoint representation, and the orbit may be degenerate.

Making use of the remaining arbitrariness in the choice of the vector, let us try to take it in such a way that the state $|0\rangle$ would be the closest to the classical one. As was shown in [67], to this end state $|0\rangle$ must be $|\mu\rangle$, where μ is a dominant weight, i.e., μ is obtained from the highest weight by means of a Weyl transformation. Then, as shown in [68], the CS minimize the invariant uncertainty relation

$$\Delta C_2=\min, \qquad \text{where} \tag{11.3.2}$$

$$C_2=\sum_j (H_j)^2+\sum_{\alpha\in R_+}(E_\alpha E_{-\alpha}+E_{-\alpha}E_\alpha) \tag{11.3.3}$$

is the quadratic Casimir operator and

$$\Delta C_2 = \langle C_2 \rangle - \left(\sum_j \langle H_j \rangle^2 + 2 \sum_{\alpha \in R_+} \langle E_\alpha \rangle \langle E_{-\alpha} \rangle \right). \tag{11.3.4}$$

As

$$\langle \lambda | E_\alpha | \lambda \rangle = 0, \quad H_j | \lambda \rangle = \lambda_j | \lambda \rangle \quad \text{then}$$

$$\Delta C_2 = \lambda^2 + 2(\lambda \varrho) - \lambda^2 = 2(\lambda \varrho), \tag{11.3.5}$$

where

$$\varrho = \tfrac{1}{2} \sum \alpha.$$

Suppose now that $T^\lambda(g)$ is a nondegenerate representation of the compact simple Lie group G with the highest weight λ, i.e., $(\lambda, \alpha) \neq 0$ for any $\alpha \in R$. We take the vector with lowest weight $|-\lambda\rangle$ as the initial vector $|0\rangle$ for the CS system. Let us consider the action of operators H_j, E_α, $E_{-\alpha} (\alpha \in R_+)$ representing Lie algebra $\mathscr{G}^c$, on this state. It can be easily seen that subalgebra $\mathscr{B}_- = \{H_j, E_{-\alpha}\}$, $\alpha \in R_+$ is the isotropy subalgebra for the vector $|-\lambda\rangle$. The corresponding group B_- is a subgroup of G^c.

Taking the lowest-weight vector $|-\lambda\rangle$ as $|0\rangle$, applying operators $T^\lambda(g)$, and using the Gaussian decomposition $g = \zeta h z$, we obtain the CS system

$$|\zeta\rangle = N T^\lambda(\zeta) |0\rangle, \quad \zeta \in Z_+, \quad N = \langle 0 | T^\lambda(g) | 0 \rangle,$$
$$|\zeta\rangle = N \exp\left(\sum_{\alpha \in R_+} \zeta_\alpha E_\alpha \right) |0\rangle, \tag{11.3.6}$$

or, in another form,

$$|\zeta\rangle = D(\xi) |0\rangle, \quad D(\xi) = \exp\left[\sum (\xi_\alpha E_\alpha - \bar{\xi}_\alpha E_{-\alpha})\right]. \tag{11.3.7}$$

Note that the unitary operators $D(\xi)$ do not form a group but their multiplication law is

$$D(\xi_1) D(\xi_2) = D(\xi_3) \exp\left(\mathrm{i} \sum_j \varphi_j \mathrm{H}_j \right). \tag{11.3.8}$$

Note also that these CS are eigenstates of operators

$$T(g) H_j T^{-1}(g) = \tilde{H}_j, \quad \tilde{H}_j |x\rangle = -\lambda_j |x\rangle. \tag{11.3.9}$$

Equations (11.3.9) determine the CS up to a phase factor $\exp(\mathrm{i}\alpha)$. The CS system constructed has all properties of a general CS system. Some of the most important are noted below.

1. Operators $T^{\lambda}(g)$ transform the CS into another one,

$$T^{\lambda}(g)|x\rangle = e^{i\varphi(x,g)}|x_g\rangle, \tag{11.3.10}$$

where $\varphi(x,g)$ is a phase shift.

2. The CS are not mutually orthogonal; the scalar product is

$$\begin{aligned}\langle\zeta_1|\zeta_2\rangle &= N_1N_2\langle 0|T^{+}(\zeta_1)T(\zeta_2)|0\rangle = N_1N_2\langle 0|T(\zeta_1^{+}\zeta_2)|0\rangle\\ &= K_{\lambda}(\zeta_1^{+}\zeta_2)[K_{\lambda}(\zeta_1^{+}\zeta_1)K_{\lambda}(\zeta_2^{+}\zeta_2)]^{-1/2},\end{aligned} \tag{11.3.11}$$

where

$$K_{\lambda}(\zeta_1^{+}\zeta_2) = \Delta_1^{\lambda_1}(\zeta_1^{+}\zeta_2)\dots\Delta_r^{\lambda_r}(\zeta_1^{+}\zeta_2)$$

and quantities Δ_j may be found from the Gaussian decomposition. For group $G=SU(n)$, $G^c=SL(n,\mathbb{C})$, quantity Δ_j is the lower angular minor of order j of the matrix $\zeta_1^{+}\zeta_2$.

3. The CS minimize the invariant uncertainty relation,

$$\Delta C_2 = \min.$$

4. The following "resolution of unity" is valid

$$\begin{aligned}&\hat{1} = \dim(T^{\lambda})\int d\mu(\zeta)|\zeta\rangle\langle\zeta|,\\ &d\mu(\zeta) = Cd^p\zeta\, d^p\bar{\zeta}\Delta_1^{-\varrho_1}\dots\Delta_r^{-\varrho_r},\quad \varrho=\tfrac{1}{2}\sum_{\alpha\in R_+}\alpha,\end{aligned} \tag{11.3.12}$$

where $p=\frac{1}{2}(\dim G - r)$.

5. Using these formulae it is possible to decompose any state $|\psi\rangle$ over a CS system

$$\begin{aligned}&|\psi\rangle = \sum_{\mu\in P(\lambda)} c_{\mu}|\lambda,\mu\rangle,\qquad |\psi\rangle = \int d\mu_{\lambda}(\zeta)\psi(\bar{\zeta})|\zeta\rangle,\\ &P(\lambda) = \{\mu : \mu = \lambda - \alpha, \alpha\in R\}.\end{aligned} \tag{11.3.13}$$

This expansion is associated with the kernel

$$K_{\lambda}(\xi,\bar{\eta}) = \sum_{\mu}\psi_{\lambda\mu}(\xi)\bar{\psi}_{\lambda\mu}(\eta) \tag{11.3.14}$$

which is an analog of the Bergmann kernel for the unbounded domain. This kernel determines the Bergmann metrics in this domain as given by the standard formulae.

6. The CS representation is appropriate for describing the operators. Namely, we may use

$$\hat{P} = \int d\mu_{\lambda}(x)P(x)|x\rangle\langle x|,\qquad Q(x) = \langle x|\hat{P}|x\rangle. \tag{11.3.15}$$

7. By decomposing the function $P(x)$ over the basis of "spherical harmonics" $Y_{lm}(x)$ on space X, we obtain the Clebsch-Gordan series

$$T_\lambda \otimes T_\lambda = \sum_l T_l,$$

$$P(x) = \sum c_{lm} Y_{lm}(x), \qquad \hat{P}_{lm}(x) = \int d\mu_\lambda(x)\, Y_{lm}(x) |x\rangle\langle x| \tag{11.3.16}$$

$$\langle \lambda v' | \hat{P}_{lm} | \lambda v \rangle = \sqrt{\frac{\dim T_l}{C}} \langle \lambda, v'; l, m | \lambda, v \rangle \langle \lambda, -\lambda; l, 0 | \lambda, -\lambda \rangle,$$

where $\langle \lambda, v'; l, m | \lambda, v \rangle$ are the so-called Clebsch-Gordan coefficients.

In the considered case of compact simple Lie groups the property of analyticity is essential for the following reasons.

i) The homogeneous space $X = G/H$ possesses a complex homogeneous structure, i.e., group G acts on X as a group of holomorphic transformations [in the previous example $G = SU(2)$, $X = G/H$ is the two-dimensional sphere isomorphic to the complex projective space $\mathbb{C}P^1$].

ii) The Hilbert sapce $\mathscr{H}$ for the unitary irreducible representation in the CS basis can be identified with the space of holomorphic functions on $X = G/H$ [for the $SU(2)$ group it is the space of polynomials in Z of a degree less than k, where k is a fixed integer].

In the general case of a compact simple Lie group there always exists a complex homogeneous (and moreover, homogeneous Kählerian) structure, and the manifold X is a homogeneous symplectic manifold with the 2-form ω. Hence it may be considered as a phase space for a classical dynamical system and this space is subject to the action of the group G, which is a group of canonical transformations, i.e., those preserving the 2-form ω.

From this point of view the coherent states may be considered as localized wave functions on the classical phase space; this result is well known for the standard CS relevant to harmonic oscillators.

The inverse problem, namely, the problem of constructing unitary irreducible representations of group G, starting from the structure of the phase space for group G (the so-called geometric quantization) is treated in [19–21].

12. Discrete Series of Representations: The General Case

This chapter, following [72, 73], introduces and investigates CS systems for certain discrete series of representations of noncompact Lie groups, namely, for groups of motion for complex homogeneous symmetric bounded domains. The CS systems considered are parametrized by the domain points; they are expressed in terms of the Bergmann kernel for this domain. Recall the simplest case, the $SU(1,1)$ group, considered in Chap. 5.

12.1 Discrete Series

Let G be a connected real semisimple Lie group with a finite center, $\mathscr{H}$ be the Hilbert space and $T(g)$ a unitary irreducible representation of group G acting in space $\mathscr{H}$, $T\colon G \to \mathscr{H}$. Let us recall a few facts from [101, 102].

Definition 1. A unitary irreducible representation $T(g)$ acting in the Hilbert space $\mathscr{H}$ is called square-integrable if for any pair of nonzero vectors $|\varphi\rangle$, $|\psi\rangle \in \mathscr{H}$ the function $\langle\psi|T(g)|\varphi\rangle$ is square-integrable with an invariant measure $d\mu(g)$ (this condition is satisfied if it holds for some pair of vectors $|\varphi\rangle$ and $|\psi\rangle$).

Square-integrable representations are in some respects similar to representations of a compact Lie group; in particular, they satisfy orthogonality relations.

Proposition 1. A constant d_T called the formal degree of representation $T(g)$ exists such that if $|\psi\rangle$, $|\varphi\rangle$, $|\psi'\rangle$, $|\varphi'\rangle \in \mathscr{H}$, then

$$\int_G \langle\psi|T(g)|\varphi\rangle\overline{\langle\psi'|T(g)|\varphi'\rangle}\, d\mu(g) = d_T^{-1}\langle\psi|\psi'\rangle\overline{\langle\varphi|\varphi'\rangle}. \tag{12.1.1}$$

Here $d\mu(g)$ is an invariant measure on G and $\langle\varphi|\psi\rangle$ stands for the scalar product of vectors $|\varphi\rangle$ and $|\psi\rangle$.

Remark 1. Obviously, d_T depends on normalization of the measure $d\mu(g)$, but if the measure is fixed, then two equivalent square-integrable representations have the same degree d_T.

Definition 2. The set of all equivalence classes of square-integrable representations [denoted by $\varepsilon_2(G)$] is called a discrete series of representations for G.

The following fundamental statement was proven in [103].

Theorem 1. A connected real semisimple Lie group has a discrete series of representations if and only if a compact Cartan subgroup exists in the group G.

Let K be the maximal compact subgroup of group G. It is well known [81] that the coset space $D=G/K$ is the symmetric space of a nonpositive curvature.

Let us suppose in addition that a complex structure exists in D, i.e., D is a complex homogeneous manifold. Then the Cartan subgroup H in G is compact and is also the Cartan subgroup in K. Note that group K has a one-dimensional center, i.e., $K=K_1\times U(1)$.

Thus G has a discrete series of representations. We shall consider only the case of discrete series realizable in space $\mathscr{F}$ of holomorphic functions on D.

For considered noncompact groups all root vectors can be divided into two sets: noncompact vectors orthogonal to the Lie algebra $\mathscr{K}$ of the compact group K, and compact vectors orthogonal to space $\mathscr{L}$, the orthogonal complement of $\mathscr{K}$ in $\mathscr{G}$: $\mathscr{G}=\mathscr{K}+\mathscr{L}$.

Proposition 2. [101, 102]. The representation of discrete series with a highest weight λ can be realized in the space $\mathscr{F}$ if and only if $(\lambda+\varrho,\alpha)<0$ for all noncompact positive roots α [here $(\lambda+\varrho,\alpha)$ is the scalar product in the root space].

There is such a normalization of the measure $d\mu(g)$ that the constant d_T in (12.1.1) is given by

$$d_T=\left|\prod_{\alpha\in R_+}\frac{(\lambda+\varrho,\alpha)}{(\varrho,\alpha)}\right|. \tag{12.1.2}$$

Remark 2. The expression in (12.1.2) is almost identical to the famous formula by Weyl for the representation dimensionality for compact Lie groups, the only difference being that the product is not positive definite, so one must take the absolute value.

12.2 Bounded Domains

This section presents information concerning the theory of Hermitian symmetric spaces. For proofs, see [81].

Let G be a connected semisimple Lie group with trivial center and K be its maximal compact subgroup. Let us consider the domain $D=G/K$ where a Hermitian structure can be introduced. The the following statement is valid.

Proposition 3. Any Hermitian symmetric space is a direct product of irreducible Hermitian symmetric spaces. (A symmetric space is called Hermitian if it admits a G-invariant complex structure.)

A classification theorem follows [81].

Theorem 2. I) Noncompact irreducible Hermitian symmetric spaces are the manifolds G/K where G is a connected noncompact simple Lie group with the center $\{e\}$ and K is the maximal compact subgroup of G with a nondiscrete center.

II) The compact irreducible Hermitian symmetric spaces are exactly the manifolds U/K, where U is a connected compact simple Lie group with center $\{e\}$ and K is the maximal connected proper subgroup of U with a nondiscrete center.

We shall use the following general theorem [81].

Theorem 3. Any bounded symmetric domain D is a Hermitian symmetric space of noncompact type.

Using these two theorems and Cartan's classification of simple Lie groups, all irreducible Hermitian spaces of compact and noncompact types are described. The following series of spaces exist (besides, there are two special spaces corresponding to exceptional Lie algebras).

Series of Hermitian symmetric spaces of classical type.

	Noncompact case	Compact case
I)	$SU(p,q)/SU(p)\times SU(q)\times U(1)$	$SU(p+q)/SU(p)\times SU(q)\times U(1)$
II)	$\mathrm{Sp}(p,\mathbb{R})/U(p)$	$\mathrm{Sp}(p)/U(p)$
III)	$SO^*(2p)/U(p)$	$SO(2p)/U(p)$
IV)	$SO_0(p,2)/SO(p)\times SO(2)$	$SO(p+2)/SO(p)\times SO(2)$.

Here $SO^*(2p)$ is a subgroup of $SO(2p,\mathbb{C})$ which preserves the form

$$z_{p+1}\bar{z}_1 - z_1\bar{z}_{p+1} + \ldots + z_{2p}\bar{z}_p - z_p\bar{z}_{2p},$$

$SO_0(p,q)$ is the connected component of unity within $SO(p,q)$, $\mathrm{Sp}(p,\mathbb{R})\in GL(2p,\mathbb{R})$ and leaves invariant the bilinear form

$$(x\wedge y) = x,\ y_{p+1} - y,\ x_{p+1} + \ldots + x_p y_{2p} - y_p\cdot x_{2p}.$$

Finally, $\mathrm{Sp}(p) = \mathrm{Sp}(p,\mathbb{C})\cap U(2p)$ is the compact symplectic group.

Bounded symmetric domains corresponding to each of the four series exist, and groups $SU(p,q)$, $\mathrm{Sp}(p,\mathbb{R})$, $SO^*(2p)$, $SO_0(p,2)$ act as analytical automorphism groups in the corresponding domains. *Cartan* [104] discovered these domains in 1935.

We use the explicit realizations for these domains.

(A) The domain D_{I} with the automorphism group $\mathrm{Aut}\,D_{\mathrm{I}} = SU(p,q)$ is realized in the space of matrices Z with p rows and q columns which satisfy the condition

$$\mathrm{I}^{(p)} - ZZ^+ > 0, \tag{12.2.1}$$

where $\mathrm{I}^{(p)}$ is the unit $p \times p$ matrix ($Z^+ = \bar{Z}'$ is the Hermitian conjugate for Z) and $A > 0$, where A is a Hermitian matrix, means that all eigenvalues of A are positive.

The group action is given by

$$Z \to Zg = (A'Z + C')(B'Z + D')^{-1}, \tag{12.2.2}$$

where A, B, C, D are $p \times p$, $p \times q$, $q \times p$ and $q \times q$ matrices, respectively, satisfying

$$AA^+ - BB^+ = \mathrm{I}^{(p)}, \qquad AC^+ = BD^+, \qquad DD^+ - CC^+ = \mathrm{I}^{(q)}.$$

(B) The domain D_{II} consists of complex symmetric $p \times p$ matrices which satisfy

$$\mathrm{I}^{(p)} - ZZ^+ > 0. \tag{12.2.3}$$

The automorphism group is $\mathrm{Sp}(p, \mathbb{R})$,

$$Z \to Zg = (A'Z + \bar{B}')(B'Z + \bar{A}')^{-1} \quad \text{and} \tag{12.2.4}$$

$$A'B = B'A, \qquad AA^+ - BB^+ = \mathrm{I}.$$

(C) The domain D_{III} consists of complex skew-symmetric $p \times p$ matrices, such that

$$\mathrm{I}^{(p)} - ZZ^+ > 0. \tag{12.2.5}$$

The group Aut D_{III} is $SO^*(2p)$ and its action is

$$Z \to Zg = (A'Z - \bar{B}')(B'Z + \bar{A}')^{-1}, \quad \text{where} \tag{12.2.6}$$

$$A'B = -B'A, \qquad A^+A - B^+B = \mathrm{I}.$$

(D) The domain D_{IV} consists of complex vectors $z = (z_1, \ldots, z_p)$ such that

$$1 + |zz'|^2 - 2\bar{z}z' > 0, \qquad |zz'| < 1. \tag{12.2.7}$$

The group Aut $(D_{\mathrm{IV}}) = SO_0(p, 2)$ consists of the following transformations

$$zg = \left\{ \left[\left(\frac{1}{2}(zz'+1), \frac{\mathrm{i}}{2}(zz'-1) \right) A' + zB' \right] \begin{pmatrix} 1 \\ \mathrm{i} \end{pmatrix} \right\}^{-1}$$

$$\times \left\{ \left(\frac{1}{2}(zz'+1), \frac{\mathrm{i}}{2}(zz'-1) \right) C' + zD' \right\}, \tag{12.2.8}$$

where A, B, C, D are real 2×2, $2\times p$, $p\times 2$, $p\times p$ matrices satisfying

$$\det\begin{bmatrix} A & B \\ C & D \end{bmatrix}=1, \qquad AA'-BB'=\mathrm{I}^{(2)}, \qquad DD'-CC'=\mathrm{I}^{(p)}, \qquad AC'=BD'.$$

Let D be an arbitrary bounded symmetric domain and G a semisimple group with the center $\{e\}$. Suppose that the elements of G are analytical automorphisms of D. It is known that the following invariant measure on D exists:

$$d\mu(z)=\varrho(z,\bar{z})\prod_{j=1}^{n} dx_j\,dy_j, \qquad z=(z_1,\ldots,z_n), \qquad z\in D,$$

$$\dim_{\mathbb{C}} D=n, \qquad z_j=x_j+\mathrm{i}y_j, \qquad \varrho(z,\bar{z})>0. \tag{12.2.9}$$

In the following we shall give explicit calculations of the Jacobian for transformation $z\to zg$ $(g\in G)$, denoted by $\mathscr{J}_g(z)$.

Lemma 1. The Jacobian can be written as

$$\mathscr{J}_g(z)=[\mathscr{J}_{g_z}(0)]^{-1}\,\mathscr{J}_{g_z g}(0). \tag{12.2.10}$$

Recall that the classical symmetric domains are complete circular domains and that they contain, in particular, the point $z=0$ (the origin). Since the measure is invariant,

$$\varrho(zg,\overline{zg})=|\mathscr{J}_g(z)|^{-2}\varrho(z,\bar{z}). \tag{12.2.11}$$

Equation (12.2.11) has a unique solution, up to a normalization factor. With the normalization condition $\varrho(0,0)=1$ we have

$$\varrho(z,\bar{z})=|\mathscr{J}_{g_z}(0)|^{-2} \tag{12.2.12}$$

and it follows from the transitivity condition that

$$\mathscr{J}_g(z)=|\mathscr{J}_{g_z}(0)|^{-1}\,\mathscr{J}_{g_z g}(0). \tag{12.2.13}$$

Therefore it is sufficient to calculate $\mathscr{J}_g(0)$ in order to find $\mathscr{J}_g(z)$ and $\varrho(z,\bar{z})$.

We shall use the explicit formulae for measures and Jacobians for all classical domains. Note that in [93] only expressions for $\varrho(z,\bar{z})$ are given. We formulate the final results in a lemma. All the necessary calculations are given in App. C.

Lemma 2. The explicit expressions for $\varrho(z,\bar{z})$ and $\mathscr{J}_g(z)$ are:
for D_{I}

$$\varrho(z,\bar{z})=[\det(\mathrm{I}-ZZ^{+})]^{-(p+q)}$$

$$\mathscr{J}_g(z)=[\det(B'Z+D')]^{-(p+q)}; \tag{12.2.14}$$

for D_{II}

$$\varrho(z,\bar{z})=[\det(\mathrm{I}-ZZ^{+})]^{-(p+1)}$$

$$\mathscr{J}_g(z)=[\det(B'Z+\bar{A}')]^{-(p+1)}; \tag{12.2.15}$$

for D_{III}

$$\varrho(z,\bar{z})=[\det(\mathrm{I}-ZZ^{+})]^{-(p-1)}$$

$$\mathscr{J}_g(z)=[\det(B'Z+\bar{A}')]^{-(p-1)}; \tag{12.2.16}$$

for D_{IV}

$$\varrho(z,\bar{z})=(1+|zz'|^2-2\,\bar{z}z')^{-p}$$

$$\mathscr{J}_g(z)=\left\{\left[\left(\frac{1}{2}(zz'+1),\frac{\mathrm{i}}{2}(zz'-1)\right)A'+zB'\right]\begin{pmatrix}1\\ \mathrm{i}\end{pmatrix}\right\}^{-p}. \tag{12.2.17}$$

12.3 Coherent States

In defining CS systems we follow [15]. Consider the transformation in the space of holomorphic functions on D:

$$T^k(g)\psi(z)=[\mathscr{J}_g(z)]^k\psi(zg),\qquad g\in G,\qquad z\in D, \tag{12.3.1}$$

where k is an integer positive number. It can be easily seen that these transformations give a representation of group G.

Let us introduce a norm in functional space, given by

$$\|\psi\|_k^2=\int|\psi(z)|^2 d\mu_k(z). \tag{12.3.2}$$

Clearly, the necessary condition for representation T^k to be unitary is

$$d\mu_k(z)=[\varrho(z,\bar{z})]^{-k}d\mu(z). \tag{12.3.3}$$

The space of all holomorphic functions in D with the finite norm $\|\psi\|_k$ forms the Hilbert space $\mathscr{F}_k$ with the scalar product

$$\langle\varphi|\psi\rangle=\int\bar{\varphi}(z)\psi(z)d\mu_k(z). \tag{12.3.4}$$

The functional representation of the lowest weight vector, $\varphi_0(z)\equiv 1$, exists

$$\varphi_g(z)=T^k(g)\varphi_0(z)=[\mathscr{J}_g(z)]^k, \tag{12.3.5}$$

and $\varphi_{g_1}=\exp(\mathrm{i}\alpha)\varphi_{g_2}$ if and only if $g_1=g_2h$, where α is a real number and $h\in H$ [H is the isotropy subgroup of vector $\varphi_0(z)$]; for details, see [73]. Therefore, every state is determined by a point $\zeta \in D$, $\varphi_g(z) = \exp(i\alpha)\psi_\zeta(z)$, and the choice of $\psi_\zeta(z)$ determines the cross section in the one-dimensional bundle with base D, and the fiber is a circle. Let $\zeta \in D$ and $g: \zeta \to 0$.

Proposition 4. Explicit expressions for CS in the case of four classical domains are

$$\psi_\zeta(z) = [B(\zeta^+, \zeta)]^{-k/2} [B(\zeta^+, z)]^k; \tag{12.3.6}$$

D_{I}

$$B(\zeta^+, z) = [\det(\mathrm{I} - \zeta^+ Z)]^{-(p+q)}; \tag{12.3.7}$$

D_{II}

$$B = [\det(\mathrm{I} - \zeta^+ Z)]^{-(p+1)}; \tag{12.3.8}$$

D_{III}

$$B(\zeta^+, z) = [\det(\mathrm{I} - \zeta^+ Z)]^{-(p-1)}; \tag{12.3.9}$$

D_{IV}

$$B(\zeta^+, z) = [1 + (\bar{\zeta}\bar{\zeta}')(zz') - 2\bar{\zeta}z']^{-p}. \tag{12.3.10}$$

These formulae follow immediately from the explicit expressions (12.2.14–17) for the Jacobians.

Expressions for the normalized measures are

$$d\mu_k^{(0)}(z) = N_k \, d\mu_k(z), \qquad \int d\mu_k^{(0)}(z) = 1.$$

Using results from [93], we obtain the expiessions for N_k:

D_{I}

$$N_k = \frac{1}{\pi^{pq}} \frac{(\lambda+p)! \cdots (\lambda+p+q-1)!}{\lambda! \cdots (\lambda+q-1)!}, \qquad \lambda = (k-1)(p+q); \tag{12.3.11}$$

D_{II}

$$N_k = \frac{1}{\pi^{p(p+1)/2}} \frac{(\lambda+p)!(2\lambda+p+1)!(2\lambda+p+2)! \cdots (2\lambda+2p-1)!}{\lambda!(2\lambda+2)!(2\lambda+4)! \cdots (2\lambda+2p-2)!},$$

$$\lambda = (k-1)(p+1); \tag{12.3.12}$$

D_{III}

$$N_k = \frac{1}{\pi^{p(p-1)/2}} \frac{(2\lambda+p-1)!(2\lambda+p)!\cdots(2\lambda+2p-3)!}{(2\lambda)!(2\lambda+2)!\cdots(2\lambda+2p-4)!},$$

$$\lambda=(k-1)(p-1); \tag{12.3.13}$$

D_{IV}

$$N_k = \frac{2^{p-1}}{\pi^p} \frac{(\lambda+p-1)!(2\lambda+p)}{\lambda!}, \qquad \lambda=(k-1)p. \tag{12.3.14}$$

Note that $N_1=[V_{\text{E}}(D)]^{-1}$, where $V_{\text{E}}(D)$ is the Euclidean volume of domain D.

If we choose measure $d\mu_k^{(0)}(z)=N_k d\mu_k(z)$, the $\psi_\zeta(z)$ are normalized:

$$\langle\psi_\zeta|\psi_\zeta\rangle = N_k \int_D |\psi_\zeta(z)|^2 d\mu_k(z)=1. \tag{12.3.15}$$

So, for all four classical domains integral (12.3.15) can be represented in explicit form

$$\langle\psi_\zeta|\psi_\zeta\rangle = N_k[B(\zeta^+,\zeta)]^{-k} \int_D [B(\zeta^+,z)]^k [B(z^+,\zeta)]^k [B(z,z^+)]^{1-k} dx\,dy=1$$

$$\int_D B_k(\zeta^+,z)\,B_k(z^+,\zeta)dx\,dy = B_k(\zeta^+,\zeta), \tag{12.3.16}$$

$x=(x_1,\ldots,x_n)$, $\quad y=(y_1,\ldots,y_n)$, $\quad B_k$ – see (12.3.18).

There is a more general relation:

$$\int_D B_k(\zeta^+,z)B_k(z^+,\zeta_1)dx\,dy = B_k(\zeta^+,\zeta_1), \qquad \text{where} \tag{12.3.17}$$

$$B_k(z^+,z_1)=N_k[B(z,z^+)B(z_1^+,z_1)]^{(1-k)/2}[B(z^+,z_1)]^k. \tag{12.3.18}$$

Relation (12.3.17) indicates that B_k is the so-called reproducing kernel. If $k=1$, then $N_1 B$ is the usual Bergmann kernel. Therefore, in general, we call B_k the generalized Bergmann kernels. These conclusions are summarized in the following important proposition.

Proposition 5. Let D be an arbitrary symmetric domain. Then the system of coherent states is defined by generalized Bergmann kernels.

Remark. The Bergmann kernel, as well as generalized kernels, exist also for the bounded homogeneous nonsymmetric domains. In all these cases CS systems are defined in terms of generalized Bergmann kernels. Unlike the case of symmetric domains, for generic domains there are no discrete subgroups Γ with a finite volume $\mu(\Gamma|D)$ of coset space $\Gamma|D$. Actually, *Siegel* [105] has proved that if

$\mu(\Gamma|D)$ is finite, the group G is unimodular, and according to *Hano* [106] this group can act transitively only in a symmetric domain.

Now let us consider general properties of the CS systems.

1. The scalar product of two states is

$$\langle\psi_{\zeta_1}|\psi_{\zeta_2}\rangle = N_k[B(\zeta_1^+,\zeta_1)B(\zeta_2^+\cdot\zeta_2)]^{-k/2}$$
$$\times\int_D [B(\zeta_1^+,z)B(z^+,\zeta_2)]^k[B(z,z^+)]^{1-k}dx\,dy$$
$$=[B(\zeta_1^+,\zeta_1)B(\zeta_2^+,\zeta_2)]^{-k/2}[B(\zeta_1^+,\zeta_2)]^k. \tag{12.3.19}$$

2. A restriction on the growth of the kernel can be obtained directly from the Schwartz inequality:

$$|B(\zeta^+,z)|\leq[B(\zeta^+,\zeta)B(z,z^+)]^{1/2}. \tag{12.3.20}$$

3. Let us consider the action of group G on the CS system:

$$T(g)\psi_\zeta(z)=T(g)\,[B(\zeta^+,\zeta)]^{-k/2}[B(\zeta^+,z)]^k$$
$$=[B(\zeta^+,\zeta)]^{-k/2}[B(\zeta^+,zg)]^k[\mathcal{J}_g(z)]^k, \tag{12.3.21}$$

$$T(g)\psi_\zeta(z)=\exp[\mathrm{i}\alpha(g,\zeta)]\psi_{\zeta g^{-1}}(z).$$

4. Let P_ζ be a projector on state $|\psi_\zeta\rangle$. Let us consider integral $\int_D P_\zeta d\mu(\zeta)$, where $d\mu(\zeta)$ is the invariant measure on D.

From Schur's lemma and from the invariance of $d\mu(\zeta)$ we obtain the resolution of unity,

$$\int P_\zeta d\mu(\zeta)=C_k\hat{\mathrm{I}}, \tag{12.3.22}$$

where $\hat{\mathrm{I}}$ is the unit operator.

Let us apply (12.3.22) to vector $|\varphi\rangle$. The result is

$$\int C(\zeta)\psi_\zeta d\mu(\zeta)=C_k\varphi,\qquad C(\zeta)=\langle\psi_\zeta|\varphi\rangle. \tag{12.3.23}$$

Substituting $|\varphi\rangle=|\psi_0\rangle$ and taking the scalar product (12.3.23) with $|\psi_0\rangle$ gives an expression for C_k

$$C_k=\int|\langle\psi_\zeta|\psi_0\rangle|^2d\mu(\zeta)=\int[B(\zeta^+,\zeta)]^{1-k}d\xi\,d\eta=N_k^{-1}. \tag{12.3.24}$$

Hence, finally

$$\int_D P_\zeta d\mu^{(0)}(\zeta)=\hat{\mathrm{I}},\qquad d\mu^{(0)}(\zeta)=N_k\,d\mu(\zeta). \tag{12.3.25}$$

5. It is easy to find an explicit expression for C_k in terms of the representation formal degree d_T. Setting $|\varphi\rangle=|\varphi'\rangle=|\psi\rangle=|\psi'\rangle=|\psi_0\rangle$ in (12.1.1) gives

$$\begin{aligned}\int |\langle\psi_0|\mathrm{T}(g)|\psi_0\rangle|^2 d\mu(g) &= \int |\langle\psi_0|T(g)|\psi_0\rangle|^2 d\mu(k)\,d\mu(\zeta)\\ &= \int |\langle\psi_0|\psi_\zeta\rangle|^2 d\mu(k)\,d\mu(\zeta)\\ &= C\int |\langle\psi_0|\psi_\zeta\rangle|^2 d\mu(\zeta)=d_T^{-1},\end{aligned} \tag{12.3.26}$$

where C is a constant.

Here $d\mu(k)$ is an invariant measure for the compact Lie group K. Now from (12.3.25) and relations

$$C_k=N_k^{-1},\qquad C_1=N_1^{-1},\qquad N_1=[V_{\mathrm{E}}(D)]^{-1}$$

we obtain $d_T=N_k/N_1$. (12.3.27)

Remark. This normalization differs from that used by *Harish-Chandra* [103]; it is more convenient for our purpose.

6. Now let us consider the transformation of the highest vector under the action of the compact subgroup $K\subset G$. Group K contains $U(1)$ which is its center, $K=U(1)\times K_0$. Let us consider a transformation $r\in U(1): z\to\exp(-\mathrm{i}\varphi)z$ (factor $\exp(-\mathrm{i}\varphi)$ corresponds to the choice of the representation with the highest weight). Then

$$T(K_0)|\psi_0\rangle=|\psi_0\rangle,\qquad T(r)|\psi_0\rangle=\exp(-ikn\varphi)|\psi_0\rangle, \tag{12.3.28}$$

where $n=\dim_{\mathbb{C}} D$, k is a number determined by the representation.

It follows from (12.3.28) that vector $|\psi_0\rangle$ transforms like a function $[P(z_1,\ldots,z_n)]^k$, where P is a homogeneous polynomial of degree n which is invariant under the group K_0. Actually, let h be an element of the Cartan subgroup $H\subset G$. Then $T(h)|\psi_0\rangle=\exp[\mathrm{i}(\lambda\varphi)]|\psi_0\rangle$, and comparing this formula with (12.3.28) we see that

$$\lambda=-k\sum_{P_+}\boldsymbol{\beta},$$

where P_+ is the set of all positive noncompact roots.

7. It is known that any function holomorphic in D could be expanded into a power series

$$\psi_\zeta(z_1,\ldots,z_n)=\sum c_m\psi_m(z_1,\ldots,z_n),$$

where ψ_m is homogeneous polynomial of degree m. For domains D_{I}, D_{II}, D_{III} the decomposition for CS can be obtained explicitly. Using identity

$$\det A=\exp(\mathrm{tr}\{\ln A\}) \tag{12.3.29}$$

and (12.3.7–10) we get

$$\psi_\zeta(z)=[B(\zeta^+,\zeta)]^{-k/2}[B(\zeta^+,z)]^k, \quad \text{where} \tag{12.3.30}$$

$$[B]^k=\exp\left[kr\sum_{m=1}^{\infty}\frac{1}{m}S_m(z_1,\ldots,z_n)\right], \quad S_m=\operatorname{tr}\{(\zeta^+Z)^m\}, \tag{12.3.31}$$

$$r=p+q \quad \textit{for} \quad D_{\rm I}, \quad r=(p+1) \quad \text{for} \quad D_{\rm II}, \quad r=(p-1) \quad \text{for} \quad D_{\rm III}.$$

13. Coherent States for Real Semisimple Lie Groups: Class-I Representations of Principal Series

This section is concerned with certain non-square-integrable CS systems. The systems in view are related to class-I representations of the principal series for groups of motion for symmetric noncompact spaces. The present exposition follows that of [107, 108].

13.1 Class-I Representations

Let G be a real connected semisimple Lie group with a finite center. It is well known [109] that such groups have series of unitary irreducible representations (UIRs) for which there exists a vector $|\psi_0\rangle$ in the Hilbert space $\mathscr{H}$ invariant under the action of the maximal compact subgroup $K \subset G$. The representations are called class-I.

For the class-I representations of the principal series we may use an explicit realization in terms of the induced representations. The constructions take the following form.

It is known [79] that the group elements have Iwasawa decomposition, $G = KAN$, where K is the maximal compact subgroup, A is an Abelian noncompact subgroup and N is the maximal nilpotent subgroup. Let M be the centralizer of A in K, i. e., the set of elements of K commuting with all elements of group A. Let B be a subgroup G containing elements of the form MAN, Ξ be the coset space $B\backslash G = M\backslash K$ and $d\mu(\xi)$ the normalized K-invariant measure in Ξ,

$$\int d\mu(\xi) = 1.$$

The class-I representations of the principal series are considered to be induced by representations of subgroup B, which are trivial on M.

Recall the construction of the induced representation. Let us consider $L_2(\Xi, d\mu)$, the space of square-integrable functions $f(\xi)$,

$$\|f\|^2 = \int |f(\xi)|^2 d\mu(\xi) < \infty. \tag{13.1.1}$$

In each coset corresponding to an element $\xi \in \Xi = B\backslash G$ we choose a group element g_ξ. Now any element of group G may be written as $g = bg_\xi$, and the action of the group on the homogeneous space $\Xi(g: \xi \to \xi_g)$ is given by the decomposition

$$g_\xi g = bg_\eta, \quad \eta = \xi_g, \quad b = nam. \tag{13.1.2}$$

The action of an operator $T(g)$ is defined as

$$T(g)f(\xi)=\alpha(\xi,g)f(\xi_g), \tag{13.1.3}$$

where $\alpha(\xi,g)$ is a function called a multiplier, and ξ_g is determined from (13.1.2).

It can be easily seen that a necessary condition for operators $T(g)$ to form a representation of group G is a functional equation for the multiplier $\alpha(\xi,g)$

$$\alpha(\xi,g_2g_1)=\alpha(\xi,g_2)\,\alpha(\xi_{g_2},g_1). \tag{13.1.4}$$

Equation (13.1.4) is fulfilled if

$$\alpha(\xi,g)=\left(\frac{d\mu(\xi_g)}{d\mu(\xi)}\right)^{1/2}\chi(a), \tag{13.1.5}$$

where $\chi(a)$ is the character of group A, and the element $a\in A$, $a=a(\xi,g)$ is determined by (13.1.2). If $\chi(a)$ is a unitary character of group A, then representation $T(g)$ is unitary and irreducible [109]. For the rank-r symmetric space $X=G/K$ the representation is defined by r real numbers $\lambda=(\lambda_1,\ldots,\lambda_r)$. (The rank of the symmetric space G/K is, by definition, the number of independent metrical invariants for a pair of its points. This number is equal to the dimensionality of subgroup A of group G [81].) This is a class-I representation. As shown in [110], all representations of principal series are irreducible. It was also proved there that for all spaces of rank-1 every class-I UIR belongs either to principal or to complementary series obtained from principal series by an analytic continuation in λ.

In other words, a K-invariant function exists in the representation space, i.e., a function $f_0(\xi)$ satisfying the functional equation

$$T(k)f_0(\xi)=\alpha(\xi,k)f_0(\xi_k)=f_0(\xi). \tag{13.1.6}$$

Let us now introduce a new function $\tilde{f}(\xi)$ defined by

$$f(\xi)=f_0(\xi)\tilde{f}(\xi). \tag{13.1.7}$$

Setting it into (13.1.3) we get

$$T(g)\tilde{f}(\xi)=\tilde{\alpha}(\xi,g)\tilde{f}(\xi_g), \tag{13.1.3$'$}$$

where the multiplier $\tilde{\alpha}(\xi,g)$ is

$$\tilde{\alpha}(\xi,g)=\frac{f_0(\xi_g)}{f_0(\xi)}\,\alpha(\xi,g) \tag{13.1.8}$$

and it evidently satisfies (13.1.4).

It can be easily seen that for all $k\in K$

$$\tilde{\alpha}(\xi,k)\equiv 1. \tag{13.1.9}$$

Hence it follows that $\tilde{\alpha}(\xi, gk) = \tilde{\alpha}(\xi, g)$, i.e., function $\tilde{\alpha}(\xi, g)$ depends only on coset $x \in X = G \backslash K$ corresponding to element g:

$$\tilde{\alpha}(\xi, g) = \beta(\xi, x(g)). \tag{13.1.10}$$

13.2 Coherent States

It follows from (13.1.3′, 9) that function $\psi_0(\xi) \equiv 1$ is invariant under transformations of subgroup K. Acting by operator $T(g)$ gives an expression for CS in the ξ representation

$$T(g)\psi_0(\xi) = \tilde{\alpha}(\xi, g) = \beta(\xi, x(g)), \qquad x(g) = \pi g. \tag{13.2.1}$$

Here $\pi: g \to x(g)$ is a mapping of elements g to corresponding cosets x. Thus, the coherent state is determined by the kernel $\beta(\xi, x)$, where $x \in X$ and $\xi \in \Xi$.

Let us now turn to properties of these kernels. First of all we shall prove the following proposition.

Proposition 1. For fixed $\xi \in \Xi$ the kernel $\langle \xi, \lambda | x \rangle = \beta(\xi, x)$ is constant on orbits of a group N_ξ which is isomorphic to the group N present in the Iwasawa decomposition $G = KAN$ and having the fixed point, ξ, $N_\xi = h_\xi N h_\xi^{-1}$.

Proof. Let us fix a point $\xi \in \Xi$ and consider function $f_\xi(g) = T^\lambda(g)\psi_0(\xi) = \tilde{\alpha}^\lambda(\xi, g)$ which depends on element g. Suppose that H_ξ is a isotropy subgroup for point ξ. Setting $\eta = \xi$ in (13.1.2) gives

$$H_\xi = g_\xi^{-1} B g_\xi, \tag{13.2.2}$$

i.e., group H_ξ is conjugated to the group $B = NAM$. Similarly, the nilpotent component N_ξ of group H_ξ is

$$N_\xi = g_\xi^{-1} N g_\xi \tag{13.2.3}$$

It can be easily seen that the point ξ is intact under transformations of this subgroup.

Let h be an element of N_ξ: $h = g_\xi^{-1} n g_\xi$, $n \in N$. The function $\tilde{\alpha}^\lambda(\xi, h)$ is completely determined by $a(\xi, g)$ present in decomposition (13.1.2). Moreover,

$$g_\xi h = g_\xi g_\xi^{-1} n g_\xi = n g_\xi. \tag{13.2.4}$$

Therefore, $a(\xi, h) = e$ (e is the unity element of G). Thus, we have proved that for $h \in N_\xi$

$$\tilde{\alpha}^\lambda(\xi, h) = 1. \tag{13.2.5}$$

Hence it follows also that for $h \in M_\xi$, $M_\xi = g_\xi^{-1} M g_\xi$

$$\tilde{\alpha}^\lambda(\xi, h) = 1. \tag{13.2.6}$$

Finally, for $h \in A_\xi = g_\xi^{-1} A g_\xi$, $h = g_\xi^{-1} a g_\xi$

$$\tilde{\alpha}^\lambda(\xi, h) = f_\xi(h) = f_\xi(a). \tag{13.2.7}$$

Next we consider an arbitrary element of G. Now $g = g_x k$, so

$$\tilde{\alpha}^\lambda(\xi, g) = \tilde{\alpha}^\lambda(\xi, g_x k) = \tilde{\alpha}^\lambda(\xi, g_x). \tag{13.2.8}$$

Introduce the notations $y = x_h$, $h \in N_\xi$ and write

$$h^{-1} g_x = g_y k_1, \quad g_x = h g_y k_1, \quad k_1 \in K \tag{13.2.9}$$

Because of the functional equation (13.1.4),

$$\tilde{\alpha}^\lambda(\xi, g_x) = \tilde{\alpha}^\lambda(\xi, h g_y k_1) = \tilde{\alpha}^\lambda(\xi, h g_y) = \tilde{\alpha}^\lambda(\xi, g_y), \quad \text{so} \tag{13.2.10}$$

$$\psi_x^\lambda(\xi) = \psi_y^\lambda(\xi), \quad y = x_h, \quad h \in N_\xi \tag{13.2.11}$$

It is remarkable that when h runs throughout the whole group N_ξ, x_h moves along the orbit of this group in space X. These orbits are called the horospheres of the maximal dimensionality in the symmetrical space X, or the horocycles. Hence, Proposition 1 can be formulated in the following equivalent form.

Proposition 1′. The kernel $\langle \xi, \lambda | x \rangle$ which describes the coherent state $|x\rangle$ is constant on horocycles of group N_ξ.

It is natural to call these kernels horospherical kernels. This relates the CS method for the considered case to the horosphere method developed by *Gel'fand* and *Graev* [111] and considered in detail for symmetric spaces by *Helgason* [112].

Let us consider in more detail how CS systems relate to horocycles in the symmetric space.

13.3 Horocycles in Symmetric Space

By definition, horospheres of maximal dimensionality in space X, also named horocycles, are orbits of subgroups conjugated to the subgroup N. Sometimes the term is also applied to horospheres of lower dimensionalities, i.e., nongeneric horospheres, but we will not consider them here. Let us consider some properties of horocycles.

Proposition 2. Let $\Omega = \{\omega\}$ be a set of horocycles, then $\Omega = G/MN$.
The proof is given in [112].

Proposition 3.
1. $K\backslash G/K = A/W$,
2. $MN\backslash G/MN = A \times W$,

where W is the Weyl group of the symmetric space: $W = N(A)/M$, $N(A)$ is the normalizer of A in K.

Any horocycle may be presented in the form

$$\omega = (ka)^{-1}\omega_0, \qquad \omega_0 = Nx_0, \qquad kx_0 = x_0$$

(x_0 is the origin of space X), while the elements k and mk determine the same horocycle and the element a is unique. Hence quantities ξ and a determine horocycle ω unambiguously. Thus, one can introduce the horospherical system of coordinates. The element $\xi \in \Xi$ is called the normal to horosphere ω, and element a is called the complex distance from the horocycle ω_0.

All symmetric spaces have been completely classified by *Cartan* (the results are presented in [81]). We consider only the important case of Hermitian symmetric spaces, i.e., symmetric spaces having a complex structure. It is known [81] that these spaces can be realized in the form of bounded domains in the n-dimensional complex space, $\mathbb{C}^n$.

An arbitrary Hermitian symmetric space is a direct product of irreducible Hermitian symmetric spaces. It is also known that the number of nonequivalent types of horospheres in such a space is equal to r (r is the rank of the space) and that this space can be realized as a bounded symmetric domain in $\mathbb{C}^n$.

First let us consider rank-1 symmetric spaces.

13.4 Rank-1 Symmetric Spaces

By definition, the rank of symmetric space G/K is the number of independent metric invariants for any pair of its points. This number equals the dimensionality of subgroup A of group G. As is well known [81], there are three series of rank-1 spaces and an exceptional rank-1 space.

I) There is the real n-dimensional hyperbolic space (Lobatschevsky space), $X_n^{\mathrm{I}} = SO(n,1)/SO(n)$, where $SO(n,1)$ and $SO(n)$ are the groups of the real unimodular matrices leaving invariant the forms $x_1^2 + \ldots + x_n^2 - x_{n+1}^2$ and $x_1^2 + \ldots + x_n^2$, respectively.

II) There is the complex hyperbolic space, of real dimensionality $2n$: $X_n^{\mathrm{II}} = SU(n,1)/SU(n) \times U(1)$, where $SU(n,1)$ and $SU(n)$ are the groups of the complex unimodular matrices leaving invariant the forms $|z_1|^2 + \ldots + |z_n|^2 - |z_{n+1}|^2$ and $|z_1|^2 + \ldots + |z_n|^2$, respectively.

III) There is the quaternion hyperbolic space of real dimensionality $4n$: $X_n^{\mathrm{III}} = \mathrm{Sp}(n,1)/\mathrm{Sp}(n) \times \mathrm{Sp}(1)$, where $\mathrm{Sp}(n,1)$ and $\mathrm{Sp}(n)$ are the groups of quaternion unimodular matrices leaving invariant the forms $|q_1|^2 + \ldots + |q_n|^2 - |q_{n+1}|^2$ and $|q_1|^2 + \ldots + |q_n|^2$, respectively. Here $|q|$ is the norm of the quaternion q.

Recall that quaternion algebra is associative but not commutative. This is the algebra over a field of real numbers whose basic elements e_0, e_1, e_2 and e_3 satisfy the following multiplication law:

$$e_0^2 = e_0, \quad e_i^2 = -e_0, \quad e_0 e_i = e_i e_0 = e_i, \quad i = 1, 2, 3$$

$$e_1 e_2 = -e_2 e_1 = e_3, \quad e_2 e_3 = -e_3 e_2 = e_1, \quad e_3 e_1 = -e_1 e_3 = e_2. \tag{13.4.1}$$

Thus, an arbitrary quaternion q is of the form $q = q^0 e_0 + \boldsymbol{q}\boldsymbol{e}$, where $\boldsymbol{q} = (q^1, q^2, q^3)$, $\boldsymbol{e} = (e_1, e_2, e_3)$ and $q^\alpha, \alpha = 0, 1, 2, 3$ are real numbers.

Let $\bar{q} = q^0 e_0 - \boldsymbol{q}\boldsymbol{e}$ be a quaternion conjugated to q. The norm of the quaternion is $|q|^2 = \bar{q}q = (q^0)^2 + \boldsymbol{q}^2$.

IV) There is a two-dimensional hyperbolic space over the algebra (nonassociative) of the Cayley numbers (octonions) of real dimensionality 16:

$$X^{\text{IV}} = F_4^{\mathbb{R}}/SO(9),$$

where $F_4^{\mathbb{R}}$ is the real form of the exceptional simple group F_4, $SO(9)$ is the group of orthogonal unimodular 9×9 matrices. (Valuable information on the algebra of Cayley numbers and the geometry of this space may be found in [113].)

Note that all three series of symmetric spaces may be realized in a unified way, namely, in all three cases one may assume that $G = \{g\}$ is the group of matrices $g = \begin{pmatrix} A & B \\ C & D \end{pmatrix}$, where A is a $n \times n$ matrix, B is a $n \times 1$ matrix, C is a $1 \times n$ matrix and D is a 1×1 matrix, and the matrix elements are real numbers, complex numbers, or quaternions, respectively. In this case the matrix g must leave invariant the form $|x_1|^2 + \ldots + |x_n|^2 - |x_{n+1}|^2$, where x_i is real, complex, or quaternion, respectively. This implies certain conditions on the matrices A, B, C and D, in particular,

$$|D|^2 - \sum_1^n |C_i|^2 = 1, \quad |D|^2 - \sum_1^n |B_i|^2 = 1.$$

Thus in all three cases the symmetric space $X = G/K$ may be treated as a set of vectors $x = (x_1, \ldots, x_n)$ satisfying the condition

$$|x_1|^2 + \ldots + |x_n|^2 < 1, \tag{13.4.2}$$

while the space $\Xi = B \backslash G$ is a set of unit vectors $\xi = (\xi_1, \ldots, \xi_n)$,

$$|\xi_1|^2 + \ldots + |\xi_n|^2 = 1. \tag{13.4.2'}$$

Group G is realized in space X (correspondingly, in space Ξ) as the group of linear-fractional transformations,

$$g: x \to x' = x_g, \quad x_i' = (A_{ij} x_j + B_i)(C_j x_j + D)^{-1}. \tag{13.4.3}$$

Space X^{IV} is the hyperbolic Cayley plane, i.e., it may be considered as a set of vectors $x=(x_1, x_2)$, where x_1 and x_2 are the Cayley numbers (octonions). Recall that octonion algebra is noncommutative and nonassociative, but alternative, that is to say, any pair of its elements generates an associative subalgebra. The basis of the Cayley algebra consists of eight elements $e_0, e_1, \ldots, e_7$. The norm of an element $x=x^0e_0+\sum_1^7 x^ie_i$ is given by $|x|^2=\bar{x}x=x\bar{x}=(x^0)^2+\sum_1^7 (x^i)^2$, where $\bar{x}=x^0e_0-\sum x^je_j$ is the element conjugated to x. Elements of the Cayley algebra may also be represented as pairs of quaternions $x=(q_1, q_2)$. Then the multiplication law is

$$(q_1, q_2)(q_1', q_2')=(q_1q_1'-\bar{q}_2'q_2,\ q_2'q_1+q_2\bar{q}_1'). \tag{13.4.4}$$

It can be easily seen that function $\alpha(\xi, g)=(d\mu(\xi_g)/d\mu(\xi))^{1/2+i\lambda'}$, where λ' is a real number, satisfies the functional equation (13.1.4) and, consequently, is the multiplier for the class-I representation of the principal series.

Calculating $d\mu(\xi_g)/d\mu(\xi)$ we get

$$\frac{d\mu(\xi_g)}{d\mu(\xi)}=|C_j\xi_j+D|^{-2\varrho}, \tag{13.4.5}$$

$$\alpha(\xi, g)=|C_j\xi_j+D|^{-\varrho+i\lambda}, \qquad \lambda=-2\varrho\lambda', \qquad \text{where} \tag{13.4.6}$$

$$2\varrho=\dim X+q'-1 \tag{13.4.7}$$

$$q'=\begin{cases} 0 & \text{for } X_n^{\mathrm{I}} \\ 1 & \text{for } X_n^{\mathrm{II}} \\ 3 & \text{for } X_n^{\mathrm{III}} \\ 7 & \text{for } X^{\mathrm{IV}} \end{cases} \tag{13.4.8}$$

The number ϱ in (13.4.5, 6) can be expressed in terms of internal characteristics of the symmetric space. To this end, let us consider the structure of space X in more detail. In view of the Iwasawa decomposition, Lie algebra $\mathscr{G}$ for group G has the form of $\mathscr{G}=\mathscr{K}+\mathscr{A}+\mathscr{N}$, where $\mathscr{K}$, $\mathscr{A}$ and $\mathscr{N}$ are Lie algebras for subgroups K, A, and N, respectively. For the case in view, $\mathscr{A}$ is one-dimensional: $\mathscr{A}=\mathbb{R}H$. It is well known that one may choose a basis X_α^i and $X_{2\alpha}^j$ in the nilpotent subalgebra $\mathscr{N}$ so that

$$\begin{aligned}&[H, X_\alpha^i]=\alpha X_\alpha^i, && i=1,2,\ldots,p\\ &[H, X_{2\alpha}^j]=2\alpha X_{2\alpha}^j, && j=1,2,\ldots,q.\end{aligned} \tag{13.4.9}$$

The elements of algebra X_α^j form a root subsapce $\mathscr{G}_\alpha$ corresponding to root α; respectively, $\mathscr{G}_{2\alpha}=\{X_{2\alpha}^j\}$. The numbers $p=\dim \mathscr{G}_\alpha$ and $q=\dim \mathscr{G}_{2\alpha}$ are called the multiplicities of roots α and 2α. Comparing (13.4.8) with the calculated numbers q for spaces X_n^{I}, X_n^{II}, X_n^{III}, and X^{IV}, one sees that $q'=q$. Moreover, (13.4.7) may be rewritten as

$$\varrho=\frac{p}{2}+q. \tag{13.4.10}$$

To conclude this section, Table 13.1 collates certain characteristics of symmetric rank-1 spaces.

13.5 Properties of Rank-1 CS Systems

To get an explicit expression for CS in the ξ representation, one can apply operator $T(g)$ to the function $f_0(\xi)\equiv 1$ and use (13.4.6). The result is

$$\langle\xi,\lambda|x\rangle=\psi_x^\lambda(\xi)=(1-|x|^2)^{(\varrho-\mathrm{i}\lambda)/2}|1-\bar{x}\xi|^{-\varrho+\mathrm{i}\lambda}, \tag{13.5.1}$$

where $\bar{x}\xi=\sum_1^n \bar{x}_i\xi_i$.

Another form is more convenient in a number of cases

$$\psi_x^\lambda(\xi)=H_x^\lambda(\xi)=|x_0-\check{x}\xi|^{-\varrho+\mathrm{i}\lambda},\qquad \text{where} \tag{13.5.2}$$

$$x_0=(1-|x|^2)^{-1/2},\qquad \check{x}=(1-|x|^2)^{-1/2}x. \tag{13.5.3}$$

In the latter form, any CS is determined by a point of hyperboloid $\{x:|x_0|^2-|\check{x}|^2=1\}$, and the function $H_x^\lambda(\xi)$ is the kernel of an integral transformation mapping functions on hyperboloid $\{x:|x_0|^2-|\check{x}|^2=1\}$, $f(x_0,\check{x})$, to functions on the cone $\{\xi:|\xi_0|^2-|\xi|^2=0\}$. One should bear in mind, however, that in cases II and III the action of groups $SU(n,1)$ and $\mathrm{Sp}\,(n,1)$ on the corresponding hyperboloids and cones is not transitive. Note also that for $\lambda\to\infty$, $\tau\to 0$ and $\lambda\tau=\mathrm{const}$ the space is flat. Then the coherent states are just the plane waves.

The CS system is overcomplete and nonorthogonal and has a number of remarkable properties, presented below.

1. Every state $|x\rangle$ is normalized to unity

$$\langle x|x\rangle=\|\psi_x^\lambda\|^2=\int|\psi_x^\lambda(\xi)|^2\,d\mu(\xi)=1. \tag{13.5.4}$$

This is true, since the representation $T^\lambda(g)$ is unitary.

2. The system $\{|x\rangle\}$ is complete. This is true since representation $T^\lambda(g)$ is irreducible.

3. The operator $T^\lambda(g)$ transforms any CS into another CS

$$T^\lambda(g)|x\rangle=|x'\rangle,\qquad x'=x_{g^{-1}}. \tag{13.5.5}$$

Table 13.1. Characteristics of rank-1 symmetric spaces

G	$\dim G$	K	$\dim K$	M	$\dim M$	$\dim X$	p	q	ϱ
$SO(n, 1)$	$n(n+1)/2$	$SO(n)$	$n(n-1)/2$	$SO(n-1)$	$(n-1)(n-2)/2$	n	$n-1$	0	$(n-1)/2$
$SU(n, 1)$	$n(n+2)$	$SU(n)\times U(1)$	n^2	$SU(n-1)\times U(1)$	$(n-1)^2$	$2n$	$2n-2$	1	n
$\mathrm{Sp}(n, 1)$	$(n+1)(2n+3)$	$\mathrm{Sp}(n)\times \mathrm{Sp}(1)$	$n(2n+1)+3$	$\mathrm{Sp}(n-1)\times \mathrm{Sp}(1)$	$(n-1)(2n-1)+3$	$4n$	$4n-4$	3	$2n+1$
F_4^{F}	52	$SO(9)$	36	$SO(7)$	21	16	8	7	11

This fact can be easily verified via a direct calculation.

4. At fixed ξ, the function $\psi_x^\lambda(\xi)$ is constant on horospheres

$$\frac{|1-\bar{x}\xi|^2}{1-|x|^2}=\text{const}, \tag{13.5.6}$$

or, equivalently, $|x_0-\check{x}\xi|^2=\text{const}$. Thus, the kernel $\langle\xi,\lambda|x\rangle$ can be called horospherical, and the CS method is closely related to the horosphere method developed in [111].

5. The kernel

$$P(x,\xi)=|\psi_x^\lambda(\xi)|^2=\left(\frac{1-|x|^2}{|1-\bar{x}\xi|^2}\right)^\varrho \tag{13.5.7}$$

is the Poisson kernel for the symmetric space $X=G/K$. Thus

$$\psi_x^\lambda(\xi)=|P(x,\xi)|^{1/2+i\lambda'}, \qquad \lambda=-2\varrho\lambda'. \tag{13.5.8}$$

6. At fixed ξ, the function $\psi_x^\lambda(\xi)$ is an eigenfunction for the Laplace-Beltrami operator Δ_x for the symmetric space $X=\{x:|x|^2<1\}$,

$$-\Delta_x\psi_x^\lambda(\xi)=(\varrho^2+\lambda^2)\,\psi_x^\lambda(\xi). \tag{13.5.9}$$

Moreover, these functions are constant on horospheres which are analogs of hyperplanes in Euclidean space. Thus, CS are a natural generalization of the plane waves $\langle \boldsymbol{n},k|\boldsymbol{r}\rangle=\exp(ik\boldsymbol{n}\boldsymbol{r})$, $|\boldsymbol{n}|=1$, for the Euclidean space.

7. The CS are not orthogonal to each other. Calculating their inner product gives

$$\langle x|y\rangle=\langle 0|T^+(g_x)T(g_y)|0\rangle=\langle 0|T(h)|0\rangle=\Phi_\lambda(\tau), \tag{13.5.10}$$

where

$$\begin{aligned} &h=g_x^{-1}g_y=k_1a(\tau)k_2\,,\\ &\cosh^2\tau=[(1-|x|^2)(1-|y|^2)]^{-1}|1-\bar{x}y|^2 \end{aligned} \tag{13.5.11}$$

and $\tau=\tau(x,y)$ determines the distance between points x and y in space X.

Using (13.5.11) it is easy to get an expression for the metric in symmetric space $X=G/K$:

$$\begin{aligned} ds^2=&\,d\tau^2+\sinh^2\tau\,[d\bar{\xi}_i d\xi_i-(\bar{\xi}_i d\xi_i)(d\bar{\xi}_j\xi_j)]\\ &+\tfrac{1}{4}\sinh^2 2\tau(\bar{\xi}_i d\xi_i)(d\bar{\xi}_j\xi_j), \end{aligned} \tag{13.5.12}$$

$$x=\xi\tanh\tau;\ |\xi|^2=1.$$

8. Function

$$\Phi_\lambda(\tau)=\langle 0|T^\lambda(g)|0\rangle, \quad g=k_1 a(\tau) k_2 \tag{13.5.13}$$

determines the inner product in the CS space. It is called a zonal spherical function and plays an important role in the theory of symmetric spaces [114]. It may be also determined by the integral

$$\Phi_\lambda(\tau)=\langle 0|x\rangle=\int \psi_x^\lambda(\xi)\,d\mu(\xi), \quad |x|=\tanh\tau. \tag{13.5.14}$$

An integral representation for the zonal spherical functions is obtained directly from (13.5.14). For space X_n^{I}

$$\Phi_\lambda(\tau)=\int_0^\pi [\cosh\tau-\sinh\tau\cos\theta]^{-\varrho+\mathrm{i}\lambda}\,d\mu(\theta), \quad \text{where} \tag{13.5.15}$$

$$d\mu(\theta)=\frac{\Gamma((p+1)/2)}{\sqrt{\pi}\,\Gamma(p/2)}(\sin\theta)^{p-1}\,d\theta. \tag{13.5.15$'$}$$

In other cases

$$\begin{aligned}\Phi_\lambda(\tau)=\int_0^{\pi/2}\int_0^{\pi} &[(\cosh\tau-\sinh\tau\cos\theta\cos\varphi)^2\\ &+\sinh^2\tau\cos^2\theta\sin^2\varphi]^{(-\varrho+\mathrm{i}\lambda)/2}\,d\mu(\theta,\varphi)\end{aligned} \tag{13.5.16}$$

where

$$d\mu(\theta,\varphi)=\frac{2\Gamma\left(\frac{p+q+1}{2}\right)}{\sqrt{\pi}\,\Gamma\left(\frac{p}{2}\right)\Gamma\left(\frac{q}{2}\right)}(\sin\theta)^{p-1}(\cos\theta)^q(\sin\varphi)^{q-1}\,d\theta\,d\varphi. \tag{13.5.16$'$}$$

These functions are even $\Phi_\lambda(-\tau)=\Phi_\lambda(\tau)$ and are normalized by condition $\Phi_\lambda(0)=1$; further, $\Phi_{-\lambda}(\tau)=\Phi_\lambda(\tau)$.

9. The zonal spherical function $\Phi_\lambda(\tau)$ is an eigenfunction for the radial part of the Laplace-Beltrami operator on symmetric space

$$-\left(\frac{d^2}{d\tau^2}+p\coth\tau\frac{d}{d\tau}+2q\coth 2\tau\frac{d}{d\tau}\right)\Phi_\lambda(\tau)=(\varrho^2+\lambda^2)\,\Phi_\lambda(\tau) \tag{13.5.17}$$

The solution of this equation is known, so one has an expression for the zonal spherical function in terms of the hypergeometric function [114]

$$\Phi_\lambda(\tau)=F(a,b,c,-\sinh^2\tau). \quad \text{Here} \tag{13.5.18}$$

$$a=\frac{\varrho+\mathrm{i}\lambda}{2}, \quad b=\frac{\varrho-\mathrm{i}\lambda}{2}, \quad c=\frac{p+q+1}{2}. \tag{13.5.19}$$

10. The Laplace-Beltrami operator is a self-adjoint operator. Therefore the zonal spherical functions are orthogonal (at $\lambda>0$ and $\lambda'>0$)

$$\int_0^\infty \bar{\Phi}_\lambda(\tau)\Phi_{\lambda'}(\tau)d\mu(\tau)=N(\lambda)\delta(\lambda-\lambda'),$$
$$d\mu(\tau)=(2\sinh\tau)^p(2\sinh 2\tau)^q d\tau. \tag{13.5.20}$$

Hence clearly also $\int_0^\infty |\Phi_\lambda(\tau)|^2 d\mu(\tau)=\infty$, i.e., the CS system considered is not square-integrable.

11. The normalizing coefficient $N(\lambda)$ in (13.5.20) is determined by the $\tau\to\infty$ asymptotic behaviour of function $\Phi_\lambda(\tau)$

$$\Phi_\lambda(\tau)\sim[c(\lambda)e^{i\lambda\tau}+c(-\lambda)e^{-i\lambda\tau}]e^{-\varrho\tau}, \quad \bar{c}(\lambda)=c(-\lambda), \quad \tau\to\infty \tag{13.5.21}$$

In view of (13.5.20, 21)

$$N(\lambda)=2\pi|c(\lambda)|^2. \tag{13.5.22}$$

Taking the asymptotics of the hypergeometric function in (13.5.18) we obtain

$$c(\lambda)=2^{\varrho-i\lambda}\Gamma\left(\frac{p+q+1}{2}\right)\frac{\Gamma(i\lambda)}{\Gamma\left(\frac{\varrho+i\lambda}{2}\right)\Gamma\left(\frac{p}{4}+\frac{1}{2}+\frac{i\lambda}{2}\right)}. \tag{13.5.23}$$

Equation (13.5.23) for coefficient $c(\lambda)$ is a particular case of a general formula by *Gindikin* and *Karpelevic* [115].

Further,

$$\mathrm{N}(\lambda)=2^{p+2q+1}\pi\left|\Gamma\left(\frac{p+q+1}{2}\right)\right|^2\frac{|\Gamma(i\lambda)|^2}{\left|\Gamma\left(\frac{\varrho}{4}+\frac{1}{2}+\frac{i\lambda}{2}\right)\right|^2\left|\Gamma\left(\frac{\varrho+i\lambda}{2}\right)\right|^2}. \tag{13.5.24}$$

12. Let us consider the space of functions $f(\tau)$ satisfying the condition

$$\int_0^\infty |f(\tau)|^2 d\mu(\tau)<\infty. \tag{13.5.25}$$

Functions $\Phi_\lambda(\tau)$ are a complete system in this space and the completeness condition is written as

$$\int_0^\infty \bar{\Phi}_\lambda(\tau)\Phi_\lambda(\tau')d\mu(\lambda)=(2\sinh\tau)^{-p}(2\sinh 2\tau)^{-q}\delta(\tau-\tau'),$$
$$d\mu(\lambda)=[2\pi|c(\lambda)|^2]^{-1}d\lambda. \tag{13.5.26}$$

13. Equalities (13.5.20, 26) enable one to write an expansion over the zonal spherical functions for any function, satisfying condition (13.5.25),

$$f(\tau)=\int \hat{f}(\lambda)\Phi_\lambda(\tau)d\mu(\lambda), \tag{13.5.27}$$

where the coefficient $\hat{f}(\lambda)$ is determined by the integral

$$\hat{f}(\lambda)=\int_0^\infty \bar{\Phi}_\lambda(\tau)f(\tau)d\mu(\tau). \tag{13.5.28}$$

14. An extension of the Plancherel formula is valid

$$\int_0^\infty |f(\tau)|^2 d\mu(\tau)=\int_0^\infty |\hat{f}(\lambda)|^2 d\mu(\lambda). \tag{13.5.29}$$

15. Functions describing the coherent states are also mutually orthogonal:

$$\int \bar{\psi}_x^\lambda(\xi)\psi_x^{\lambda'}(\xi')d\mu(x)=N(\lambda)\delta(\lambda-\lambda')\delta(\xi,\xi'), \tag{13.5.30}$$

where

$$x=\hat{x}\tanh\tau, \quad |\hat{x}|=1, \quad d\mu(x)=d\mu(\tau)d\mu(\hat{x}), \quad \int d\mu(\hat{x})=1. \tag{13.5.31}$$

Besides,

$$\int \bar{\psi}_x^\lambda(\xi)\psi_{x'}^\lambda(\xi)d\mu(\xi)\, d\mu(\lambda)=\delta(x,x'). \tag{13.5.32}$$

The delta functions $\delta(\xi,\xi')$ and $\delta(x,x')$ here are nonzero only for $\xi'=\xi$ and $x'=x$, and satisfy

$$\int \delta(\xi,\xi')d\mu(\xi')=1, \qquad \int \delta(x,x')d\mu(x')=1. \tag{13.5.33}$$

Equality (13.5.32) follows directly from the identity

$$\int \bar{\psi}_x^\lambda(\xi)\psi_{x'}^\lambda(\xi)d\mu(\xi)=\Phi_\lambda(\tau(x,x')) \tag{13.5.34}$$

combined with completeness for the zonal spherical functions (13.5.26).

The completeness (13.5.30) stems from Schur‘s lemma. Note, by the way, that by integrating (13.5.30) over $d\mu(\xi')$ we get the orthogonality (13.5.20).

16. With (13.5.30, 32) one is able to expand an arbitrary function $f(x)$, satisfying the condition

$$\int |f(x)|^2 d\mu(x)<\infty, \tag{13.5.35}$$

over the CS system. The result is

$$f(x)=\int \hat{f}(\xi,\lambda)\psi_x^\lambda(\xi)d\mu(\xi)d\mu(\lambda) \tag{13.5.36}$$

$$\hat{f}(\xi, \lambda) = \int \bar{\psi}_x^\lambda(\xi) f(x) d\mu(x) \tag{13.5.37}$$

and the Plancherel formula holds

$$\int |f(x)|^2 d\mu(x) = \int |\hat{f}(\xi, \lambda)|^2 d\mu(\xi) d\mu(\lambda). \tag{13.5.38}$$

17. The zonal spherical functions are closely related to the quantal scattering problem in a potential $V(\tau)$. To make this fact evident we substitute a new function

$$\Psi_\lambda(\tau) = (2 \sinh \tau)^{p/2} (2 \sinh 2\tau)^{q/2} \Phi_\lambda(\tau) \tag{13.5.39}$$

into (13.5.17). For function $\Psi_\lambda(\tau)$ we get an equation which is the Schrödinger equation with potential $V(\tau)$

$$\left[-\frac{d^2}{d\tau^2} + V(\tau) \right] \Psi_\lambda(\tau) = \lambda^2 \Psi_\lambda(\tau), \quad \text{where} \tag{13.5.40}$$

$$V(\tau) = \frac{a}{\sinh^2 \tau} + \frac{b}{\sinh^2 2\tau},$$

$$a = \left[\left(\frac{p+q-1}{2} \right)^2 - \left(\frac{q-1}{2} \right)^2 \right], \quad b = [(q-1)^2 - 1]. \tag{13.5.41}$$

The corresponding asymptotic behavior of function $\Psi_\lambda(\tau)$ as τ goes to infinity is

$$\Psi_\lambda(\tau) \sim [c(\lambda) e^{i\lambda\tau} + c(-\lambda) e^{-i\lambda\tau}]. \tag{13.5.42}$$

18. The scattering in potential $V(\tau)$ is determined by the so-called scattering matrix [116]. Here

$$S(\lambda) = -c(\lambda)/c(-\lambda). \tag{13.5.43}$$

Using expression (13.5.23) for $c(\lambda)$ then

$$S(\lambda) = -2^{-2i\lambda} \frac{\Gamma(i\lambda) \Gamma\left(\frac{\varrho - i\lambda}{2}\right) \Gamma\left(\frac{p}{4} + \frac{1}{2} - \frac{i\lambda}{2}\right)}{\Gamma(-i\lambda) \Gamma\left(\frac{\varrho + i\lambda}{2}\right) \Gamma\left(\frac{p}{4} + \frac{1}{2} + \frac{i\lambda}{2}\right)}. \tag{13.5.44}$$

It is remarkable that functions $c(\lambda)$ and $S(\lambda)$ are meromorphic and that $c(\lambda)$ have no zeros (and no poles) in the lower half-plane of λ. This fact means that there are no bound states (no discrete spectrum) in this problem. In upper half-plane, function $c(\lambda)$ [and respectively the scattering matrix $S(\lambda)$] has zeros and poles on the imaginary semiaxis $\lambda = i\kappa$, $\kappa > 0$. These poles arise since potential $V(\tau)$ decreases too slowly at $\tau \to \infty$ (as an exponential), and no bound states correspond to them. As for the zeros of $c(\lambda)$ in the upper half-plane, in all cases except for $SO(n,1)$ for odd n they correspond to bound states (discrete spectrum) for compact symmetric space dual to space X, according to *Cartan* [104].

19. Since representation $T^{\lambda}(g)$ is restricted to the maximal compact subgroup K, it can be decomposed over irreducible representations of this group.

Let $\hat{K}$ be a set of all UIRs of group K, and $\hat{K}_0$ be a set of representations involved in the decomposition of representation $T^{\lambda}(g)$. It is known that any representation which belongs to $\hat{K}_0$ is present in this decomposition only once. From the Frobenius reciprocity theorem [109] it follows that $\hat{K}_0$ contains only those representations of group K which contain the identity representation, being restricted on subgroup M.

It is seen from Table 13.1 that the problem is reduced to a consideration of the group restrictions: $SO(n)\to SO(n-1)$, $SU(n)\to SU(n-1)$, $\mathrm{Sp}(n)\to\mathrm{Sp}(n-1)$ and $SO(9)\to SO(7)$, respectively. Recall that any representation of a rank-ν compact simple group is determined by ν integer nonnegative numbers $l_1,\ldots,l_\nu$. Making use of the result presented in [70], we get the following results:

$$\begin{aligned}
&\text{In case I} && \hat{K}_0=\{(l,0,\ldots,0)\}\,;\\
&\text{In case II} && \hat{K}_0=\{(l_1,0,\ldots,0,l_2)\}\,;\\
&\text{In case III} && \hat{K}_0=\{(l_1,l_2,0,\ldots,0)\}\,;\\
&\text{In case IV} && \hat{K}_0=\{(l_1,l_2,0,0)\}.
\end{aligned} \tag{13.5.45}$$

20. Any coherent state $|x\rangle$ can be decomposed over representations which belong to $\hat{K}_0$. If we use notation $|l,m\rangle$ for the basis function of representation D_l present in $\hat{K}_0$, $\langle\xi|l,m\rangle=Y_{lm}(\xi)$, where l is the set of the numbers characterizing the representation and m is the set of the numbers enumerating the functions transformed by representations D_l, then we get an analog of the plane-wave decomposition,

$$|x\rangle=\sum_{lm}\Phi_{\lambda lm}(x)|l,m\rangle, \tag{13.5.46}$$

$$\psi_x^{\lambda}(\xi)=\sum_{lm}\Phi_{\lambda lm}(x)\,Y_{lm}(\xi), \tag{13.5.47}$$

$$\Phi_{\lambda lm}(x)=\langle l,m|T^{\lambda}(g)|0\rangle. \tag{13.5.48}$$

From the normalization condition for $|x\rangle$ one gets the condition

$$\sum_{lm}|\Phi_{\lambda lm}(x)|^2=1,\qquad l\in\hat{K}_0\,. \tag{13.5.49}$$

21. Note that (13.5.48) is equivalent to the integral representation

$$\Phi_{\lambda lm}(x)=\int\bar{Y}_{lm}(\xi)\psi_x^{\lambda}(\xi)d\mu(\xi). \tag{13.5.50}$$

It may also be shown that

$$\Phi_{\lambda lm}(x)=\Phi_{\lambda l}(\tau)\,\bar{Y}_{lm}(\hat{x}), \qquad x=\hat{x}\tanh\tau. \tag{13.5.51}$$

Expansion (13.5.47) now takes the form

$$\psi_x^\lambda(\xi)=\sum_{lm}\Phi_{\lambda l}(\tau)\,\bar{Y}_{lm}(\hat{x})\,Y_{lm}(\xi)$$

$$=\sum_{lm}\Phi_{\lambda l}(\tau)c_l\,Y_{l0}(|\hat{x}\bar{\xi}|), \qquad c_l=d_l/Y_{l0}(1), \tag{13.5.47'}$$

$$\int \bar{Y}_{lm}(\xi)\,Y_{lm'}(\xi)d\mu(\xi)=\delta_{mm'}, \qquad \int d\mu(\xi)=1.$$

where $x=\hat{x}\tanh\tau$ and d_l is the dimensionality of representation T_l of group K. The explicit expression for d_l is given by the well-known Weyl formula.

22. Functions $\Phi_{\lambda lm}(x)$, which are the matrix elements of the operator $T^\lambda(g)$, are eigenfunctions of the Laplace-Beltrami operator Δ_x for the symmetric space

$$-\Delta_x\Phi_{\lambda lm}(x)=(\varrho^2+\lambda^2)\Phi_{\lambda lm}(x), \tag{13.5.52}$$

$$\Delta_x=(1-|x|^2)\left(\frac{\partial^2}{\partial x_i\partial\bar{x}_i}-x_i\bar{x}_j\,\frac{\partial}{\partial\bar{x}_i}\,\frac{\partial}{\partial x_j}\right). \tag{13.5.53}$$

23. Functions $\Phi_{\lambda l}(\tau)$ (the so-called associated spherical functions) are orthogonal to each other

$$\int \bar{\Phi}_{\lambda l}(\tau)\Phi_{\lambda' l}(\tau)d\mu(\tau)=N_l(\lambda)\delta(\lambda-\lambda'). \tag{13.5.54}$$

24. Coefficient $N_l(\lambda)$ in (13.5.54) is determined by the asymptotics of $\Phi_{\lambda l}(\tau)$ at $\tau\to\infty$

$$\Phi_{\lambda l}(\tau)\sim[C_l(\lambda)e^{i\lambda\tau}+C_l(-\lambda)e^{-i\lambda\tau}]e^{-\varrho\tau}. \tag{13.5.55}$$

Here $C_l(\lambda)=\sqrt{d_l}\,e^{i\delta_l(\lambda)}c(\lambda)$, $N_l(\lambda)=2\pi|C_l(\lambda)|^2$, where d_l is the representation dimensionality.

25. Functions $\Phi_{\lambda l}(\tau)$ are a complete system. The condition of completeness is

$$\int \bar{\Phi}_{\lambda l}(\tau)\Phi_{\lambda l}(\tau')d\mu_l(\lambda)=\frac{1}{(2\sinh\tau)^p(2\sinh 2\tau)^q}\,\delta(\tau-\tau'),$$

$$d\mu_l(\lambda)=\frac{d\lambda}{2\pi|C_l(\lambda)|^2}\,. \tag{13.5.56}$$

26. An arbitrary square-integrable function $f(\tau)$

$$\int_0^\infty |f(\tau)|^2 d\mu(\tau)<\infty$$

can be expanded over the associated spherical functions $\Phi_{\lambda l}(\tau)$,

$$f(\tau)=\int \hat{f}_l(\lambda)\Phi_{\lambda l}(\tau)\,d\mu_l(\lambda), \quad \text{where} \tag{13.5.57}$$

$$\hat{f}_l(\lambda)=\int \bar{\Phi}_{\lambda l}(\tau) f(\tau)\,d\mu(\tau) \quad \text{and} \tag{13.5.58}$$

$$\int |f(\tau)|^2 d\mu(\tau)=\int |\hat{f}_l(\lambda)|^2 d\mu_l(\lambda). \tag{13.5.59}$$

27. Function

$$\Psi_{\lambda l}(\tau)=(2\sinh\tau)^{p/2}(2\sinh 2\tau)^{q/2}\,\Phi_{\lambda l}(\tau) \tag{13.5.60}$$

satisfies the Schrödinger equation with the potential

$$V_l(\tau)=\frac{a_l}{\sinh^2\tau}+\frac{b_l}{\sinh^2 2\tau}$$

$$a_l=\left[\left(l_1+\frac{p+q-1}{2}\right)^2-\left(l_2+\frac{q-1}{2}\right)^2\right],$$

$$b_l=4\left[\left(l_2+\frac{q-1}{2}\right)^2-\frac{1}{4}\right]. \tag{13.5.61}$$

Here coefficients a_l and b_l are eigenvalues of some operators $-\Delta_1$ and $-\Delta_2$ acting on the sphere $|x|=\text{const}$.

Hence it follows that the asymptotics of function $\Phi_{\lambda l}(\tau)$ have the form of (13.5.55), and $\Phi_{\lambda l}(\tau)$ and $C_l(\lambda)$ are obtained from $\Phi_\lambda(\tau)$ and $c(\lambda)$ by substituting

$$p\to p+2(l_1-l_2), \qquad q\to q+2l_2. \tag{13.5.62}$$

In conclusion, note that some results similar tio those presented here can be also obtained for symmetric spaces of arbitrary rank, considered in Sect. 13.6.

13.6 Complex Homogeneous Bounded Domains

As already mentioned in Sects. 13.3, 4, any irreducible Hermitian noncompact symmetric space can be realized as a complex homogeneous symmetric domain.

It is known [93] that there exist four series of such domains (the so-called classical domains) and two exceptional domains which will not be considered here. The classical domains are the coset spaces G/K:

$$D^{\mathrm{I}}_{p,q}, \quad p\geq q, \quad G=SU(p,q), \quad K=SU(p)\times SU(q)\times U(1); \tag{13.6.1}$$

$$D^{\mathrm{II}}_p \quad G=\mathrm{Sp}(p,\mathbb{R}), \quad K=U(p); \tag{13.6.2}$$

$$D^{\mathrm{III}}_p \quad G=SO^*(2p), \quad K=U(p), \tag{13.6.3}$$

where $SO^*(2p)$ is a subgroup of $SO(2p, \mathbb{C})$ group leaving invariant the Hermitian form

$$\bar{z}_1 z_{p+1} - \bar{z}_{p+1} z_1 + \ldots;$$

$$D_p^{\mathrm{IV}} \qquad G = SO_0(p,2), \qquad K = SO(p) \times SO(2), \tag{13.6.4}$$

where $SO_0(p,2)$ stands for the connected component of the identity transformation within the $SO(p,2)$ group.

All these domains are irreducible Hermitian symmetric spaces except for the type-IV domain at $p=2$. Table 13.2 presents the main characteristics of the classical domains. Note that the dimensionalities of spaces D and Ξ are related by $\dim \Xi = \dim D - r$, which follows from $\dim(G/K) = \dim[G/(MN)]$, here r is the rank of D (recall that space Ξ is defined in Sect. 13.1).

Table 13.2. The main characteristics of classical domains

D	$\dim D$	$\dim \Delta$	r
D^{I}	$2pq$	$2pq-q$	q
D^{II}	p^2+p	p^2	p
D^{III}	p^2-p	$p^2-p-[p/2]$	$[p/2]$
D^{IN}	$2p$	$2p-2$	2

Two realizations of classical domains are known: a bounded realization (like a unit disk) and an unbounded realization (like the upper half-plane).

The bounded realization has been described in detail [93]. For our purpose the unbounded realization is more convenient. We consider only the most important class, namely, the tube domains.

Recall that a homogeneous domain D is called the tube domain if it can be represented in the form

$$D = \{Z : Z = X + \mathrm{i}Y, X \in \mathbb{R}^n, Y \in V^n\},$$

where $V^n \subset \mathbb{R}^n$ is a convex self-adjoint homogeneous cone.

The tube domains belong to the four classical series, and in addition there is an exceptional domain:

$$\text{I)} \quad D_{pp}^{\mathrm{I}} = SU(p,p)/SU(p) \times SU(p) \times U(1); \tag{13.6.1'}$$

$$\text{II)} \quad D_p^{\mathrm{II}} = \mathrm{Sp}(p, \mathbb{R})/U(p); \tag{13.6.2'}$$

$$\text{III)} \quad D_{2p}^{\mathrm{III}} = SO^*(4p)/U(2p); \tag{13.6.3'}$$

$$\text{IV)} \quad D_p^{\mathrm{IV}} = SO_0(p,2)/SO(p) \times SO(2); \tag{13.6.4'}$$

$$\text{V)} \quad D^{\mathrm{V}} = E_7^{\mathbb{R}}/E_6^{\mathbb{C}} \times SO(2). \tag{13.6.5}$$

Here $E_7^{\mathbb{R}}$ is a real form of the exceptional group E_7, $E_6^{\mathbb{C}}$ is the compact form of the exceptional group E_6, $\dim_{\mathbb{C}} D^{\mathrm{V}}=27$, and the rank of domain D^{V} is equal to 3.

It is known that the possibility of representing a domain D in tube form is determined by the structure of its root diagram. Domain D has tube form if its root diagram is of type C_p and cannot be realized as a tube if the root diagram is of type BC_p [117].

To obtain explicit formulae for the kernels describing CS, one should find the character $\chi^\lambda(a)$ of representation T^λ. Note the geometrical meaning of element $a \in A$; it characterizes the complex distance between parallel horocycles.

Proposition 4. [112]. Let x_0 be the origin in D, and $\omega_0 = Nx_0$ be a horocycle parallel to ω, and $x \in \omega$. Then element $a \in A$ defining the complex distance between horocycle ω_0 and ω may be found from

$$x = a(x)n(x)x_0, \qquad x \in \omega. \tag{13.6.6}$$

Gel'fand and *Graev* [111] proved this statement for the coset space of a complex group. Their proof can be also applied to our case, since the family of horocycles is transitive with respect to the action of group G.

We turn now to explicit calculation of the complex distances and relevant subgroups for the four series of the classical tube domains. Some general results will elucidate the structure of the calculations.

Proposition 5. Any analytical automorphism of a tube domain with fixed infinity point has the form

$$Z \to AZ + B, \tag{13.6.7}$$

where A is an affine transformation of the cone V on itself, and B is a real vector.

Group G_1 of the affine transformations of cone V acts as follows

$$g\colon Y \to Y' = AYA^+, \qquad Y, Y' \in V, \qquad g \in G_1, \tag{13.6.8}$$

where A belongs, correspondingly, to (notations taken from [81])

1. $SL(p, \mathbb{C}) \times \mathbb{R}^+$ for D^{I};
2. $SL(p, \mathbb{R}) \times \mathbb{R}^+$ for D^{II}; (13.6.9)
3. $SU^*(2p, \mathbb{R}) \times \mathbb{R}^+$ for D^{III};

where $SU^*(2p, \mathbb{R}) \times \mathbb{R}^+$ is the group of the real quaternionic matrices, $\mathbb{R}^+$ is the multiplicative group of positive real numbers. Further,

4. $SO(p-1, 1) \times \mathbb{R}^+$ for D^{IV}.

Note that any element $Y \in V$ can be represented in the form

$$Y = AY_0A^+, \tag{13.6.10}$$

where $A \in N_1$, N_1 is the maximal nilpotent subgroup of G_1, and Y_0 is a diagonal matrix with positive elements.

Note that the cones associated with the domains of the types D^{I}, D^{II}, D^{III}, D^{IV} can be described as:

I) the cone of positive-definite complex Hermitian matrices for D^{I};
II) the cone of positive-definite real symmetric matrices for D^{II};
III) the cone of positive-definite Hermitian-quaternion matrices for D^{III};
IV) the cone of positive-definite "Hermitian" 3×3 matrices containing Cayley numbers (octonions) for D^{V}; i.e., matrices of the form

$$\begin{pmatrix} \alpha & c & b \\ \bar{c} & \beta & a \\ \bar{b} & \bar{a} & \gamma \end{pmatrix},$$

where α, β, γ are real numbers, a, b, c are octonions, and $\bar{a}$, $\bar{b}$, $\bar{c}$ are conjugated octonions.

Finally, note that for the D^{IV} the cone is the set of vectors

$$\{y = (y_1, \ldots, y_p) \colon y_1^2 - y_2^2 - \ldots - y_p^2 > 0, y_1 > 0\}.$$

Let us consider classical tube domains in more detail. We shall use the following notations: A^+: the matrix Hermitian conjugated to matrix A, $A > 0$: for a Hermitian matrix A means that all its eigenvalues are positive; $A^{(p,q)}$: a matrix with p lines and q columns; $A^{(p)}$: a matrix of order p.

13.6.1 Type-I Tube Domains

We will study these domains in some detail. They are defined as

$$D_p^{\mathrm{I}} = SU(p,p)/SU(p) \times SU(p) \times U(1) \quad \text{or}$$

$$D_p^{\mathrm{I}} = \{Z \colon Z = X + \mathrm{i}Y, X^+ = X, Y^+ = Y, Y > 0\},$$

where $G = SU(p,p) = \{g\}$ acts on D as a group of fractional linear transformations:

$$g = \begin{pmatrix} A & B \\ C & D \end{pmatrix} \colon Z \to Z' = (AZ + B)(CZ + D)^{-1}, \quad \det g = 1, \tag{13.6.11}$$

where A, B, C, D are matrices of order p.

Some useful facts concerning the structure of the group G and its Lie algebra $\mathscr{G}$ are given below.

1. The maximal compact subgroup $K=\{k\}$ is isomorphic to $SU(p)\times SU(p) \times U(1)$. Its Lie algebra $\mathscr{K}$ consists of the matrices

$$\begin{pmatrix} \tilde{A} & \tilde{B} \\ -\tilde{B} & \tilde{A} \end{pmatrix}, \tag{13.6.12}$$

where $\tilde{A}^+ = -\tilde{A}$, $\tilde{B}^+ = \tilde{B}$, $\operatorname{tr}\tilde{A}=0$.

2. Exponential mapping of the matrix in (13.6.12) leads to an expression for a matrix $k\in K$

$$k=\frac{1}{2}\begin{pmatrix} (A+D) & \pm\mathrm{i}(A-D) \\ \mp\mathrm{i}(A-D) & (A+D) \end{pmatrix}, \tag{13.6.13}$$

where $A^+A=D^+D=1$, $\det A \det D=1$.

3. Subgroup A consists of matrices

$$a=\begin{pmatrix} A & 0 \\ 0 & A^{-1} \end{pmatrix}, \tag{13.6.14}$$

where A is a diagonal matrix with nonnegative elements.

4. Subgroup M (the centralizer of A in K) consists of the following matrices:

$$m=\begin{pmatrix} T & 0 \\ 0 & T \end{pmatrix}, \tag{13.6.15}$$

where T is a diagonal unitary matrix with $\det T=\pm 1$. Correspondingly, $M = U(1)\times \ldots \times U(1)$ (($p-1$) times).

5. Subalgebra $\mathscr{N}$ consists of matrices

$$n=\begin{pmatrix} A & B \\ 0 & -A^+ \end{pmatrix}, \tag{13.6.16}$$

where A is an upper triangular matrix with zeros on the diagonal and B is an Hermitian matrix.

6. Group N is produced as the exponential mapping of the Lie algebra $\mathscr{N}$

$$N=\left\{\begin{pmatrix} A & AB_1 \\ 0 & (A^+)^{-1} \end{pmatrix}\right\}, \tag{13.6.17}$$

where A is an upper triangular matrix with unities on the diagonal, B_1 is an Hermitian matrix; N is the semidirect product of two groups N_1 and N_2, where

$$N_1=\left\{\begin{pmatrix} A & 0 \\ 0 & (A^+)^{-1} \end{pmatrix}\right\}, \quad N_2=\left\{\begin{pmatrix} \mathrm{I} & B_1 \\ 0 & \mathrm{I} \end{pmatrix}\right\}. \tag{13.6.18}$$

Group N_1 acts on the cone V in the natural way

$$n_1: Y \to Y' = AYA^+. \tag{13.6.19}$$

Its action on the cone is not transitive. It seems natural to call the orbits of this group the horocycles ω_c in the cone. Group N_2 acts as the translation group.

So, horocycle ω of the whole symmetric space can be represented by

$$\omega = \{X + iY,\ Y \in \omega_c,\ X \in S\}, \qquad S = \{X; X^+ = X\}. \tag{13.6.20}$$

7. Let $Z = X + iY$ be a generic point in domain D. It can be easily seen that we can transfer point Z into point iY by a transformation belonging to group N_2 and by applying a transformation of the group AN_1 transfer it into point $Z_0 = iI$. Therefore $Z = X + iAA^+$, $Y = AA^+$. Matrix Y is positive definite and it can be represented unambiguously in the form AY_0A^+, where A is a complex matrix with unities on the diagonal, Y_0 is a diagonal matrix with $y_{jj} > 0$. The elements of matrix Y_0 determine the complex distance between matrix Y and the standard horocycle passing through point $Z_0 = iI$. It is not difficult to obtain the following expression for the elements y_{jj}:

$$y_{jj} = \frac{\Delta_{p-j+1}(Y)}{\Delta_{p-j}(Y)}, \tag{13.6.21}$$

where $\Delta_j(Y)$ is the principal lower angle minor of the jth order, $\Delta_0(Y) = 1$. The values y_{jj} are positive according to the Sylvester criterion of positive definiteness for matrix Y, and the cone V is defined by inequalities

$$y_{11} > 0, \ldots, y_{pp} > 0. \tag{13.6.22}$$

13.6.2 Type-II Tube Domains

The approach we use is essentially the same as for type-I domains with some evident modifications. So it is sufficient to describe the matrix realization of domain D^{II} and to formulate the final results

1. $D_p^{\mathrm{II}} = \mathrm{Sp}(p, \mathbb{R})/U(p)$

and in the tube realization

$$D_p^{\mathrm{II}} = \{Z: Z = X + iY\},$$

where X and Y are the real symmetric matrices of order p, and Y is positive definite. Group $G = \mathrm{Sp}(p, \mathbb{R})$ acts as a group of the fractional linear transformations (13.6.11). Matrices A, B, C and D satisfy the following conditions:

$$A'C = C'A, \qquad B'D = D'B, \qquad A'D - C'B = I. \tag{13.6.23}$$

2. Let $Z_0 = \mathrm{i}I$, $Z = X + \mathrm{i}Y$. As in item (7) of the previous section, $Y = AA'$ (A is a real upper triangular matrix with positive elements on the diagonal). Alternatively, $Y = \tilde{A} Y_0 \tilde{A}'$, where $\tilde{A}$ is the upper triangular matrix with unities on the diagonal, and Y_0 is the diagonal matrix with elements $y_{11}, \ldots, y_{pp}$. Then

$$y_{jj} = \frac{\Delta_{p-j+1}(Y)}{\Delta_{p-j}(Y)}, \qquad \Delta_0(Y) = 1.$$

13.6.3 Type-III Tube Domains

Domain D_{2p}^{III} of type-III, see (13.6.3), is a tube. Its realization is

$$D_{2p}^{\mathrm{III}} = \{Z\colon Z = X + \mathrm{i}Y, X^+ = X, Y^+ = Y, Y > 0\},$$

where X and Y are the matrices of order $2p$ satisfying the supplementary condition

$$ZJ = J'Z, \qquad \text{where} \tag{13.6.24}$$

$$J = \begin{pmatrix} j & 0 & & \\ 0 & j & & \\ & & \ddots & \\ & & & j \end{pmatrix}, \qquad j = \begin{pmatrix} 0 & -1 \\ 1 & 0 \end{pmatrix}. \tag{13.6.25}$$

It is convenient to construct all the $2p \times 2p$ matrices of p^2 blocks of 2×2 matrices representing quaternions

$$q = q^0 \tau_0 + q^1 \tau_1 + q^2 \tau_2 + q^3 \tau_3, \qquad \text{where} \tag{13.6.26}$$

$$\tau_0 = \begin{pmatrix} 1 & 0 \\ 0 & 1 \end{pmatrix}, \quad \tau_1 = \begin{pmatrix} 0 & -\mathrm{i} \\ -\mathrm{i} & 0 \end{pmatrix}, \quad \tau_2 = \begin{pmatrix} 0 & -1 \\ 1 & 0 \end{pmatrix}, \quad \tau_3 = \begin{pmatrix} -\mathrm{i} & 0 \\ 0 & \mathrm{i} \end{pmatrix},$$

$$\tau_k^+ = -\tau_k, \qquad k = 1, 2, 3. \tag{13.6.27}$$

Elements τ_0, τ_k $(k = 1, 2, 3)$ provide the familiar basis for noncommutative associative algebra with division over the field of real numbers.

As usual, we can define the conjugate quaternion $\bar{q} = q^0 \tau_0 - q^k \tau_k$ and the norm $N(q) = \bar{q}q = q\bar{q} = (q^0)^2 + q^k q^k$. It is easy to verify that

$$jq'j^{-1} = \bar{q}. \tag{13.6.28}$$

The complex-conjugate quaternion is defined as $q_c = q_c^0 + q_c^k \tau_k$, where q_c^k is the number complex conjugate to q^k. Now $q^+ = q_c^0 \tau_0 - q_c^k \tau_k$.

Let $X = (x_{kl})$ be the $p \times p$ quaternionic Hermitian matrix, satisfying condition (13.6.24). Then

$$\tilde{X} = JX'J^{-1} = (\bar{x}_{lk}). \tag{13.6.29}$$

Since $X^+ = X$, then $x^+_{lk} = \bar{x}_{lk} = x_{kl}$ and consequently, both matrices X and Y are real quaternionic and Hermitian.

Positive definite matrices $Y (Y > 0)$ form a cone V in the $p(2p-1)$-dimensional real linear space. The affine transformations of V are

$$Y \to Y_1 = AYA^+. \tag{13.6.30}$$

As Y belongs to the cone V, the following condition applies:

$$\begin{aligned} AYA^+ &= JY_1' J^{-1} = J(AYA^+)' J^{-1} \\ &= J\bar{A} Y' A' J^{-1} = J\bar{A} J^{-1} YJA' J^{-1}. \end{aligned}$$

Hence A must satisfy

$$JAJ^{-1} = \tilde{A} = A^+, \tag{13.6.31}$$

i.e., it is real quaternionic but not necessarily Hermitian. The real quaternionic matrices form a group $SU^*(2p) \times R^+$, which is the group of affine transformations of the cone V.

We will not present here explicitly the properties of the domains, since they are similar to those of domains of types I and II.

13.6.4 Type-IV Domains

All type-IV domains are tubes. The domain can be realized in the form

$$D_p^{\mathrm{IV}} = \{Z\colon Z = X + \mathrm{i}Y, X \in \mathbb{R}^p, Y \in V\},$$

where the cone V is fixed by inequalities

$$y_1^2 - y_2^2 - \ldots - y_p^2 > 0, \qquad y_1 > 0, \tag{13.6.32}$$

or, with another metric:

$$2y_1 y_p - y_2^2 - \ldots - y_{p-1}^2 > 0, \qquad y_1 > 0. \tag{13.6.32$'$}$$

The affine transformations of this cone are

$$y \to Ay. \tag{13.6.33}$$

Here A is a real $p \times p$ matrix, satisfying $A'EA = \lambda E$ (λ is a positive number), where matrix E is

$$E = \begin{pmatrix} 1 & 0 & & & \\ 0 & -1 & & & \\ & & \ddots & & \\ & & & \ddots & \\ & & & & -1 \end{pmatrix} \tag{13.6.34}$$

for the cone (13.6.32) and

$$E=\begin{pmatrix} 0 & & & & 1 \\ & -1 & & & \\ & & \ddots & & \\ & & & -1 & \\ 1 & & & & 0 \end{pmatrix} \tag{13.6.34'}$$

for the cone (13.6.32').

Evidently, the group of affine transformations of the cone is $SO(p-1,1) \times R^+$. The group A is a two-dimensional group. It contains dilatation $y_j \to \lambda y_j$ and the hyperbolic rotation of the $(1,p)$ plane

$$y_1' = y_1 \cosh \tau + y_p \sinh \tau$$

$$y_p' = y_1 \sinh \tau + y_p \cosh \tau.$$

In the metric of (13.6.32')

$$x^2 = 2(x_1 x_{p+1} + x_2 x_{p+2}) + x_3^2 + \ldots + x_p^2.$$

Elements of group A are matrices of the form

$$a=\begin{pmatrix} \begin{matrix} \lambda_1 & 0 \\ 0 & \lambda_2 \end{matrix} & 0 & 0 \\ 0 & \mathrm{I}_q & 0 \\ 0 & 0 & \begin{matrix} \lambda_1^{-1} & 0 \\ 0 & \lambda_2^{-1} \end{matrix} \end{pmatrix} \tag{13.6.35}$$

The group transformations are

$$x_1 \to \lambda_1 x_1, \qquad x_{p+1} \to \lambda_1^{-1} x_{p+1}$$

$$x_2 \to \lambda_2 x_2, \qquad x_{p+2} \to \lambda_2^{-1} x_{p+2}.$$

Lie algebra $\mathscr{A} = \{a\}$ is represented by the diagonal matrices

$$\begin{pmatrix} \alpha & 0 & 0 \\ 0 & 0_q & 0 \\ 0 & 0 & -\alpha \end{pmatrix}, \tag{13.6.36}$$

where α is a real diagonal 2×2 matrix.

To describe the Lie algebra $\mathscr{G}$ of group G, we use an equation for its elements:

$$A'E + EA = 0. \tag{13.6.37}$$

It is suitable to represent A in block form

$$A=\begin{pmatrix} A_{11} & A_{12} & A_{13} \\ A_{21} & A_{22} & A_{23} \\ A_{31} & A_{32} & A_{33} \end{pmatrix}\begin{matrix} \}2 \\ \}q, \\ \}2 \end{matrix} \qquad (13.6.38)$$

with column blocks of widths 2, q, 2.

Hence, condition (13.6.37) is rewritten as follows:

$$A'_{31}=-A_{31}, \quad A'_{22}=-A_{22}, \quad A'_{13}=-A_{13}$$

$$A_{33}=-A'_{11}, \quad A_{32}=-A'_{21}, \quad A_{23}=-A'_{12}. \qquad (13.6.39)$$

13.6.5 The Exceptional Domain D^{V}

Besides the "classical" tube domains, an exceptional domain exists in $\mathbb{C}^{27}$. It is $D^{\mathrm{V}}=E_7^{\mathrm{R}}/E_6^{\mathrm{C}}\times SO(2)$, where E_7^{R} is the real form of the exceptional simple group E_7. This space can be represented as

$$D^{\mathrm{V}}=\{Z\colon Z=X+\mathrm{i}Y\},$$

where X and Y are Hermitian 3×3 matrices whose elements are Cayley numbers, and Y is positive definite ($Y>0$). Matrices Y form a convex self-adjoint cone in space $\mathbb{R}^{27}$. The rank of D^{V} is 3.

13.7 Properties of the Coherent States

As shown above, the coherent state is given by the horospherical kernel

$$\psi_x^\lambda(\xi)=\langle\xi,\lambda|x\rangle, \qquad x\in X, \qquad \xi\in\Xi.$$

For tube domains, it is suitable to use the unbounded realization of space $X=G/K$.

Let $\psi_0^\lambda(\xi)$ be a K-invariant function on Ξ, N a maximal nilpotent subgroup leaving invariant the point $\xi_0=\{\infty\}$, and $\boldsymbol{\varrho}=(\varrho_1,\ldots,\varrho_p)=\frac{1}{2}\sum\boldsymbol{\alpha}$, where α are positive roots of the symmetric space. As shown in [108],

$$\langle\xi_0,\lambda|z\rangle=\psi_z^\lambda(\xi_0)=\prod_{j=1}^{p} y_j^{-\varrho_j+\mathrm{i}\lambda_j}\psi_0^\lambda(\xi_0). \qquad (13.7.1)$$

Here p is the rank of domain D, y_j are diagonal elements of matrix Y_0, which were given explicitly above.

Using the group-theoretical properties of the multiplicator $\tilde{\alpha}^{\lambda}(z,\xi)$ one can obtain an expression for the function $\langle\xi,\lambda|z\rangle$ for an arbitrary ξ [108].

Here are some properties of a CS system.

1. The horospherical kernels $\psi_x^{\lambda}(\xi)$ are the eigenfunctions of the Laplace-Beltrami operator on symmetric space $X=G/K$

$$-\Delta_x\psi_x^{\lambda}(\xi)=(\lambda^2+\varrho^2)\psi_x^{\lambda}(\xi). \tag{13.7.2}$$

The kernels are constant on the horocycles of space X. Therefore these states are a natural analog of the plane-waves system for the symmetric space which has a nonzero curvature.

2. The natural orthonormal basis in the space of the unitary irreducible representation of class-I consists of zonal spherical functions on X.

3. Function $|\psi_z^{\lambda}(\xi)|^2$ is the Poisson kernel for space X.

4. The CS system $\{|x\rangle\}$ is complete but not orthogonal

$$\langle x|y\rangle=\Phi^{\lambda}(\tau), \tag{13.7.3}$$

where $\tau=\tau(x,y)$ is the complex distance between points x and y in space X. Function $\Phi_{\lambda}(\tau)$ is

$$\Phi_{\lambda}(\tau)=\langle 0|T^{\lambda}(g)|0\rangle \tag{13.7.4}$$

or, equivalently,

$$\Phi_{\lambda}(\tau)=\langle 0|x\rangle=\int\psi_x^{\lambda}(\xi)d\mu(\xi). \tag{13.7.4$'$}$$

5. The explicit expression for this function is known only for symmetric spaces of rank-I [114]. However, the asymptotic behavior at $\tau\to\infty$ is known for the general case. Namely,

$$\Phi_{\lambda}(\tau)\sim\sum_{s\in W}c(s\lambda)\exp[(\mathrm{i}s\lambda-\varrho)(\tau)],\qquad \tau\to\infty \tag{13.7.5}$$

where W is the Weyl group of space X, and function $c(\lambda)$ is defined by

$$c(\lambda)=\int_N\exp[(-\varrho+\mathrm{i}\lambda)H(\bar{n})]d\mu(\bar{n}). \tag{13.7.6}$$

Here measure $d\mu(\bar{n})$ is normalized so that

$$\int_{\bar{N}}\exp[-2\varrho H(\bar{n})]d\mu(\bar{n})=1,\ \text{i.e.,}\ c(-\mathrm{i}\varrho)=1. \tag{13.7.7}$$

Function $H(g)$ is determined from the Iwasawa decomposition

$$g=n(g)\exp(H(g))k(g),\qquad n\in N,\qquad H\in A,\qquad k\in K. \tag{13.7.8}$$

It follows from the *Bruhat* decomposition [112] that space $\bar{N}MAN$ is dense in G, and element $\bar{n}$ in (13.7.6, 7) is determined by g unambiguously, as $g=\bar{n}amn$.

6. The explicit expression for function $c(\lambda)$ was obtained by *Gindikin* and *Karpelevich* [115]:

$$c(\lambda)=\frac{\mathrm{I}(\mathrm{i}\lambda)}{\mathrm{I}(\varrho)}, \qquad \mathrm{I}(\lambda)=\prod_{\alpha\in R^+} B\left(\frac{m_\alpha}{2}, \frac{m_{\alpha/2}}{4}+\frac{(\lambda,\alpha)}{(\alpha,\alpha)}\right). \tag{13.7.9}$$

Here $B(\mu, \nu)=\dfrac{\Gamma(\mu)\Gamma(\nu)}{\Gamma(\mu+\nu)}$ is the beta function, R^+ is the set of positive roots of symmetric space X, m_α is the multiplicity of root α, and (α, β) is the standard scalar product in the root space.

7. The set of functions corresponding to the CS system is complete and orthogonal,

$$\int \bar{\psi}_x^\lambda(\xi)\psi_x^{\lambda'}(\xi')\,d\mu(x)=N\delta(\lambda-\lambda')\delta(\xi,\xi') \tag{13.7.10}$$

$$\int \bar{\psi}_x^\lambda(\xi)\psi_{x'}^\lambda(\xi)\,d\mu(\xi)\,d\mu(\lambda)=N_1\delta(x,x'). \tag{13.7.11}$$

Here functions $\delta(x,x')$, $\delta(\xi,\xi')$ are normalized to give identities

$$\int \delta(x,x')f(x')\,d\mu(x')=f(x),$$

$$\int \delta(\xi,\xi')f(\xi')\,d\mu(\xi')=f(\xi), \tag{13.7.12}$$

$$d\mu(\lambda)=|c(\lambda)|^2 d\lambda, \qquad N_1=|c(\lambda)|^{-2}N,$$

and $c(\lambda)$ is defined in (13.7.9).

The proof of (13.7.10) stems from a generalization of Schur's lemma [109]. Relation (13.7.11) follows from completeness of the system of spherical functions.

8. Let us now turn to decomposition of representations $T^\lambda(g)$ into irreducible components when restricted to the maximal compact subgroup K. It follows from the Frobenius reciprocity theorem [109] that the decomposition contains those and only those representations of group K which, being restricted to subgroup M, contain the identity representation. Thereby the problem reduces to that for compact groups.

We should note in conclusion that the results and calculation methods relevant to classical domains of the tube type are also valid for other complex homogeneous domains.

14. Coherent States and Discrete Subgroups: The Case of $SU(1,1)$

This Chapter considers the properties of completeness for subsystems of the CS system which are related to the discrete series of representations of the group $G=SU(1,1)/Z_2$, where $Z_2=\{I, -I\}$ is the center of $SU(1,1)$. The presentation follows a previous work [118].

14.1 Preliminaries

Group $G=SU(1,1)/Z_2$ is the group of motions of the Lobachevsky plane and the coherent states considered correspond to the points of this plane, as shown in Chap. 5.

An appropriate realization of the Lobachevsky plane is the unit disk $D=\{\zeta:|\zeta|<1\}$, and the CS system relevant to the representation $T^k(g)$ of discrete series of group $SU(1,1)$ is the set of functions (Chap. 5).

$$\psi_\zeta(z)=(1-|\zeta|^2)^k(1-\bar{\zeta}z)^{-2k} \tag{14.1.1}$$

These functions belong to space $\mathscr{F}_k$ for representation $T^k(g)$, i.e., the space of functions analytic inside the unit disk $D=\{z:|z|<1\}$, and satisfy condition

$$\|f\|_k<\infty, \quad \text{where} \tag{14.1.2}$$

$$\|f\|_k^2=\int_D |f(z)|^2\, d\mu_k(z),$$

$$d\mu_k(z)=\frac{2k-1}{\pi}(1-|z|^2)^{2k-2}dx\,dy, \quad z=x+iy. \tag{14.1.3}$$

It can be easily verified that functions

$$f_n(z)=\sqrt{\frac{\Gamma(n+2k)}{\Gamma(n+1)\Gamma(2k)}}\,z^n, \quad n=0,1,2,\dots \tag{14.1.4}$$

constitute an orthonormalized basis in space $\mathscr{F}_k$. The expansion of a CS in terms of the orthonormalized basis is

$$\psi_\zeta(z)=(1-|\zeta|^2)^k\sum_{n=0}^{\infty}\overline{f_n(\zeta)}f_n(z). \tag{14.1.5}$$

The CS system is overcomplete and a question arises as to what are complete subsystems of this system. To answer this question let us try the following trick.

Let $|\psi\rangle$ be an arbitrary vector in the Hilbert space corresponding to the function $\psi(\zeta)$,

$$\langle\psi|\zeta\rangle = \int \bar{\psi}(z)\psi_\zeta(z)\,d\mu_k(z) = (1-|\zeta|^2)^k\psi(\zeta). \tag{14.1.6}$$

In view of (14.1.5), $\psi(\zeta)\in\mathscr{F}_k$, i.e., it is analytic in the unit disk $D=\{\zeta : |\zeta|<1\}$, and

$$\|\psi(\zeta)\|_k^2 = \int d\mu_k(\zeta)|\psi(\zeta)|^2 = \|\psi\|^2 < \infty. \tag{14.1.7}$$

Hence a subsystem of states $\{|\zeta_n\rangle\}$ is complete if the set $\{\zeta_n\}$ has a limiting point within the disk D. Such a subsystem remains complete even after removal of a finite number of CS.

The following proposition is also useful.

Proposition 1. The subsystem $\{|\zeta_n\rangle\}$ is not complete if and only if $\{\zeta_n\}$ is the set of zeros of a function $f(\zeta)\in\mathscr{F}_k$, which is not identically zero.

Indeed, if a system $\{|\zeta_n\rangle\}$ is not complete, a vector $|\psi\rangle$ exists in the Hilbert space which is orthogonal to all vectors of the system. The function corresponding to it, $\psi(\zeta)\in\mathscr{F}_k$, vanishes at the points ζ_n. Inversely, if $\psi(\zeta)\in\mathscr{F}_k$ and $\psi(\zeta_n)=0$, then the vector

$$|\psi\rangle = \frac{2k-1}{\pi}\int d\mu(\zeta)\,(1-|\zeta|^2)^k\psi(\zeta)|\zeta\rangle \tag{14.1.8}$$

belongs to the Hilbert space and is orthogonal to all the vectors $|\zeta_n\rangle$.

The simplest subsystems $\{|\zeta_n\rangle\}$ are related to discrete subgroups of group $G=SU(1,1)/\mathbb{Z}_2$. Let $\Gamma=\{\gamma_n\}$ be a discrete subgroup of G and let ζ_0 be some point in the domain D.

Definition. The set of states $\{|\zeta_n\rangle\}$, where $\zeta_n=\zeta_0\cdot\gamma_n$, is called a subsystem of coherent states related to subgroup Γ.

We now turn to the question of completeness of such subsystems.

14.2 Incompleteness Criterion for CS Subsystems Related to Discrete Subgroups

To prove the criterion of incompleteness for the subsystems in view we will make use of the theory of automorphic forms with respect to the discrete subgroup Γ. [A detailed treatment of the properties if discrete subgroups of group $SU(1,1)$ [or of the group $SL(2,R)$ isomorphic to it] and the corresponding automorphic forms can be found in [119–121] cf. Sect. 14.5.] We restrict ourselves to those discrete subgroups for which the fundamental domain $\Gamma\backslash D$ has a finite area (in general, $\Gamma\backslash D$ is not necessarily compact). It is well known that in this case the fundamental domain can be taken as a polygon with a finite number of sides which are segments of geodesics. Vertices of the polygon lying on the boundary of the domain $D=\{z:|z|=1\}$ are called parabolic vertices. Let us denote by $\mathscr{P}$ the set of the parabolic vertices and let $D^+=D\cup\mathscr{P}$.

Definition. An automorphic form of weight m (m is an integer) is called a function $f_m(z)$, analytic within the disk D, satisfying the functional equation

$$f_m(z\gamma_n)=(\beta_n z+\bar{\alpha}_n)^{2m}f_m(z), \qquad \gamma\in\Gamma, \tag{14.2.1}$$

and regular in the domain D^+ (this means that it must have a definite limit at every parabolic vertex z_p,

$$\lim\,(z-z_p)^{2m}f_m(z)$$

as $z\to z_p$ from the interior of domain $\Gamma\backslash D$).

An automorphic form $f_m(z)$ is called parabolic if it vanishes at all the parabolic vertices.

The set of automorphic forms of a weight m is a finite-dimensional vector space. We denote by $d_m(\Gamma)$ and $d_m^+(\Gamma)$ the dimensionalities of spaces of automorphic and parabolic forms, respectively. Let m_0 be the least value of m for which $d_m(\Gamma)\geq 2$ (respectively, m_0^+ for $d_m^+(\Gamma)\geq 2$). If for this Γ the domain $\Gamma\backslash D$ is compact, then $d_m(\Gamma)=d_m^+(\Gamma)$, $m_0=m_0^+$, and every automorphic form can be considered as parabolic.

Theorem 1. Suppose $\{|\zeta\rangle\}$ is a CS system of the type $(T^k, |0\rangle)$. A subsystem $\{|n\rangle\}$, $|n\rangle=|\zeta_n\rangle$, $\zeta_n=\zeta_0\cdot\gamma_n$, $\gamma_n\in\Gamma$ related to a discrete subgroup Γ is incomplete if $k>m_0^+ +1/2$.

Proof. Note, first of all, that in this case there exists a parabolic form $f_{m_0^+}(z)$ of a weight m_0^+ which vanishes at an arbitrary point ζ_0 of domain D and thus vanishes at all points ζ_n.

In accordance with Proposition 1, we can consider the case where the function $f_{m_0^+}(\zeta)$ belongs to space $\mathscr{F}_k$. For arbitrary m function $|f_m(\zeta)|^2$ satisfies

$$|f_m(\zeta\gamma_n)|^2=|\beta_n\zeta+\bar{\alpha}_n|^{4m}|f_m(\zeta)|^2. \tag{14.2.2}$$

In view of identity

$$|\beta\zeta+\bar{\alpha}|^2=(1-|\zeta|^2)/(1-|\zeta g|^2), \tag{14.2.3}$$

function $|f_m(\zeta)|^2$ can be represented in the form

$$|f_m(\zeta)|^2=(1-|\zeta|^2)^{-2m}F_m(\zeta,\bar{\zeta}), \tag{14.2.4}$$

where $F_m(\zeta,\bar{\zeta})$ is a nonnegative Γ-invariant function: $F_m(\zeta\gamma_n, \overline{\zeta\gamma_n}) = F_m(\zeta,\bar{\zeta})$. So, if the form $f_m(\zeta)$ is parabolic, function $F_m(\zeta,\bar{\zeta})$ is bounded within the fundamental domain, and, therefore, throughout the whole disk D.

The norm of function $f_{m_0^+}(\zeta)$ is given by integral

$$\|f_{m_0^+}(\zeta)\|_k^2 = \frac{2k-1}{\pi} \int_D d^2\zeta(1-|\zeta|^2)^{2k-2-2m_0^+} F_{m_0^+}(\zeta,\bar{\zeta}). \tag{14.2.5}$$

It can be easily seen that this integral converges when $k > (m_0^+ + 1/2)$. Thus, the proof of Theorem 1 is completed.

Remark 1. If $k < (m_0^+ + 1/2)$, the integral in (14.2.5) diverges. The boundary case, $k = (m_0^+ + 1/2)$, requires a more careful analysis. Now

$$\begin{aligned} \frac{\pi}{2k-1} \|f_{m_0^+}\|_k^2 &= \int_D d^2\zeta(1-|\zeta|^2)^{-1} F_{m_0^+}(\zeta,\bar{\zeta}) \\ &= \int_{\Gamma\backslash D} d^2\zeta(1-|\zeta|^2)^{-2} F_{m_0^+}(\zeta,\bar{\zeta}) \sum_n (1-|\zeta_n|^2). \end{aligned} \tag{14.2.6}$$

It is known, however [Ref. 121, p. 181], that if area S_Γ of the fundamental domain $\Gamma\backslash D$ is finite, the series $\sum_n (1-|\zeta_n|^2)$, where $\zeta_n = \zeta_0\gamma_n$, diverges. [The proof of this statement is quite simple if the fundamental domain $\Gamma\backslash D$ is compact. If it is noncompact but has a finite area one can use *Hedlund's* results [122] on the distribution of the points ζ_n near the boundary.] Thus, for $k \le (m_0^+ + 1/2)$ the integral in (14.2.5) diverges.

Remark 2. Suppose group Γ is given so that the area of the fundamental domain $\Gamma\backslash D$ is infinite. Then the system of states $\{|\zeta_n\rangle\}$, $\zeta_n = \zeta_0\gamma_n$ is incomplete. Actually, one can show [121] that the series $\sum_n (1-|\zeta_n|^2)$ converges for any $\zeta_0 \in D$ and that a function exists analytic and bounded in D, which has zeros at the points ζ_n.

Remark 3. The method used here to prove the incompleteness of the subsystem of CS is applicable also to CS subsystems for the groups of motion of homogeneous domains, Chap. 15.

14.3 Growth of a Function Analytical in a Disk Related to the Distribution of Its Zeros

To prove the criterion of completeness of subsystems $\{|\zeta_n\rangle\}$ we use a theorem concerning the lower bound of the growth of functions analytic in the disk $\{z : |z| < 1\}$. An analogous theorem for entire analytic functions is well known [123].

Let $M(r)$ be the maximum modulus of function $f(z)$ on the circle $|z|=r$, $M_1(r)=\left[\frac{1}{2\pi}\int|f(re^{i\varphi})|^2 d\varphi\right]^{1/2}$, and $n(r)$ be the number of zeros of $f(z)$ in the disk $\{z:|z|<r\}$. Suppose that the limit $v=\underline{\lim}_{r\to 1}(1-r^2)n(r)$ exists and that $v\neq 0$. The numbers $\tau=\overline{\lim}_{r\to 1}[\ln M(r)]/\ln[(1-r^2)^{-1}]$ and $\tau_1=\overline{\lim}_{r\to 1}\frac{\ln M_1(r)}{\ln[(1-r^2)^{-1}]}$ characterizing the growth of function $f(z)$ as $|z|\to 1$ are called *genralized types of function* $f(z)$.

Theorem 2. The following inequalities are valid for $v>0$:

$$\tau\geq\frac{v}{2},\qquad \tau_1\geq\frac{v}{2}. \tag{14.3.1}$$

Proof. Dividing $f(z)$ by az^n, if necessary, we get a function $\tilde{f}(z)$, that is $\tilde{f}(0)=1$. From the Jensen formula

$$\frac{1}{2\pi}\int_0^{2\pi}\ln|\tilde{f}(re^{i\varphi})|d\varphi=\int_0^r\frac{n(t)}{t}\,dt \tag{14.3.2}$$

one gets

$$\ln M(r)\geq\int_0^r\frac{n(t)}{t}\,dt. \tag{14.3.3}$$

On the other hand, because of the generalized inequality between the arithmetic and geometric means,

$$\begin{aligned}\ln M_1(r)&=\frac{1}{2}\ln\left[\frac{1}{2\pi}\int_0^{2\pi}|\tilde{f}(re^{i\varphi})|^2 d\varphi\right]\\ &\geq\frac{1}{2\pi}\int_0^{2\pi}\ln|\tilde{f}(re^{i\varphi})|d\varphi=\int_0^r\frac{n(t)}{t}\,dt.\end{aligned} \tag{14.3.4}$$

Because of the definitions of the quantities v, τ and τ_1, for any numbers $\varepsilon>0$, $\varepsilon_1>0$, and $\delta>0$, a radius $r_0<1$ exists such that

$$\ln M(r)\leq(\tau+\varepsilon)\ln\left(\frac{1}{1-r^2}\right),\quad \ln M_1(r)\leq(\tau_1+\varepsilon_1)\ln\left(\frac{1}{1-r^2}\right),$$

$$n(r)\geq\frac{v-\delta}{1-r^2}\,r^2 \tag{14.3.5}$$

for every $r > r_0$. We can now rewrite inequalities (14.3.3, 4) as

$$(\tau+\varepsilon)\ln\left(\frac{1}{1-r^2}\right) \geq \ln M(r) \geq \int_0^{r_0} \frac{n(t)}{t}\,dt + \frac{\nu-\delta}{2}\ln\left(\frac{1-r_0^2}{1-r^2}\right) \tag{14.3.6}$$

$$(\tau_1+\varepsilon_1)\ln\left(\frac{1}{1-r^2}\right) \geq \ln M_1(r) \geq \int_0^{r_0} \frac{n(t)}{t}\,dt + \frac{\nu-\delta}{2}\ln\left(\frac{1-r_0^2}{1-r^2}\right). \tag{14.3.7}$$

If r tends to 1 in (14.3.6, 7) and $\nu > 0$, then (14.3.1) results.

14.4 Completeness Criterion for CS Subsystems

Consider a subsystem of coherent states $\{|\zeta_n\rangle\}$, $\zeta_n = \zeta_0\gamma_n$, $\gamma_n \in \Gamma$ of the type $\{T^k, |0\rangle\}$. Denote by $n(r)$ the number of points ζ_n within the disk of radius r, whose non Euclidean area is $S(r) = \pi r^2/(1-r^2)$. We assume that a lower limit for the density of points ζ_n exists, i.e., that

$$\varliminf_{r\to 1} \frac{n(r)}{S(r)} = \frac{1}{\pi} \varliminf_{r\to 1} r^{-2}(1-r^2)n(r) = \frac{\nu}{\pi}. \tag{14.4.1}$$

Theorem 3. If $k < (\nu+1)/2$, the CS system $\{|\zeta_n\rangle\}$ of the type $\{T^k, |0\rangle\}$ is complete.

Proof. As already shown, it is sufficient to prove that integral

$$I_k = \int_D |f(\zeta)|^2 (1-|\zeta|^2)^{2k-2} d\mu(\zeta) = 2\pi \int_0^1 M_1^2(r)(1-r^2)^{2k-2} r\,dr$$

diverges for any function $f(\zeta)$ with zeros at points ζ_n. According to Theorem 2, in this case the generalized type of function $f(z)$ is $\tau_1 \geq \nu/2$. Hence for any $\delta > 0$ there exists $r_0 < 1$ such that $M_1^2(r) \geq (1-r^2)^{-(\nu-\delta)}$, if $1 > r > r_0$, and therefore the integral I_k diverges for $k < (\nu+1)/2$.

For subsystems of coherent states related to discrete subgroups with compact fundamental domains, one can use the results of [124]. As was shown, limits do exist for the sequence $\zeta_n = \zeta_0\gamma_n$,

$$\frac{\nu}{\pi} = \varliminf_{r\to 1} \frac{n(r)}{S(r)} = \varlimsup_{r\to 1} \frac{n(r)}{S(r)} = \frac{1}{S_\Gamma}, \tag{14.4.2}$$

where S_Γ is the area of the fundamental domain. Equation (14.4.2) remains valid also for a noncompact fundamental domain with $S_\Gamma \neq \infty$, following from the results of [124, 125]. (Thus, Theorem 4 holds also for that case.) Theorem 3 and (14.4.2) lead to Theorem 4.

Theorem 4. Subsystem $\{|\zeta_n\rangle\}$ of the type $\{T^k, |0\rangle\}$ is complete if $S_\Gamma < \pi/(2k-1)$. The proof is evident.

Remark 4. Thus, the criterion of completeness suggested in [15] for subsystems of coherent states related to discrete subgroups is justified in the present case.

When group Γ admits an automorphic form with one zero in the fundamental domain, one can choose the automorphic form $f_{m_0}(\zeta)$ ($m_0=\pi/2S_\Gamma$) as $f(\zeta)$. It turns out that 21 such groups exist; the next section is devoted to finding these groups.

14.5 Discrete Subgroups of $SU(1,1)$ and Automorphic Forms

First, we present necessary information concerning discrete subgroups of group $SU(1,1)/\mathbb{Z}_2$ [119–121]. We restrict ourselves to groups with a finite area of their quotient space (the fundamental domain) $\Gamma\backslash D$. It is known that in this case the fundamental domain is a polygon with an even number of sides $2n$: the sides, grouped in pairs, are equivalent with respect to the action of group transformations. The vertices of the polygon are joined in cycles of vertices which are equivalent to each other. The sum of the polygon angles at the vertices of a given cycle is $2\pi/l$, where l is either a positive integer or ∞. If $l=1$, the cycle is called accidental, if $l=\infty$ the vertices of the cycle lie on the boundary of domain D and the cycle is called parabolic, while in all other cases the cycle is called elliptic and l is the order of the cycle. Let c be the total number of cycles. If equivalent sides and vertices are considered to be the same, a Riemannian surface is formed. The genre p of this surface is found by

$$2p=1+n-c. \tag{14.5.1}$$

The set of numbers $(p,c;l_1,l_2,\ldots,l_c)$ is called the signature of group Γ; the area of the fundamental domain is completely determined by the signature of the group, and with our choice of the invariant measure

$$d\mu(\zeta)=(1-|\zeta|^2)^{-2}d\xi\,d\eta, \qquad \zeta=\xi+i\eta$$

the area is

$$S_\Gamma=\pi\left[p-1+\frac{1}{2}\sum_{i=1}^{c}\left(1-\frac{1}{l_i}\right)\right]. \tag{14.5.2}$$

It is noteworthy that this invariant measure differs by a factor from the standard invariant measure $d\mu_0(z)=y^{-2}dx\,dy$, $z=x+iy$ in the upper half-plane $\{z:z=x+iy,\ y>0\}$, $d\mu\frac{1}{4}\,d\mu_0$.

It is seen in (14.5.2) that the area S_Γ cannot be arbitrarily close to zero and it is possible to show [126] that the minimal value of $S_\Gamma=\pi/84$ is achieved for a group with signature $(0,3;2,3,7)$. If the fundamental domain is not compact, i.e., group Γ contains parabolic elements, then $S_\Gamma\geq\pi/12$; $S_\Gamma=\pi/12$ is relevant to the modular group $\Gamma=(0,3;2,3,\infty)$.

It is also known that the signature of the group can be arbitrary if $p \geq 2$. If $p = 1$, condition $c \geq 1$ must hold. For $p = 0$, one of the following conditions must be satisfied: i) $c \geq 5$; ii) $c = 4$ and $\sum l_j^{-1} < 2$; iii) $c = 3$, $\sum l_j^{-1} < 1$. Let us consider now the automorphic forms defined by (14.2.1). Dimensionalities of spaces of automorphic forms of weight m are

$$d_m(\Gamma) = \begin{cases} 0 & \text{for} \quad m < 0 \\ 1 & \text{for} \quad m = 0 \\ g_1 & \text{for} \quad m = 1 \\ (2m-1)(p-1) + \sum_{j=1}^{c} \left[m\left(1 - \frac{1}{l_j}\right)\right] & \text{for} \quad m \geq 2. \end{cases} \tag{14.5.3}$$

Here p is the genre of the fundamental domain, $[m]$ is the integral part of number m, and $g_1 \geq p$ is the number of holomorphic differentials on the Riemannian surface $\Gamma \backslash D$. With this, the number of zeros of $f_m(z)$ inside the fundamental domain is given by *Poincaré's* formula [127] (here presented slightly differently)

$$N = 2mS_\Gamma/\pi. \tag{14.5.4}$$

Note that if some elliptic or parabolic vertices are present, this number is not necessarily integral.

Further, comparing (14.5.3, 2) we find that

$$N \geq d_m + p - 1 \tag{14.5.5}$$

and the equality takes place only when numbers m/l_i are integers, including zero.

In the following we are interested in automorphic forms for which $d_m(\Gamma) \geq 2$. Let m_0 be the minimal weight of the forms. Which values are allowed for this number?

I) For $p \geq 2$, $m_0 = 1$, as seen from (14.5.3).
II) For $p = 1$ the result depends on C_2, the number of parabolic cycles

a) $C_2 \geq 2$, then $m_0 = 1$,
b) $C_2 = 1$, then $m_0 = 2$,
c) $C_2 = 0$ and $\Gamma = (1, 1; 2)$, then $m_0 = 4$
d) $C_2 = 0$ and $\Gamma = (1, 1; l)$, $l \geq 3$, then $m_0 = 3$,
e) $C_2 = 0$, $c \geq 2$, then $m_0 = 2$.

III) For $p = 0$

$$m_0 \geq m_1 = \frac{\pi}{2S_\Gamma} = \left[\sum_{j=1}^{c}\left(1 - \frac{1}{l_j}\right) - 2\right]^{-1} = \left(c - 2 - \sum_{1}^{c} \frac{1}{l_j}\right)^{-1}. \tag{14.5.6}$$

Using the notation

$$N_0 = 2m_0 S_\Gamma / \pi,$$

one can see, following (14.5.5), that $N_0 \geq p+1$, so N_0 can be equal to unity only for $p=0$. Hence $d_{m_0}=2$ and the value of m_0 is determined by (14.5.4):

$$m_0 = \frac{\pi}{2S_\Gamma} = \left[\sum_1^c \left(1 - \frac{1}{l_j}\right) - 2\right]^{-1}. \tag{14.5.7}$$

Let l be the least common multiple of all finite l_j. Then (14.5.7) can be rewritten as $m_0 = l\left[\sum_j \left(l - \frac{l}{l_j}\right) - 2l\right]^{-1}$ so that $m_0 \leq l$. However, m_0 must be divisible by all finite l_j, so it must coincide with l, $m_0 = l$.

Table 14.1. Signatures of discrete subgroups satisfying (14.5.8)

m_0	Signature of Γ
1	$(0,3;\ \infty,\infty,\infty)$
2	$(0,3;\ 2,\infty,\infty)$, $(0,4;\ 2,2,2,\infty)$, $(0,5;2,2,2,2,2)$
3	$(0,3;\ 3,3,\infty)$
4	$(0,3;\ 4,4,4)$, $(0,3;2,4,\infty)$, $(0,4;2,2,2,4)$
6	$(0,3;\ 2,3,\infty)$, $(0,3;3,3,6)$ $(0,3;2,6,6)$, $(0,4;2,2,2,3)$
8	$(0,3;\ 2,4,8)$
10	$(0,3;\ 2,5,5)$
12	$(0,3;\ 3,3,4)$, $(0,3;2,3,12)$, $(0,3;2,4,6)$
18	$(0,3;\ 2,3,9)$
20	$(0,3;\ 2,4,5)$
24	$(0,3;\ 2,3,8)$
42	$(0,3;\ 2,3,7)$

Thus Proposition 2 has been prooved.

Proposition 2. If a group Γ of signature $(0, c; l_1, \ldots, l_{c_1}, \infty, \ldots, \infty)$ admits an automorphic form $f_{m_0}(z)$ with a single zero in the fundamental domain, then m_0 is the least common multiple of numbers $l_1, l_2, \ldots, l_c$ and, moreover, it must satisfy condition

$$m_0(c-2) - \sum_1^{c_1} \frac{m_0}{l_j} = 1. \tag{14.5.8}$$

It is not difficult to show that (14.5.8) has solutions only for $c=3$, 4 and 5 and that the number of solutions is finite. Correspondingly, there are 21 discrete subgroups Γ, all of which are listed in Table 14.1.

15. Coherent States for Discrete Series and Discrete Subgroups: General Case

This chapter considers the property of completeness for those subsystems of the CS system considered in Chap. 12, which are relevant to discrete subgroups of the groups of motion for bounded homogeneous domains. It is helpful to consider first some properties of automorphic forms for discrete subgroups.

15.1 Automorphic Forms

Let us consider a bounded symmetric domain D (Chap. 12), a discrete subgroup Γ of automorphisms of this domain and suppose that the quotient (denoted by $\Gamma\backslash D$) is compact. We suppose that Γ acts on D effectively and there are no fixed points ($\gamma x = x$ for some $x \in D$, $\gamma \in \Gamma$, only if $\gamma = e$). In this case $\Gamma\backslash D$ is a compact complex algebraic manifold. It has been proved in [128] that discrete groups with this property do indeed exist. Recall the definition of the automorphic form.

Definition 1. A holomorphic function $f: D \to \mathbb{C}$ ($\mathbb{C}$ is the field of complex numbers) is an automorphic form of weight m with respect to Γ if for any pair $z \in D$, $\gamma \in \Gamma$

$$[J_\gamma(z)]^m f(\gamma z) - f(z), \tag{15.1.1}$$

where $J_\gamma(z)$ is the Jacobian of mapping $z \to \gamma z$ at point z.

Automorphic forms of weight m constitute a vector space over $\mathbb{C}$ which is usually denoted by $H^\circ(D, \Gamma, m)$. The dimensionality of $H^\circ(D, \Gamma, m)$ in our case was calculated by *Hirzebruch* [129]. The result is

$$\dim H^\circ(D, \Gamma, m) = \begin{cases} 0, & m < 0 \\ 1, & m = 0 \\ g_n, & m = 1 \\ (-1)^n \dim T^{\lambda_m} \chi(\Gamma\backslash D), & m \geq 2. \end{cases} \tag{15.1.2}$$

Here $n = \dim_{\mathbb{C}} D$ is the complex dimensionality of domain D, g_l is the dimensionality of the space of holomorphic differentials of degree l on $\Gamma\backslash D$; T^λ is the finite-dimensional representation of group U with the highest weight λ, $\lambda_m = (m-1) \sum_{P_+} \beta$, where $P_+ = \{\beta\}$ is the set of all positive noncompact roots of Lie algebra $\mathscr{G}^c$. Further, U is the maximal compact subgroup of G^c, and G^c is the simply-connected complex group corresponding to Lie algebra $\mathscr{G}^c$. The so-called arithmetical genre of the manifold $\Gamma\backslash D$ is $\chi(\Gamma\backslash D)=\sum_{l=0}^{n}(-1)^l g_l$ (note that $\chi(\Gamma\backslash D)\neq 0$).

Using Weyl's formula for the dimensionality of representation T^λ gives explicit formulae for $\dim H^\circ(D,\Gamma,m)$ at $m\geq 2$ for all four types of domains:

I) D^{I}, $\dim H^\circ(D,\Gamma,m)=(-1)^{pq}\chi(\Gamma\backslash D)\prod_{i,j}\frac{m(p+q)-i-j}{p+q-i-j}$

$$0\leq i\leq p-1,\qquad 1\leq j\leq q; \tag{15.1.3}$$

II) D^{II}, $\dim H^\circ(D,\Gamma,m)$

$$=(-1)^{p(p+1)/2}\chi(\Gamma\backslash D)\prod_{0\leq i\leq j\leq p}\frac{2(m-1)(p+1)+i+j}{i+j}; \tag{15.1.4}$$

III) D^{III}, $\dim H^\circ(D,\Gamma,m)$

$$=(-1)^{p(p-1)/2}\chi(\Gamma\backslash D)\prod_{0\leq i\leq j\leq p-1}\frac{2(m-1)(p-1)+i+j}{i+j}; \tag{15.1.5}$$

IV) D^{IV}, $\dim H^\circ(D,\Gamma,m)=(-1)^p\chi(\Gamma\backslash D)\,[C^p_{mp-1}+C^p_{mp}]$. (15.1.6)

15.2 Completeness of Some CS Subsystems

Let us consider a subsystem of coherent states associated with a discrete subgroup Γ of group G which satisfy the conditions listed in the previous section, that is, subgroup Γ acts effectively in domain D and there are no fixed points, while the quotient space $\Gamma\backslash D$ is compact.

Let $\zeta_0\in D$ be a given point in D. Let us consider the CS subsystem $\{|\psi_m\rangle\}$

$$|\psi_m\rangle\equiv|\psi_{\zeta_m}\rangle,\qquad \zeta_m=\gamma_m\zeta,\qquad \gamma_m\in\Gamma. \tag{15.2.1}$$

The next lemma indicates whether this subsystem is incomplete.

Lemma [73]. Subsystem $\{|\psi_m\rangle\}$ of coherent states of the type $\{T^k,|0\rangle\}$ is incomplete if and only if a holomorphic function $\psi(\zeta)\in\mathscr{F}_k$ exists such that $\psi(\zeta_n)=0$.

Let $|\psi\rangle$ be a vector which belongs to the representation space $\mathscr{F}_k$. Let us consider function $\langle\psi|\psi_\zeta\rangle$. It is evident from (12.3.6) that function $\langle\psi|\psi_\zeta\rangle$ can be written as

$$\langle\psi|\psi_\zeta\rangle=[B(\zeta,\zeta^+)]^{-k/2}\psi(\zeta), \tag{15.2.2}$$

where function $\psi(\zeta)$ is holomorphic in D. The norm of vector $|\psi\rangle$ is

$$\|\psi\|^2 = \|\psi(\zeta)\|_k^2 = N_k \int [B(\zeta, \zeta^+)]^{1-k} |\psi(\zeta)|^2 \, d\xi \, d\eta,$$

so $\psi(\zeta) \in \mathscr{F}_k$.

If the system is incomplete, then a nonzero vector $|\psi\rangle$ exists orthogonal to all vectors $\{|\psi_m\rangle\}$; meanwhile it follows from (15.2.2) that $\psi(\zeta_n) = 0$. The inverse statement can be proved as easily as the former.

Now we will construct function $\psi(\zeta)$ for which $\psi(\zeta_n) = 0$, at $\zeta_n = \gamma_n \zeta_0$, $\gamma_n \in \Gamma$. Suppose $\psi(\zeta)$ is an automorphic form with respect to discrete subgroup Γ (Sect. 15.1) and let $\psi(\zeta_0) = 0$.

If $\dim H^\circ(D, \Gamma, m) \geq 2$, then such a form does exist. Indeed, let $\psi_1(\zeta)$ and $\psi_2(\zeta)$ be the automorphic forms of weight m. Then $\psi(\zeta) = c_1 \psi_1(\zeta) + c_2 \psi_2(\zeta)$ is an automorphic form and it is possible to choose the constants c_1 and c_2 so that $\psi(\zeta_n) = 0$.

Consider an arbitrary automorphic form of weight m, $f(z)$. The norm of f is

$$\|f\|_k^2 = \mathrm{N}_k \int_D |f(z)|^2 \varrho(z, \bar{z})^{1-k} \, dx \, dy. \tag{15.2.3}$$

Because of (15.1.1)

$$|f(\gamma z)|^2 = |J_\gamma(z)|^{-2m} |f(z)|^2$$

so $|f(z)|^2 = |F|^2 |\varrho(z, \bar{z})|^m$, where $\varrho(z, \bar{z})$ is defined in Sect. 12.2, $\varrho(z, \bar{z}) > 0$ and $F(\gamma z, \overline{\gamma z}) = F(z, \bar{z})$.

Hence, the norm of function f is equal to

$$\|f\|_k^2 = N_k \int \varrho(z, \bar{z})^{m-k+1} |F(z, \bar{z})|^2 \, dx \, dy. \tag{15.2.4}$$

This integral converges if $m \leq k-1$, so here the function $f(z)$, $f(\zeta_n) = 0$, $f(z) \neq 0$ does actually exist. Hence the system under consideration is incomplete.

Let m_0 be the least m for which $\dim H^\circ(D, \Gamma, m) \geq 2$. It follows from (15.1.3–6) that $\dim H^\circ(D, \Gamma, m) \geq 2$, if $m \geq 2$. On the other hand, $g_n \geq 1$, so that $m_0 = 1$, if $g_n > 1$, and $m = 2$ if $g_n = 1$.

We are now in a position to formulate the final result.

Theorem. The subsystem of coherent states $\{|\psi_n^k\rangle\}$ is incomplete if $k \geq m_0 + 1$. It is seen from this theorem that the subsystem of CS is incomplete, as a rule. The sufficient condition is $k \geq 3$.

16. Coherent States and Berezin's Quantization

A quantization method proposed by *Berezin* [22, 130, 131] is described in this chapter. The method can be applied to homogeneous Kähler manifolds. Its advantage as compared with the standard method of geometric quantization is that it incorporates the correspondence principle.

Quantization is a procedure of constructing a quantum system starting from a classical mechanical system. It is required that the quantum system obtained must go over into the original classical one in the limit $\hbar \to 0$, where $\hbar$ is Planck's constant. This requirement is called the correspondence principle. Evidently, many quantizations satisfying this requirement may exist.

This quantization concept is interesting from the point of view of pure mathematics since it is a fruitful source of important ideas and constructions. For example, the method is exploited in the so-called geometrical quantization approach developed by *Kirillov* and *Kostant* (see [19–21] and references therein) and gave rise to some new results in the theory of group representations.

Quantization is well known for the case of the standard phase space of a classical system with n degrees of freedom; the phase space is the linear real space $\mathbb{R}^{2n} = \{(p, q)\}$, $p = (p_1, \ldots, p_n)$ is the momentum vector, $q = (q_1, \ldots, q_n)$ is the coordinate vector.

The Hilbert space of states of the quantum system is the space of square-integrable functions $\{\psi(x)\}$ of n *real* variables, $x = (x_1, \ldots, x_n)$. Operators $\hat{p}_j$ and $\hat{q}_k$ corresponding to the classical momenta p_j and coordinates q_k are represented as

$$\hat{q}_k \psi(x) = x_k \psi(x), \qquad \hat{p}_k \psi(x) = -i\hbar \frac{\partial \psi}{\partial x_k}.$$

Every classical quantity $f(p, q)$ corresponds to a "quantum observable", i.e., an operator $f(\hat{p}, \hat{q})$ obtained by replacing the classical variables p_j and q_k by operators $\hat{p}_j$ and $\hat{q}_k$. However, due to noncommutativity of operators $\hat{p}_j$ and $\hat{q}_k$ one should, in general, fix the ordering of these operators in $f(p, q)$. The ordering proposed by *Weyl* [14] has a number of remarkable features.

The conventional quantization procedure is applicable only to classical systems with standard flat phase space with the canonical Cartesian coordinates p_j, q_k. The approach does not work, however, for systems with curved phase spaces, for example, for a rigid body rotating around a fixed point.

An apparently adequate quantization procedure was proposed and developed by *Berezin* [22, 130] for the case where the phase space is the homogeneous Kähler manifold. This construction can be represented suitably in terms of the CS related to this phase space.

16.1 Classical Mechanics

Classical mechanics is introduced as a pair $(\mathcal{M}, \omega)$, where $\mathcal{M}$ is a manifold and ω is a skew-symmetric tensor field of rank two whose components $\omega^{jk}(x)$ written in the local coordinates x^j must satisfy the condition

$$\omega^{jk}\frac{\partial\omega^{lm}}{\partial x^k}+\omega^{lk}\frac{\partial\omega^{mj}}{\partial x^k}+\omega^{mk}\frac{\partial\omega^{jl}}{\partial x^k}=0. \tag{16.1.1}$$

(Hereafter we use tensor notations. In particular, summation is implied over repeated indices.) Let us consider the set of differentiable functions on $\mathcal{M}$, $\mathcal{F}(\mathcal{M})$. It is a commutative and associative algebra with respect to the usual addition and multiplication; besides, this set is the Lie algebra determined by the Poisson brackets

$$\{f,g\}=\omega^{ij}(x)\frac{\partial f}{\partial x^i}\frac{\partial g}{\partial x^j}. \tag{16.1.2}$$

The fact that the Poisson brackets (16.1.2) define the Lie algebra, i.e., that the Jacobi identity

$$\{f,\{g,h\}\}+\{g\{h,f\}\}+\{h\{f,g\}\}=0 \tag{16.1.3}$$

is valid, is equivalent to condition (16.1.1). This can be easily checked by a direct calculation. Hence it follows that condition (16.1.1) is independent of the choice of a coordinate system in M.

If the tensor field $\omega(x)$ is nondegenerate, i.e., $\det|\omega^{jk}|\neq 0$ at all x, then it is possible to consider the inverse matrix ω_{jk} and the corresponding external 2-form

$$\omega=\omega_{jk}(x)dx^j\wedge dx^k. \tag{16.1.4}$$

It is not difficult to verify that condition (16.1.1) is equivalent to the statement that form ω is closed. Thus, the manifold $\mathcal{M}$ is provided with a symplectic structure. One should bear in mind, however, that some important mechanical systems do not fulfill the requirement that the field ω^{jk} be nondegenerate.

We present here several examples of symplectic manifolds $\mathcal{M}$.

1. $\mathcal{M}=\mathbb{R}^2$, the two-dimensional plane with coordinates p, q;

$$\omega^{12}=-\omega^{21}=1, \qquad \omega^{11}=\omega^{22}=0$$

$$\{f,g\}=\frac{\partial f}{\partial p}\frac{\partial g}{\partial q}-\frac{\partial f}{\partial q}\frac{\partial g}{\partial p}. \tag{16.1.5}$$

2. $\mathcal{M}=S^1\times\mathbb{R}^1$, the two-dimensional cylinder. The coordinates will be denoted by p, q, where $0\leq q<2\pi$, $-\infty<p<\infty$. The differentiable functions on $\mathcal{M}$ are periodic in q with the period of 2π. Tensor ω and the Poisson brackets are the same as in the previous case.

3. $\mathcal{M}=S^1\times S^1$, the two-dimensional torus with coordinates $0\leq q<2\pi$, $0\leq p<2\pi$. The differentiable functions on $\mathcal{M}$ are periodic in both variables with the period of 2π. The Poisson bracket is given by (16.1.5).

4. $\mathcal{M}=S^2$, the two-dimensional sphere. The measure on S^2, invariant under rotations, is

$$d\mu(\theta,\varphi)=r^2\sin\theta\, d\theta\wedge d\varphi,$$

where r is the radius of the sphere. The same expression may be taken for the 2-form ω. Thus, one gets the following expression for the Poisson brackets in the spherical coordinate system:

$$\{f,g\}=\frac{1}{r^2\sin\theta}\left(\frac{\partial f}{\partial\varphi}\frac{\partial g}{\partial\theta}-\frac{\partial f}{\partial\theta}\frac{\partial g}{\partial\varphi}\right). \tag{16.1.6}$$

It is more convenient, however, to work with complex coordinates on S^2, instead of spherical angles. The complex structure is introduced via the stereographic projection

$$z=x+\mathrm{i}y=r\cot\frac{\theta}{2}\,\mathrm{e}^{\mathrm{i}\varphi}. \tag{16.1.7}$$

The external form ω is now written as

$$\omega=\frac{1}{2\mathrm{i}}\left(1+\frac{|z|^2}{r^2}\right)^{-2} dz\wedge d\bar{z} \tag{16.1.8}$$

and the Poisson bracket is

$$\{f,g\}=2\mathrm{i}\left(1+\frac{|z|^2}{r^2}\right)^2\left(\frac{\partial f}{\partial z}\frac{\partial g}{\partial\bar{z}}-\frac{\partial f}{\partial\bar{z}}\frac{\partial g}{\partial z}\right). \tag{16.1.9}$$

5. $\mathcal{M}=\mathcal{L}^2$, the Lobachevsky plane given by a disk of radius r with its center at the origin of the complex z plane. The Lobachevsky plane has an external closed 2-form which is invariant under the group of its motions; this form

coincides with the invariant measure,

$$\omega = \frac{1}{2\mathrm{i}}\left(1 - \frac{|z|^2}{r^2}\right)^{-2} dz \wedge d\bar{z}. \tag{16.1.10}$$

The corresponding Poisson brackets are

$$\{f, g\} = 2\mathrm{i}\left(1 - \frac{|z|^2}{r^2}\right)^2 \left(\frac{\partial f}{\partial z}\frac{\partial g}{\partial \bar{z}} - \frac{\partial f}{\partial \bar{z}}\frac{\partial g}{\partial z}\right). \tag{16.1.11}$$

6. Suppose that $\mathscr{G}$ is an arbitrary Lie algebra, C_k^{ij} are its structure constants, $i, j, k = 1, \ldots, n$. The manifold to be considered is the space $\mathscr{G}^*$ dual to $\mathscr{G}$: $\mathscr{M} = \mathscr{G}^*$. The Poisson brackets are defined by the tensor

$$\omega^{ij} = C_k^{ij} x^k. \tag{16.1.12}$$

Condition (16.1.1) follows from the Jacobi identity for the Lie algebra [132]. This general construction incorporates examples 1, 4, 5 above.

a) Let Lie algebra $\mathscr{G}$ be the familiar Heisenberg-Weyl algebra. The standard basis in $\mathscr{G}$ is e_1, e_2, e_0, while

$$[e_1, e_2] = e_0, \qquad [e_1, e_0] = [e_2, e_0] = 0. \tag{16.1.13}$$

The coordinates in $\mathbb{R}^3$ corresponding to e_0, e_1 and e_2 will be denoted by r, p, q so

$$\omega^{12} = -\omega^{21} = r.$$

The Poisson brackets have the classical form

$$\{f, g\} = r\left(\frac{\partial f}{\partial p}\frac{\partial g}{\partial q} - \frac{\partial f}{\partial q}\frac{\partial g}{\partial p}\right). \tag{16.1.14}$$

b) Let us consider Lie algebra of the $SO(3)$ group and introduce the standard basis e_1, e_2, e_3 with the commutation relations

$$[e_i, e_j] = \varepsilon_{ijk} e_k, \tag{16.1.15}$$

where ε_{ijk} is the usual totally skew-symmetrical tensor. Now the Poisson brackets are

$$\{f, g\} = \varepsilon_{ijk} x^i \partial_j f \partial_k g, \qquad \partial_j f = \frac{\partial f}{\partial x^j}. \tag{16.1.16}$$

With the spherical coordinates

$$x^1 = r \sin\theta \cos\varphi, \qquad x^2 = r \sin\theta \sin\varphi, \qquad x^3 = r\cos\theta$$

we obtain

$$\{f, g\} = \frac{1}{r \sin\theta}\left(\frac{\partial f}{\partial \theta}\frac{\partial g}{\partial \varphi} - \frac{\partial f}{\partial \varphi}\frac{\partial g}{\partial \theta}\right).$$

Thus, at fixed r the Poisson brackets (16.1.16) for functions f and g coincide with (16.1.6) up to a factor. The Lie algebra of a three-dimensional Lorentz group similarly leads to the Poisson brackets for the Lobachevsky plane (16.1.11).

16.2 Quantization

The following general definition was given by *Berezin* [22].

Quantization of classical mechanics $(\mathcal{M}, \omega)$ is defined as an associative algebra $\mathcal{a}$ with involution, possessing the following properties.

1. A family A_h of associative algebras exists with involution such that

a) parameter h belongs to a set E on the positive real axis with O as a limiting point (O does not belong to E),

b) algebra $\mathcal{a}$ consists of functions $f(h)$ taking values in A_h, $h \in E$. Involution and multiplication in algebras $\mathcal{a}$ and A_h are in the usual correspondence:

i) $(f^{\tilde{\sigma}})(h) = [f(h)]^{\sigma}$,

where $\tilde{\sigma}$ and σ are involutions in $\mathcal{a}$ and A_h, respectively,

ii) $(f_1 \tilde{*} f_2)(h) = f_1(h) * f_2(h)$,

where $\tilde{*}$ and $*$ are multiplication operations in $\mathcal{a}$ and A_h, respectively. In the following multiplication and involution in algebras $\mathcal{a}$ and A_h are denoted by the same symbols.

2. A homomorphism φ exists of algebra $\mathcal{a}$ onto the algebra $A(\mathcal{M})$ of the differentiable functions on $\mathcal{M}$ with the usual operations of addition and multiplication. The homomorphism must have the following properties:

a) for any two points $x_1, x_2 \in \mathcal{M}$ there is a function

$f(x) \in \varphi(\mathcal{a})$ such that $f(x_1) + f(x_2)$

b) for any functions $f(x)$, $g(x)$

$$\varphi\left(\frac{1}{h}(f*g - g*f)\right) = \mathrm{i}\{\varphi(\mathrm{f}), \varphi(\mathrm{g})\},$$

where $*$ stands for multiplication in $\mathcal{a}$ and $\{,\}$ are the Poisson brackets in $A(\mathcal{M})$.

c) $\varphi(f^{\sigma}) = \overline{\varphi(f)}$,

where $f \to f^\sigma$ stands for involution in $\mathfrak{a}$ and the bar is the complex conjugation. Parameter h is called Planck's constant.

Let us consider a particular but important case of quantization.

Special quantization. This is a quantization which has some additional properties.

3. Algebra A_h consists of differentiable functions $f(x)$, $x \in \mathcal{M}$.
4. Algebra $\mathfrak{a}$ consists of functions $f(h, x)$, such that $f(h, x) \in A_h$ for fixed h.
5. Homomorphism φ: $\mathfrak{a} \to A(\mathcal{M})$ is given by

$$\varphi(f) = \lim_{h \to 0} f(h, x).$$

A consistent theory is now available only for special quantization. Note also that all the special quantizations investigated up to now have the following additional properties.

6. Algebra A_h contains the unity that is the function $f_0(h, x) \equiv 1$.
7. There is a trace operation in algebra A_h

$$\operatorname{tr}(f) = C \int f(x)\, d\mu(x),$$

where $d\mu(x)$ is a measure in the manifold $\mathcal{M}$, and C is a number.

Note that if the tensor field $\omega^{ij}(x)$ is nondegenerate, i.e., the closed external 2-form exists on $\mathcal{M}$, then on $\mathcal{M}$ there exists also a natural measure

$$d\mu(x) = C\omega^{n/2}.$$

With slight over-simplification one can say that the quantization procedure is a correspondence between functions $f(x)$ on the phase space $\mathcal{M}$ of the classical system and operators $\hat{F}$ in Hilbert space $\mathcal{H}$, and this correspondence must satisfy the correspondence principle.

A possible way to solve the problem is as follows.

Suppose we have an overcomplete system of states $\{|x\rangle\}$ parametrized by points x of manifold $\mathcal{M}$, and the system satisfies condition

$$\int |x\rangle\langle x|\, d\mu(x) = \hat{1},$$

where $|x\rangle\langle x|$ is the projection operator on state $|x\rangle$, $d\mu(x)$ is a measure in manifold $\mathcal{M}$, and $\hat{1}$ is the identity operator. A natural way to determine an operator corresponding to function $f(x)$ is to define it as an integral $\hat{F} = \int |x\rangle\langle x| f(x)\, d\mu(x)$. The problem is to insert a constant h in such a manner that the correspondence principle is satisfied.

The problem has been solved at present for the case where phase space $\mathcal{M}$ is a homogeneous Kähler manifold.

The group of motions G of manifold $\mathcal{M}$ has a discrete series of representations $T^k(g)$, where k is a parameter. As shown in Chap. 12 a CS system $\{|x\rangle\}$

exists related to the representation of view. Let $d\mu(x)$ be the G-invariant measure on manifold $\mathscr{M}$ and $h=k^{-1}$.

It was found that the construction considered above satisfied the correspondence principle at $h\to 0$, and, therefore does solve the quantization problem. For simple cases we will consider the construction in more detail.

16.3 Quantization on the Lobachevsky Plane

Take the model of the Lobachevsky plane as a unit disk

$$D=\{z:|z|<1\}$$

in the complex z plane. The space of functions analytical in domain D is provided with the scalar product

$$(f,g)=\left(\frac{1}{h}-1\right)\int \bar{f}(z)g(z)(1-|z|^2)^{1/h}d\mu(z,\bar{z}), \tag{16.3.1}$$

where

$$d\mu(z,\bar{z})=\frac{1}{2\pi i}\frac{dz\wedge d\bar{z}}{(1-|z|^2)^2}$$

is the invariant measure in the Lobachevsky plane. All the functions with a finite norm

$$\|f\|^2=(f,f)<\infty \tag{16.3.2}$$

are elements of the functional space $\mathscr{F}_h$, which we consider now. [Note that factor $\left(\frac{1}{h}-1\right)$ in (16.3.1) is related to the normalization condition $(f_0,f_0)=1$ for $f_0(z)\equiv 1$.] Functions

$$f_l(z)=(l!)^{-1/2}\left[\left(\frac{1}{h}\right)\cdots\left(\frac{1}{h}-1+l\right)\right]^{1/2}z^l \tag{16.3.3}$$

constitute an orthonormalized basis in space $\mathscr{F}_h$. The kernel defined as an infinite series

$$K_h(\bar{\zeta},z)=\sum_l f_l(z)\overline{f_l(\zeta)} \tag{16.3.4}$$

reduces, after a simple calculation, to

$$K_h(\bar{\zeta},z)=(1-\bar{\zeta}z)^{-1/h}. \tag{16.3.5}$$

It is easy to see that at a fixed $\bar{\zeta}$, this function belongs to space $\mathscr{F}_h$. Let us denote by $|\bar{\zeta}\rangle$ the state in the Hilbert space corresponding to this function. For an arbitrary state $|f\rangle$ determined by a function $f(z) \in \mathscr{F}_h$,

$$\langle\bar{\zeta}|f\rangle = f(\zeta). \tag{16.3.6}$$

Comparing with the results presented in Sect. 5.2, one sees that the system obtained coincides with the CS system for the discrete series of representations $T^k(g)$, $k=(2h)^{-1}$.

16.3.1 Description of Operators

Let A be a bounded linear operator in $\mathscr{F}_h$. The corresponding function exists

$$A(\zeta,\bar{\zeta}) = \frac{\langle\bar{\zeta}|A|\bar{\zeta}\rangle}{\langle\bar{\zeta}|\bar{\zeta}\rangle}, \tag{16.3.7}$$

which is called the symbol of this operator. Note that function

$$A(\xi,\bar{\eta}) = \frac{\langle\bar{\xi}|A|\bar{\eta}\rangle}{\langle\bar{\xi}|\bar{\eta}\rangle} \tag{16.3.8}$$

depends on variables ξ and $\bar{\eta}$ analytically and coincides with $A(\xi,\bar{\xi})$ at $\eta=\xi$. Therefore function $A(\xi,\bar{\eta})$ is the analytic continuation of $A(\xi,\bar{\xi})$ and is determined by it completely. Thus clearly a one-to-one correspondence exists between the operators and their symbols. In particular, the symbol $A(z,\bar{z})=a$ $=$const corresponds to operator $\hat{A}=a\hat{1}$, where $\hat{1}$ is identity in $\mathscr{F}_h$.

The action of operator $\hat{A}$ on a function is given by the following integral involving the symbols

$$(\hat{A}f)(z) = \left(\frac{1}{h}-1\right)\int A(z,\bar{\zeta})\,f(\zeta)\left(\frac{1-|\zeta|^2}{1-z\bar{\zeta}}\right)^{1/h} d\mu(\zeta,\bar{\zeta}). \tag{16.3.9}$$

There is an integral representation for operator $\hat{A}$

$$\hat{A} = \int \mathring{A}(\zeta,\bar{\zeta})|\zeta\rangle\langle\zeta|\,d\mu(\zeta), \quad \text{and} \tag{16.3.10}$$

$$A(z,\bar{z}) = \left(\frac{1}{h}-1\right)\int \mathring{A}(\zeta,\bar{\zeta})\left(\frac{(1-|\zeta|^2)(1-|z|^2)}{(1-z\bar{\zeta})(1-\bar{z}\zeta)}\right)^{1/h} d\mu(\zeta,\bar{\zeta}). \tag{16.3.11}$$

Multiplication is defined as follows. If $\hat{A}=\hat{A}_1\cdot\hat{A}_2$, the symbol of $\hat{A}$ is given by the integral

$$A(z,\bar{z}) = \left(\frac{1}{h}-1\right)\int A_1(z,\bar{\zeta})A_2(\zeta,\bar{z})\left(\frac{(1-|z|^2)(1-|\zeta|^2)}{(1-\zeta\bar{z})(1-\bar{\zeta}z)}\right)^{1/h} d\mu(\zeta). \tag{16.3.12}$$

16.3.2 The Correspondence Principle

Equations (16.3.11, 12) imply the significance of the operator T_h, whose action in functional space is defined by integral

$$(T_h f)(z,\bar{z})=\left(\frac{1}{h}-1\right)\int f(\zeta,\bar{\zeta})\left(\frac{(1-z\bar{z})(1-\zeta\bar{\zeta})}{(1-z\bar{\zeta})(1-\zeta\bar{z})}\right)^{1/h} d\mu(\zeta). \tag{16.3.13}$$

Let us investigate the $h\to 0$ asymptotics of function

$$\varphi_h(z,\bar{z})=(T_h f)(z,\bar{z}),$$

where $f(z,\bar{z})$ is assumed to be a continuously differentiable function.

We start from the point $z=0$,

$$(T_h f)(0,0)=\left(\frac{1}{h}-1\right)\int f(\zeta,\bar{\zeta})(1-\zeta\bar{\zeta})^{1/h} d\mu(\zeta,\bar{\zeta}). \tag{16.3.14}$$

Function $(1-\zeta\bar{\zeta})^{1/h}$ is localized near the origin at $h\to 0$. Since

$$\left(\frac{1}{h}-1\right)\int (1-\zeta\bar{\zeta})^{1/h} d\mu(\zeta,\bar{\zeta})=\left(\frac{1}{h}-1\right)\int (1-x)^{[(1/h)-2]}dx=1,$$

the result is

$$(T_h f)(0,0)=f(0,0)+h\left.\frac{\partial^2 f}{\partial z\partial\bar{z}}\right|_{z=0}+O(h). \tag{16.3.15}$$

To obtain the asymptotic behavior of function $(T_h f)$ at an arbitrary point z of domain D it is helpful to change the variables

$$\zeta\to\eta,\qquad \zeta=\frac{\eta-z}{1-\bar{z}\eta}. \tag{16.3.16}$$

Transformation (16.3.16) is a motion in the Lobachevsky plane, so the measure $d\mu(\zeta,\bar{\zeta})$ is conserved. Therefore

$$(T_h f)(z,\bar{z})=\left(\frac{1}{h}-1\right)\int f_1(\eta,\bar{\eta})(1-\eta\bar{\eta})^{1/h} d\mu(\eta,\bar{\eta}), \tag{16.3.17}$$

where

$$f_1(\zeta,\bar{\zeta})=f\left(\frac{\zeta-z}{1-\bar{z}\zeta},\frac{\bar{\zeta}-\bar{z}}{1-z\bar{\zeta}}\right).$$

Note that

$$\left.\frac{\partial^2 f_1}{\partial\zeta\partial\bar{\zeta}}\right|_{\zeta=\bar{\zeta}=0}=(1-z\bar{z})^2\frac{\partial^2 f}{\partial z\partial\bar{z}},$$

while the operator

$$\Delta=(1-z\bar{z})^2\frac{\partial^2}{\partial z\partial\bar{z}}$$

is just the Laplace-Beltrami operator for the Lobachevsky plane. Thus, using (16.3.15) yields the $h\to 0$ asymptotics for function $T_h f$

$$(T_h f)(z,\bar{z})=f(z,\bar{z})+h\Delta f(z,\bar{z})+o(h). \tag{16.3.18}$$

The correspondence principle follows from (16.3.18): substituting $A_1(z,\bar{\zeta})A_2(\zeta,\bar{z})=f(\zeta,\bar{\zeta})$ in (16.3.12) gives

$$A(z,\bar{z})=A_1(z,\bar{z})A_2(z,\bar{z})+h(1-z\bar{z})^2\frac{\partial A_1}{\partial\bar{z}}\frac{\partial A_2}{\partial z}+o(h). \tag{16.3.19}$$

Hence we have verified that the correspondence principle does hold; first,

$$\lim_{h\to 0}(A_1 * A_2)=A_1(z,\bar{z})A_2(z,\bar{z}), \tag{16.3.20}$$

and the second term leads to the commutator

$$\begin{aligned}\lim_{h\to 0}\frac{1}{h}(A_1 * A_2-A_2 * A_1)&=(1-z\bar{z})^2\left(\frac{\partial A_1}{\partial\bar{z}}\frac{\partial A_2}{\partial z}-\frac{\partial A_1}{\partial z}\frac{\partial A_2}{\partial\bar{z}}\right)\\&=\mathrm{i}\{A_1,A_2\}.\end{aligned} \tag{16.3.21}$$

16.3.3 Operator T_h in Terms of the Laplacian

Operator T_h commutes with the group transformations

$$f(z,\bar{z})\to f(gz,\overline{gz}),$$

where g is a motion in the Lobachevsky plane. Hence it must be a function of the Laplace-Beltrami operator Δ.

Explicit calculations in [130] lead to the following expression for T_h:

$$T_h=\prod_{l=0}^{\infty}\left(1-h^2\frac{\Delta}{(1+lh)\,[1+(l-1)h]}\right)^{-1}. \tag{16.3.22}$$

16.3.4 Representation of Group of Motions of the Lobachevsky Plane in Space $\mathscr{F}_h$

Transformation of the group elements in the space of symbols has the form

$$(\mathscr{T}_g A)(z,\bar{z}) = A(gz,\overline{gz}), \qquad g\in G. \tag{16.3.23}$$

It can be easily seen that this transformation is an automorphism of the operator algebra A_h. Since it is known that all automorphisms of the algebra of bounded operators in the Hilbert space are internal, a bounded operator U_g must exist which generates the automorphism of (16.3.23),

$$\frac{\langle\bar{z}|U_g A U_{g^{-1}}|\bar{z}\rangle}{\langle\bar{z}|\bar{z}\rangle} = \frac{\langle\overline{gz}|A|\overline{gz}\rangle}{\langle\overline{gz}|\overline{gz}\rangle} = A(gz,\overline{gz}). \tag{16.3.24}$$

It can be easily seen that U_g is a unitary operator times a factor.

Now we will show that the mapping $g\to U_g$ determines an irreducible representation of group G. Let A be a bounded operator, commuting with all the group operators U_g. Because of (16.3.24) its symbol $A(gz,\overline{gz})$ is independent of g and since the action of group G in the Lobachevsky plane is transitive, it must be a constant

$$A(z,\bar{z}) = a = \text{const.}$$

Therefore, the operator must be identity times a factor

$$\hat{A} = a\hat{1}.$$

Thus, we have got a proof that the representation $g\to U_g$ is irreducible.

Using (16.3.9) we can also get an explicit expression for the action of operator U_g:

$$(U_g f)(z) = \varepsilon f\left(\frac{\alpha z+\beta}{\bar{\beta}z+\bar{\alpha}}\right)(\bar{\beta}z+\bar{\alpha})^{-1/h},$$

$$g = \begin{pmatrix}\alpha & \beta\\ \bar{\beta} & \bar{\alpha}\end{pmatrix}, \quad |\alpha|^2-|\beta|^2=1, \quad |\varepsilon|=1. \tag{16.3.25}$$

The unitarity of transformation (16.3.25) can be verified directly. To get an explicit expression for the symbol of the operator $U_g(z,\bar{z})$, (16.3.1, 6) are combined. The matrix element is

$$\langle\bar{z}|U_g|\bar{z}\rangle = \varepsilon(\alpha+\beta\bar{z}+\bar{\beta}z+\bar{\alpha}z\bar{z})^{-1/h} \quad \text{so} \tag{16.3.26}$$

$$U_g(z,\bar{z}) = \frac{\langle\bar{z}|U_g|z\rangle}{\langle\bar{z}|\bar{z}\rangle} = \varepsilon\left(\frac{1-z\bar{z}}{\alpha+\beta\bar{z}+\bar{\beta}z+\bar{\alpha}z\bar{z}}\right)^{1/h}. \tag{16.3.27}$$

16.3.5 Quantization by Inversions: Analog to Weyl Quantization

Some elements of group G can be represented as point inversions. If a point ζ is the inversion center, the complex plane transformation is

$$g(\zeta,\bar{\zeta})z=\frac{-(1+\zeta\bar{\zeta})z+2\zeta}{-2\bar{\zeta}z+1+\zeta\bar{\zeta}} \tag{16.3.28}$$

so the matrix of the group element is of the form

$$g=\frac{1}{1-\zeta\bar{\zeta}}\begin{pmatrix}-(1+\zeta\bar{\zeta}) & 2\zeta\\ -2\bar{\zeta} & (1+\zeta\bar{\zeta})\end{pmatrix}.$$

The symbol of the inversion operator denoted by $U_{\zeta,\bar{\zeta}}$ obtained by (16.3.27) is

$$U_h(\zeta,\bar{\zeta};z,\bar{z})=\left(\frac{(1-z\bar{z})(1-\zeta\bar{\zeta})}{(1-z\bar{\zeta})(1-\bar{z}\zeta)}\right)^{1/h}\left(1+\frac{z-\zeta}{1-\bar{z}\zeta}\,\frac{\bar{z}-\bar{\zeta}}{1-z\bar{\zeta}}\right)^{-1/h} \tag{16.3.29}$$

times a constant factor ε, $|\varepsilon|=1$. This factor will be fixed in such a way that the symbol is just given by (16.3.29).

Function

$$a(z,\bar{z})=\operatorname{tr}\{\hat{A}\,U_{z,\bar{z}}\} \tag{16.3.30}$$

is called the Weyl symbol of operator $\hat{A}$. If the operator can be represented by means of the integral

$$\hat{A}=\left(\frac{1}{h}-1\right)\int \mathring{a}(z,\bar{z})\,U_{z,\bar{z}}\,d\mu(z,\bar{z}), \tag{16.3.31}$$

function $\mathring{a}(z,\bar{z})$ presented here is called the covariant symbol of operator $\hat{A}$. The relationships between symbols $\mathring{A}$ and a, and $\mathring{a}$ and A are integral transformations,

$$a(z,\bar{z})=(S_h\mathring{A})(z,\bar{z}) \tag{16.3.32}$$

$$A(z,\bar{z})=(S_h\mathring{a})(z,\bar{z}),\quad \text{where} \tag{16.3.33}$$

$$(S_h f)(z,\bar{z})=\left(\frac{1}{h}-1\right)\int f(\zeta,\bar{\zeta})\,U_h(\zeta,\bar{\zeta};z,\bar{z})\,d\mu(\zeta,\bar{\zeta}). \tag{16.3.34}$$

Equation (16.3.32) is obtained from the integral for the trace of the product

$$\operatorname{tr}\{\hat{A}\hat{B}\}=\int A(x)\,\mathring{B}(x)\,d\mu(x)$$

and from the symmetry relation

$$U_h(\zeta,\bar{\zeta};z,\bar{z})=U_h(z,\bar{z};\zeta,\bar{\zeta}).$$

To relate symbols a and $\mathring{a}$ one has to know the operator S_h as a function of the Laplace-Beltrami operator Δ. (In fact, operator S_h commutes with all operators of the group representation, so it is a function of Δ.)

Berezin's calculation [130] leads to the explicit expression

$$S_h=\prod_{l=0}^{\infty}\left(1-h^2\frac{\Delta}{(1+2lh)\,[1+(2l-1)h]}\right)^{-1}. \tag{16.3.35}$$

Comparing (16.3.22, 35) yields

$$T_h=S_h S_h', \quad \text{where} \tag{16.3.36}$$

$$S_h'=\prod_{l=0}^{\infty}\left(1-h^2\frac{\Delta}{[1+(2l+1)h]\,(1+2lh)}\right)^{-1}. \tag{16.3.37}$$

With (16.3.36, 11)

$$\mathring{a}=S_h'\mathring{A}=S_h'S_h^{-1}a. \tag{16.3.38}$$

16.4 Quantization on a Sphere

If $l=h^{-1}$ is an integer, quantization on a sphere is closely analogous to that for the Lobachevsky plane; this is discussed here.

It is appropriate to map the sphere on the complex z plane by means of stereographical projection. The invariant measure on the sphere is replaced by the measure on the z plane

$$d\mu(z,\bar{z})=(1+z\bar{z})^{-2}\frac{dz\wedge d\bar{z}}{2\pi\mathrm{i}}. \tag{16.4.1}$$

The functional space $\mathscr{F}_h$ to be considered is that of holomorphic functions $f(z)$ with the scalar product

$$(f,g)=\left(\frac{1}{h}+1\right)\int \bar{f}(z)\,g(z)\,(1+z\bar{z})^{-1/h}d\mu(z,\bar{z}). \tag{16.4.2}$$

Note that function $f(z)\in\mathscr{F}_h$ must be a polynomial of a degree less than h^{-1}; otherwise, the integral defining the norm $\|f\|^2=(f,f)$ diverges.

Therefore, dimensionality of space $\mathscr{F}_h$ is $\left[1+\frac{1}{h}\right]$. Functions

$$f_{\bar{\zeta}}(z)=(1+\bar{\zeta}z)^{1/h}, \tag{16.4.3}$$

which are the elements of the functional space $\mathscr{F}_h$, correspond to vectors $|\bar{\zeta}\rangle$ of the Hilbert space with the norm $\langle\bar{\zeta}|\bar{\zeta}\rangle=(1+\zeta\bar{\zeta})^{1/h}$. Thus, we get a CS system for the rotation group of the three-dimensional Euclidean space discussed in Chap. 4. Note that if the quantity h^{-1} is not an integer, function $f_{\bar{\zeta}}(z)$ is multivalued and, therefore, it does not belong to space $\mathscr{F}_h$.

We present now some more formulae for the symbols of operators in space $\mathscr{F}_h$. The covariant symbol of operator $\hat{A}$ is the function

$$A(z,\bar{\zeta})=\frac{\langle\bar{z}|\hat{A}|\bar{\zeta}\rangle}{\langle\bar{z}|\bar{\zeta}\rangle} \tag{16.4.4}$$

taken at $\zeta=z$. Function (16.4.4) is the analytic continuation of the symbol and, therefore, there is a one-to-one correspondence between covariant symbols and the operators.

The operator action in the functional space is given by the integral

$$(\hat{A}f)(z)=\left(\frac{1}{h}+1\right)\int A(z,\bar{\zeta})\left(\frac{1+z\bar{\zeta}}{1+|\zeta|^2}\right)^{1/h}d\mu(\zeta,\bar{\zeta}). \tag{16.4.5}$$

Hence the symbol for the operator product $\hat{A}=\hat{A}_1\hat{A}_2$ is

$$A(z,\bar{\zeta})=\left(\frac{1}{h}+1\right)\int A_1(z,\bar{\zeta})A_2(\zeta,\bar{z})\left[\frac{(1+z\bar{\zeta})(1+\bar{z}\zeta)}{(1+|z|^2)(1+|\zeta|^2)}\right]^{1/h}d\mu(\zeta,\bar{\zeta}). \tag{16.4.6}$$

The relation between the different types of symbols is given by the integral transformation

$$\begin{aligned}A(z,\bar{z})&=(T_h\mathring{A})(z,\bar{z})\\&=\left(\frac{1}{h}+1\right)\int \mathring{A}(\zeta,\bar{\zeta})\left(\frac{(1+z\bar{\zeta})(1+\zeta\bar{z})}{(1+|z|^2)(1+|\zeta|^2)}\right)^{1/h}d\mu(\zeta,\bar{\zeta}).\end{aligned} \tag{16.4.7}$$

As for (16.3.22), operator T_h can be expressed in terms of the Laplace-Beltrami operator

$$T_h=\prod_{k=1}^{\infty}\left(1+h^2\frac{\Delta}{(1+kh)\,[1+(k+1)h]}\right). \tag{16.4.8}$$

The asymptotic formula, like (16.3.18) for the Lobachevsky plane, is valid also for the sphere. Hence one verifies the correspondence principle.

The relationship using the representation theory for the groups $SO(3)$ and $SU(2)$ is established by analogy to the case of the Lobachevsky plane. The action of operator U_g in the functional space is given by

$$(U_gf)(z)=f\left(\frac{\alpha z-\bar{\beta}}{\beta z+\bar{\alpha}}\right)(\bar{\alpha}+\beta z)^{1/h},\quad g=\begin{pmatrix}\alpha & \beta\\ -\bar{\beta} & \bar{\alpha}\end{pmatrix},\quad |\alpha|^2+|\beta|^2=1, \tag{16.4.9}$$

where h^{-1} is an integer.

It is well known that the whole set of unitary irreducible representations of the $SU(2)$ group is exhausted by these representations, up to equivalence transformations.

16.5 Quantization on Homogeneous Kähler Manifolds

The construction of quantization considered in Sects. 16.3, 4 for the simplest manifolds is extended to a more general class of manifolds, namely, to the homogeneous Kähler manifolds [22].

The main relevant definitions are as follows.

Let $\mathcal{M}$ be a complex manifold and z^j are local coordinates of a point on this manifold. Suppose a metric exists on $\mathcal{M}$,

$$ds^2 = g_{j\bar{k}} dz^j d\bar{z}^k, \quad j, k = 1, \ldots, n, \tag{16.5.1}$$

such that $\det(g_{j\bar{k}}) \neq 0$ and the 2-form corresponding to this metric

$$\omega = g_{j\bar{k}} dz^j \wedge d\bar{z}^k \tag{16.5.2}$$

is closed,

$$d\omega = 0. \tag{16.5.3}$$

The manifold possessing this property is called Kählerian and metric (16.5.1) is called the Kählerian metric.

Clearly, the Kähler manifold is symplectic. Therefore, a classical mechanics given by $(\mathcal{M}, \omega)$ is established on the manifold. In the local coordinates the Poisson bracket is

$$\{f, g\} = \mathrm{i}\, g^{j\bar{k}} \left(\frac{\partial f}{\partial z^j} \frac{\partial g}{\partial \bar{z}^k} - \frac{\partial f}{\partial \bar{z}^j} \frac{\partial g}{\partial z^k} \right). \tag{16.5.4}$$

Evidently, since the 2-form is closed, a function $F(z, \bar{z})$ (the potential of the metric) must exist such that

$$g_{j\bar{k}} = \frac{\partial^2 F}{\partial z^j \partial \bar{z}^k}. \tag{16.5.5}$$

Let us consider the Hilbert space $\mathscr{F}_h$, the space of functions analytical in $\mathcal{M}$ with the scalar product defined by

$$(f, g) = c(h) \int \bar{f}(z) g(z) \exp\left[-\frac{1}{h} F(z, \bar{z}) \right] d\mu(z, \bar{z}), \tag{16.5.6}$$

where

$$d\mu(z,\bar{z})=\omega^n=\det(g_{j\bar{k}})\prod_{k=1}^{n}\frac{dz^k\wedge d\bar{z}^k}{2\pi i}. \qquad (16.5.7)$$

Constant h in (16.5.6, 7) and in the following plays the role of Planck's constant, function $c(h)$ is to be determined below.

Suppose we have an orthonormalized basis on space $\mathscr{F}_h$, $\{f_k(z)\}$. Then the following theorem holds.

Theorem 1 [22]. 1) In each coordinate vicinity the series

$$L_h(z,\bar{z})=\sum_k f_k(z)\overline{f_k(z)} \qquad (16.5.8)$$

converges absolutely and uniformly.

2) Function $L_h(z,\bar{z})$ is independent of the choice of the basis $\{f_k(z)\}$.

An important estimate,

$$|L_h(z,\bar{\zeta})|^2\leq L_h(z,\bar{z})\,L_h(\zeta,\bar{\zeta}), \qquad (16.5.9)$$

is a consequence of the Cauchy inequality. Using another notation,

$$\Phi_{\bar{\zeta}}(z)=L_h(z,\bar{\zeta}), \qquad (16.5.10)$$

one has with (16.5.6)

$$c(h)\int|\Phi_{\bar{\zeta}}(z)|^2\exp\left[-\frac{1}{h}F(z,\bar{z})\right]d\mu(z,\bar{z})$$
$$=\sum_k f_k(\zeta)\overline{f_k(\zeta)}=L_h(\zeta,\bar{\zeta}). \qquad (16.5.11)$$

Hence $\Phi_{\bar{\zeta}}(z)\in\mathscr{F}_h$. Because of (16.5.10, 8),

$$(f,\Phi_{\bar{\zeta}})=f(\zeta) \qquad (16.5.12)$$

for any $f(z)\in\mathscr{F}_h$. Therefore the set $\{\Phi_{\bar{\zeta}}(z)\}$ is an overcomplete system of vectors in space $\mathscr{F}_h$.

Using this system we define the covariant symbol $A(z,\bar{z})$ of operator $\hat{A}$ as the diagonal value ($z=\zeta$) of function

$$A(z,\bar{\zeta})=\frac{\langle\Phi_{\bar{z}}|\hat{A}|\Phi_{\bar{\zeta}}\rangle}{\langle\Phi_{\bar{z}}|\Phi_{\bar{\zeta}}\rangle}, \qquad (16.5.13)$$

which is defined in space $\mathscr{M}\times\mathscr{M}$.

By analogy with Sects. 16.3, 4,

$$(\hat{A}f)(z)=c(h)\int A(z,\bar{\zeta})f(\zeta)L_h(z,\bar{\zeta})\exp\left[-\frac{1}{h}F(\zeta,\bar{\zeta})\right]d\mu(\zeta,\bar{\zeta}) \quad (16.5.14)$$

$$(A_1*A_2)(z,\bar{z})=c(h)\int A_1(z,\bar{\zeta})A_2(\zeta,\bar{z})\times\frac{L_h(z,\bar{\zeta})\,L_h(\zeta,\bar{z})}{L_h(z,\bar{z})\,L_h(\zeta,\bar{\zeta})}$$

$$\times\left\{L_h(\zeta,\bar{\zeta})\exp\left[-\frac{1}{h}F(\zeta,\bar{\zeta})\right]\right\}d\mu(\zeta,\bar{\zeta}) \quad (16.5.15)$$

$$\operatorname{tr}\hat{A}=c(h)\int A(z,\bar{z})\left\{L_h(z,\bar{z})\exp\left[-\frac{1}{h}F(Z,\bar{z})\right]\right\}d\mu(z,\bar{z}). \quad (16.5.16)$$

Note that according to (16.5.14) operator $\hat{A}$ is reconstructed from $A(z,\bar{\zeta})$ which is an analytical continuation of $A(z,\bar{z})$, i.e., the covariant symbol of the operator. Therefore there is a one-to-one correspondence between operators and their symbols. Thus, algebra A_h can be the algebra of covariant symbols of bounded operators.

The correspondence principle can be proved only under an additional assumption that the manifold $\mathcal{M}$ is homogeneous. So we suppose that $\mathcal{M}$ is a homogeneous Kähler manifold, G is the group of its motions and $F(z,\bar{z})$ is the potential of the Kähler metric on $\mathcal{M}$ which is invariant with respect to G. It is supposed that the potential $F(z,\bar{z})$ exists globally on a set $\tilde{\mathcal{M}}$ obtained when a submanifold $\mathcal{M}'$ of a smaller dimensionality is removed from $\mathcal{M}$. Respectively, functions $f(z)\in\mathcal{F}_h$ are also defined on $\tilde{\mathcal{M}}$. Since the metric generated by potential $F(z,\bar{z})$ is invariant under the group transformations,

$$F(gz,\overline{gz})=F(z,\bar{z})+\psi(g,z)+\overline{\psi(g,z)}, \quad (16.5.17)$$

where ψ(g, z) is an analytic function of z at fixed g, defined on $\tilde{\mathcal{M}}\cap g\tilde{\mathcal{M}}$. Equation (16.5.17) determines function $\psi(g,z)$ up to an imaginary part: a more detailed treatment may be found in [22]. It was also shown there that with function $\psi(g,z)$ one can construct a unitary projective representation of group G. The representation operators are given by their action in the functional space $\mathcal{F}_h$,

$$(T_gf)(z)-\exp\left[-\frac{1}{h}\psi(g^{-1},z)\right]f(g^{-1}z). \quad (16.5.18)$$

These operators are unitary and form the projective representation of the local Lie group U_G, i.e., the representation in a vicinity of the unity element of group G. This representation can be continued to the unitary projective representation of the whole group G.

The existence of this representation is essentially used in the proof that the quantization considered satisfies the requirements formulated in Sect. 16.2.

The case considered here is the manifold $\mathcal{M}=D$, a complex homogeneous bounded domain in $\mathbb{C}^n$. Here D is the Kähler manifold relative to metric ds^2 with potential $F(z,\bar{z})=\ln K(z,\bar{z})$, where $K(z,\bar{\zeta})$ is the Bergman kernel of this domain (Chap. 12). Now, the elements of the functional space $\mathcal{F}_h$ are functions analytical in D, and the scalar product is

$$(f,f)=c(h)\int_D |f(z)|^2[K(z,\bar{z})]^{-1/h}d\mu(z,\bar{z}), \tag{16.5.19}$$

where

$$d\mu(z,\bar{z})=\det\left(\frac{\partial^2 \ln K}{\partial z^i \partial \bar{z}^j}\right)\frac{d\mu_L(z,\bar{z})}{\pi^h}; \tag{16.5.20}$$

$d\mu_L(z,\bar{z})$ is the Lebesque measure of domain D. Functions $L_h(z,\bar{\zeta})$ and $\Phi_{\bar{\zeta}}(z)=L_h(z,\bar{\zeta})$ are defined by (16.5.8). Algebra A_h is the algebra of covariant symbols of bounded operators acting in $\mathcal{F}_h$; algebra $\mathcal{A}$ consists of functions $f(h|z,\bar{z}), 0<h\leq 1$, which are elements of A_h (at fixed h) and are continuous on the whole set of variables $h, z, \bar{z}$.

Theorem 2 [22]. Algebra $\mathcal{A}$ is the quantization which satisfies the correspondence principle.

The proof of this theorem is not presented here. Note only that in the process of proving it, the following properties of the complex homogeneous bounded domain are essentially used

$$\text{1)}\ \det\left\|\frac{\partial^2 \ln K}{\partial z^i \partial \bar{z}^j}\right\|=\lambda K(z,\bar{z}); \tag{16.5.21}$$

$$\text{2)}\ L_h(z,\bar{z})=\mu[K(z,\bar{z})]^{1/h}, \tag{16.5.22}$$

where $\lambda=\lambda(D)$, $\mu=\mu(D)$ are constants depending only on the domain D.

In concluding this section, explicit formulae for the operator T_h which relates covariant and contravariant symbols are presented. The operator acts in space $\mathcal{F}_h$ as given by (16.3.13) (for the explicit form of operator T_h, see [22]). The following theorems have been proved in [131].

Theorem 3. For the complex homogeneous bounded symmetric domains, Laplace-Beltrami operators Δ_{2k} of order $2k$, $k=1,2,\ldots,l$ exist, whose eigenvalues which correspond to the unitary irreducible representations $T(g)$ of calss-I, are given by

$$S_{2k}=\sum_{i=1}^{l} x_i^{2k}. \tag{16.5.23}$$

Here l is the rank of the domain, and x_j are parameters which specify the representation [if the representation $T(g)$ is finite-dimensional, the quantities x_j are components of vector $(\lambda+\varrho)$, where 2ϱ is the sum of positive roots of the Lie algebra, and λ is the representation highest weight]. In the case of a bounded domain, $T(g)=T^{\lambda}(g)$ is a unitary representation of principal series, and x_j are arbitrary real numbers.

Theorem 4. The eigenvalues of operator T_h, corresponding to the irreducible representation of class-I for spaces of noncompact type are

$$t(\lambda; x_1, \ldots, x_l)=\frac{\prod_{k=1}^{l} \Gamma\left(\lambda-\frac{v-1}{2}+\mathrm{i}x_k\right)\Gamma\left(\lambda-\frac{v-1}{2}-\mathrm{i}x_k\right)}{\prod_{k=1}^{l} \Gamma\left(\lambda-\frac{v-1}{2}+\varrho_k\right)\Gamma\left(\lambda-\frac{v-1}{2}-\varrho_k\right)}, \tag{16.5.24}$$

where $\lambda=v/h$, $\Gamma(z)$ is the Euler gamma function, ϱ_j are the components of ϱ which are:

$$\begin{aligned}
&\text{for } D^{\mathrm{I}}_{p,q}, && \varrho_j=\frac{p+q+1}{2}-j, && 1\le j\le p;\ q\le p;\\
&\text{for } D^{\mathrm{II}}_{p}, && \varrho_j=\frac{p}{2}-\frac{j-1}{2}, && 1\le j\le p;\\
&\text{for } D^{\mathrm{III}}_{p}, && \varrho_j=\frac{2p+1}{2}-2j, && 1\le j\le \left[\frac{p}{2}\right];\\
&\text{for } D^{\mathrm{IV}}_{p}, && \varrho_j=\frac{p-1}{2},\ \varrho_2=\frac{1}{2}.
\end{aligned} \tag{16.5.25}$$

The numbers v are

$$v=\begin{cases} p+q & \text{for } D^{\mathrm{I}}_{p,q}\\ p+1 & \text{for } D^{\mathrm{II}}_{p}\\ 2(p-1) & \text{for } D^{\mathrm{III}}_{p}\\ p & \text{for } D^{\mathrm{IV}}_{p}\end{cases} \tag{16.5.26}$$

Theorem 5. The eigenvalues of operator T_h for the manifold $\mathscr{M}$ of the compact type are

$$t(-\lambda; \mathrm{i}x_1, \ldots, \mathrm{i}x_l), \tag{16.5.27}$$

where $t(\lambda; x_1, \ldots x_l)$ is the function given in (16.5.24), which corresponds to the dual (in the sense of *Cartan*) noncompact space, $\lambda>0$ is an integer, and $x_k=m_k+\varrho_k$, m_k are the components of the representation highest weight.

Note that Theorems 4 and 5 enable one to express operator T_h in terms of the Laplace-Beltrami operators Δ_{2k}. For symmetric spaces of rank-1, see [131].

Part III

Physical Applications

17. Preliminaries

The purpose of this part is to show the facilities of the CS method for some concrete examples. This chapter presents basis principles in order to explain the relevance of CS to the description of certain physical systems.

As noted in the Introduction, the CS method is now widely exploited in different domains of physics, and it has helped to solve a lot of problems. Since space is limited, I am, of course, not able to give an exhaustive exposition, so present instead some selected problems, which illustrate the CS method. Some of these problems could be solved by other means, yet the following shows that the CS method considerably simplifies the solution. Another advantage of the CS method is that it enables solutions of problems quite different in origin and statement to be unified. Note, that a number of applications of Glauber's coherent states [7, 8] have been considered in [9, 10, 13].

Some problems, beyond the scope of this exposition, have been solved using the CS method. These are listed below.

1. The superfluidity of weakly nonideal Bose gases was derived originally by *Bogolubov* [133]. The problem can also be solved easily with the CS method [12, 134, 135]. The superconductivity of weakly nonideal Fermi gases can be considered along the same lines [136].

2. A soft photon cloud around a charge particle has been described and the infrared divergences eliminated from quantum electrodynamics [137–139].

3. Spin waves in the Heisenberg model of ferromagnets have been described [140].

4. An approximate quantum description of localized field configurations (solitons) has been proposed [141–143].

5. The $SU(N)$ gauge lattice field theories were formulated [144].

6. States similar to the lattice CS (Sect. 1.5) have been used [145] to describe the ground states of a two-dimensional electron in a periodical magnetic field.

7. Long-range collisions of classical charged particles scattered on hydrogen atoms have been described [146, 147].

8. The Dicke model of two-level molecules and the transition to the superradiative phase have been treated [148–151]. This approach has been also extended to n-level molecules [152, 153].

9. Generating functions for invariants of $SU(N)$ were obtained [154].

10. An algebraic classification of dynamical systems where the coherence of generalized CS is preserved has been given [155].

11. It is shown [156] that a crucial concept in theory of solitons, the τ-functions, is related to CS for an infinite-dimensional Lie algebra $GL(\infty)$ which can be realized in terms of bilinear products of fermion operators.

12. The CS method has been applied to calculation of partition functions by means of the graph methods [157].

13. Basic ideas of using generalized CS in theoretical nuclear physics have been presented [158].

Some authors have also investigated overcomplete systems of states different from those I consider, e.g., the work by *Barut* and *Girardello* [16], and the series of works by *Nieto* (references may be found in [159]). In [160–162], overcomplete systems, which are related to the Fermi operators and the Grassmann variables, have been investigated. These systems differ from those considered in Chap. 9. Some CS systems related to superalgebras have been also considered [163].

We now briefly review applications of the CS method to cases where the system Hamiltonian has a dynamical symmetry group G.

The simplest case important for applications concerns the systems where the Hamiltonian H is a linear combination of generators X_k of a unitary irreducible representation of the corresponding Lie algebra:

$$H=\sum_k h^k X_k, \qquad X_k^+=X_k, \qquad \bar{h}^k=h^k. \tag{17.1}$$

Here the operators X_k satisfy the standard commutation relations

$$[X_k, X_l]=C_{kl}^m X_m, \tag{17.2}$$

where C_{kl}^m are the structure constants of the Lie algebra $\mathcal{G}$ of the corresponding group G; the coefficients h^k in (17.1) are, in general, time dependent: $h^k=h^k(t)$.

Such a situation takes place in a number of interesting physical systems, some of which being treated in Chaps. 18–26. Here a general description of such systems is presented.

The starting point is the fact that a unitary irreducible representation $T(g)$ of the group G corresponding to the Lie algebra of (17.2) acts in the Hilbert space of the system states. Respectively, generators X_k are transformed by the adjoint representation of G,

$$T(g)X_k T^{-1}(g)=A_k^l(g)X_l. \tag{17.3}$$

This transformation does not change the structure of the operator H,

$$\tilde{H}=T(g)HT^{-1}(g)=\tilde{h}^k X_k, \qquad \text{where}$$

$$\tilde{h}^k=A_l^k(g)h^l. \tag{17.4}$$

The problem in view is the nonstationary Schrödinger equation

$$i\hbar \frac{d}{dt}|\Psi(t)\rangle = H(t)|\Psi(0)\rangle. \tag{17.5}$$

An important property of the system stems from (17.1)

$$|\Psi(t)\rangle = T(g(t))|\Psi(0)\rangle. \tag{17.6}$$

Clearly, if at a moment t_0 the system state $|\Psi(t_0)\rangle$ was CS and related to the representation $T(g)$, it remains CS at any moment. This fact makes it possible to solve a number of problems relevant to (17.5). Here we consider problems of three types.

1. The Hamiltonian (17.1) is time independent. We must find its spectrum and the eigenfunctions.

One can use the unitary transformation (17.4) and fit a group element g to set a simpler form of H. For instance, one can make H an element of the Cartan subalgebra. For a compact Lie algebra this is always possible.

Now the spectrum $\{\tilde{E}_n\}$ and the eigenfunctions $\{\tilde{\Psi}_n\}$ can be found at once. The spectrum $\{E_n\}$ coincides with the spectrum $\{\tilde{E}_n\}$, and the eigenfunctions $\{\Psi_n\}$ are related to $\{\tilde{\Psi}_n\}$ by a unitary transformation

$$|\Psi_n\rangle = T^{-1}(g)|\tilde{\Psi}_n\rangle. \tag{17.7}$$

Hence it is seen that if when constructing a CS system one chooses $|\tilde{\Psi}_0\rangle$ (the ground state of Hamiltonian $\tilde{H}$) as an initial vector in the Hilbert space, then the ground state of the Hamiltonian (17.1) is a generalized CS.

2. The Hamiltonian (17.1) is time dependent but at $t \to \pm\infty$ it tends to some limits sufficiently rapidly so that asymptotic states $|\Psi_\pm\rangle$ exist.

Now the evolution operator $U(t, t_0)$ is

$$U(t, t_0) = T(g(t, t_0)), \tag{17.8}$$

and an S matrix exists

$$S = U(\infty, -\infty) = T(g_0). \tag{17.9}$$

The probability of transition from a state $|m\rangle$ at $t \to -\infty$ to a state $|n\rangle$ at $t \to +\infty$ is the matrix element T_{nm} squared,

$$W_{nm} = |T_{nm}(g_0)|^2 = |\langle n|T(g_0)|m\rangle|^2. \tag{17.10}$$

3. The Hamiltonian (17.1) is a periodical function of time,

$$H(t+\tau) = H(t). \tag{17.11}$$

Now there are states for which

$$|\Psi_\varepsilon(t+\tau)\rangle = \exp\left(-\mathrm{i}\frac{\varepsilon\tau}{\hbar}\right)|\Psi_\varepsilon(t)\rangle, \tag{17.12}$$

the so-called states with a definite quasienergy; their properties have been considered in detail in [164–167]. The evolution operator of the system $U(t, t_0)$ can be written as

$$U(t_0+\tau, t_0) = T(g_0) = \exp(-\mathrm{i}\tau\tilde{\mathscr{H}}/\hbar), \tag{17.13}$$

where the operator $\tilde{\mathscr{H}}$ has the form (17.1). Thus, the spectrum of the operator $\tilde{\mathscr{H}}$ gives the quasienergy spectrum for the problem in view.

Let us now turn to concrete examples.

18. Quantum Oscillators

Some nonstationary problems for quantum oscillators are solved quite simply by means of the CS method. In the cases considered the quantum problems are reduced exactly to the corresponding classical problems.

18.1 Quantum Oscillator Acted on by a Variable External Force

This problem has been solved by *Feynman* [56] and *Schwinger* [55]; the relation of their approaches to the CS method will be explained below.

The time evolution of the system concerned is governed by the Schrödinger equation

$$\mathrm{i}\,\frac{d}{dt}\,|\psi(t)\rangle=(H_0+H_1)|\psi(t)\rangle, \qquad \text{where} \tag{18.1.1}$$

$$H_0=\tfrac{1}{2}\,(p^2+\omega^2q^2)=\omega(a^+a+\tfrac{1}{2}) \tag{18.1.2}$$

is the free quantum oscillator Hamiltonian, and

$$H_1=-f(t)q=-f(t)\,\frac{1}{\sqrt{2\omega(t)}}\,(a+a^+) \tag{18.1.3}$$

describes the action of the external force $f(t)$. Here p and q are momentum and coordinate operators, a and a^+ are bosonic annihilation and creation operators. We set $\hbar=m=1$.

First, employing the unitary transformation

$$|\psi(t)\rangle=\exp\,(-\mathrm{i}H_0t)|\tilde{\psi}(t)\rangle,$$

we reduce the Schrödinger equation (18.1.1), and the equation for $|\tilde{\psi}(t)\rangle$ is

$$\mathrm{i}\,\frac{d}{dt}|\tilde{\psi}(t)\rangle=\tilde{H}_1|\tilde{\psi}(t)\rangle, \qquad \text{where} \tag{18.1.4}$$

$$\begin{aligned}\tilde{H}_1(t)&=\exp\,(\mathrm{i}H_0t)H_1\,\exp\,(-\mathrm{i}H_0t)\\&=-f(t)\,\frac{1}{\sqrt{2\omega}}\,[a\,\exp\,(-\mathrm{i}\omega t)+a^+\,\exp\,(\mathrm{i}\omega t)].\end{aligned} \tag{18.1.5}$$

It is convenient to rewrite (18.1.4) as

$$\frac{d}{dt}|\tilde{\psi}(t)\rangle=[\beta(t)a^{+}-\bar{\beta}(t)a]|\tilde{\psi}(t)\rangle, \qquad \text{where} \tag{18.1.6}$$

$$\beta(t)=\frac{\mathrm{i}}{\sqrt{2\omega}}f(t)\,\mathrm{e}^{\mathrm{i}\omega t} \tag{18.1.7}$$

Since the Hamiltonian $\tilde{H}_1(t)$ is a linear combination of operators of the Lie algebra W_1 (the Heisenberg-Weyl algebra), the evolution operator $\tilde{S}(t)$, defined by $|\tilde{\psi}(t)\rangle=\tilde{S}(t)|\tilde{\psi}(0)\rangle$, is an element of the group representation $T(g)$

$$\tilde{S}(t)=T(g(t))=\exp[-\mathrm{i}\varphi(t)]D(\gamma(t)). \tag{18.1.8}$$

In particular, it follows that if the initial state was a CS of group W_1, it remains coherent for any moment of time. Thus, the solution $|\tilde{\psi}(t)\rangle$ is

$$|\tilde{\psi}(t)\rangle=\exp[-\mathrm{i}\varphi(t)]|\alpha(t)\rangle. \tag{18.1.9}$$

In this state the expectation value of the operator a is

$$\langle\tilde{\psi}|a|\tilde{\psi}\rangle=\alpha(t). \tag{18.1.10}$$

The t derivative of the expectation value is calculated from (18.1.6)

$$\dot{\alpha}=\beta, \quad \alpha(t)=\alpha_0+\int_0^t\beta(t')dt'. \tag{18.1.11}$$

For small $t=\Delta t$, (18.1.6) gives

$$|\tilde{\psi}(t+\Delta t)\rangle\sim D(\beta(t),\,\Delta t)|\tilde{\psi}(t)\rangle. \tag{18.1.12}$$

Using $T(t,\alpha)|\beta\rangle=\mathrm{e}^{\mathrm{i}\varphi}|\beta+\alpha\rangle$, one gets from (18.1.9)

$$\dot{\varphi}=\mathrm{Im}\,\{\bar{\beta}\alpha\}=\mathrm{Im}\,\{\dot{\bar{\alpha}}\alpha\}. \tag{18.1.13}$$

Clearly, (18.1.11) is the classical equation of motion in presence of the external force, and (18.1.13) shows that $\varphi(t)$ is equal to the doubled area encircled by the radius vector as the point moves in the phase plane, namely,

$$\varphi(t)=\frac{1}{\hbar}\int_{q_0}^{q}p\,dq,$$

and the phase in (18.1.9) has a simple semiclassical meaning.

The result is especially simple if the force $f(t)$ tends to zero rapidly enough as $t\to\pm\infty$. Then the limits $\alpha_{\pm}$ and $\varphi_{\pm}$ exist, and a useful concept is the transition probability from a state $|m\rangle$ at $t\to-\infty$ to a state $|n\rangle$ at $t\to+\infty$. The probability is given by

$$W_{mn}=|\langle m|S|n\rangle|^2=|\langle m|D(\gamma)|n\rangle|^2 \tag{18.1.14}$$

and writing here the explicit expression for D_{mn} yields

$$W_{mn}=\frac{n_<!}{n_>!}|\gamma|^{2|m-n|}\exp(-|\gamma|^2)|L_{n_<}^{|m-n|}(|\gamma|^2)|^2 \tag{18.1.15}$$

where

$$\gamma=\int_{-\infty}^{\infty}\beta(t')dt'=\frac{\mathrm{i}}{\sqrt{2\omega}}\int_{-\infty}^{\infty}f(t)\exp(\mathrm{i}\omega t)dt, \tag{18.1.16}$$

it is the Fourier component of the force $f(t)$, times a factor.

In particular, if $|\psi(t)\rangle\to|0\rangle$ as $t\to-\infty$ (i.e., the oscillator was in its ground state from the very beginning), then the force $f(t)$ induces the transition to a coherent state, and the excitation probability to the nth level is given by the Poisson distribution

$$W_n=\mathrm{e}^{-\varrho}\frac{\varrho^n}{n!},\qquad \varrho=|\gamma|^2. \tag{18.1.17}$$

The second example is as follows.

18.2 Parametric Excitation of a Quantum Oscillator

The parametric excitation of a quantum oscillator is the excitation of an oscillator resulting from a change of its parameters, the mass $m=m(t)$ and for the frequency $\omega=\omega(t)$. The generic case of variable $m(t)$ and $\omega(t)$ is easily reduced to $m=\text{const}$, as $t'=\int[dt/m(t)]$ and $\omega'=m\omega$, yielding a quantum oscillator with variable frequency [168–170].

The problem we are concerned with has been considered in detail in [171–174]. Here we propose a solution using the CS system for a discrete series of representations of $SU(1,1)$ (Chap. 5). This method is quite straightforward and illuminating.

The Schrödinger equation for a quantum oscillator with variable frequency is

$$\mathrm{i}\frac{d}{dt}|\psi(t)\rangle=H(t)|\psi(t)\rangle,\qquad \text{where} \tag{18.2.1}$$

$$H(t)=\frac{p^2}{2}+\frac{\omega^2(t)}{2}q^2,\qquad p=-\mathrm{i}\frac{\partial}{\partial q} \tag{18.2.2}$$

and we set $\hbar=m=1$. Let us rewrite it as

$$\mathrm{i}\,\frac{d}{dt}\,|\psi(t)\rangle=(\boldsymbol{\Omega K})|\psi(t)\rangle, \quad \text{where} \tag{18.2.3}$$

$$\boldsymbol{\Omega K}=\Omega_0 K_0-\Omega_1 K_1-\Omega_2 K_2=\mathrm{i}\,(\beta K_+-\bar{\beta}K_- -\mathrm{i}\gamma K_0)$$

$$K_{0,1}=\frac{1}{4}\left(\frac{p^2}{\omega_0}\pm\omega_0 q^2\right), \quad K_2=\frac{1}{4}\,(pq+qp) \tag{18.2.4}$$

$$\Omega_{0,1}=\omega_0\left[\left(\frac{\omega(t)}{\omega_0}\right)^2\pm 1\right], \quad \Omega_2=0, \quad K_\pm=K_1\pm\mathrm{i}K_2\,. \tag{18.2.5}$$

Evidently, the operators K_0, K_1 and K_2 satisfy the commutation relations

$$[K_1,K_2]=-\mathrm{i}K_0, \quad [K_2,K_0]=\mathrm{i}K_1, \quad [K_0,K_1]=\mathrm{i}K_2 \tag{18.2.6}$$

(for an arbitrary value for the frequency ω_0 present in the definition of operators K_j), and according to Chap. 5 are generators of representations of a discrete series of the $SU(1,1)$ group with $k=1/4$ and $k=3/4$.

Thus, the Hamiltonian is linear in the generators of the Lie algebra of the $SU(1,1)$ group, and the solution of the Schrödinger equation is

$$\psi(t)=\exp\left[-\mathrm{i}\varphi(t)\right]|\zeta(t)\rangle, \quad |\zeta|<1, \tag{18.2.7}$$

where $|\zeta(t)\rangle$ is a CS relevant to the representation of discrete series of the $SU(1,1)$ group, with $k=1/4$ or $k=3/4$.

As in the preceding example, we substitute (18.2.7) into the Schrödinger equation (18.2.3) and get differential equations for ζ and φ:

$$\dot{\zeta}=\beta-\bar{\beta}\zeta^2-\mathrm{i}\gamma\zeta \tag{18.2.8}$$

$$\mathrm{i}\dot{\varphi}=k(\beta\bar{\zeta}-\bar{\beta}\zeta+\mathrm{i}\gamma). \tag{18.2.9}$$

Note that the ζ plane, which is the phase plane for the problem, is the Lobatschevsky plane here. Equation (18.2.8) describes the motion of a classical system (oscillator) on the phase plane. The quantum state $|\zeta(t)\rangle$ is just governed by the classical motion. The phase shift $\varphi(t)$ is exactly equal to the area (in Lobatschevsky metrics) encircled by the radius vector $\zeta(t)$. Both these facts arise since the "semiclassical approximation" gives the exact answer in this case.

Though this result is valid for an arbitrary time dependence of $\omega(t)$, two cases are of special physical interest.

A) The quantity $\omega(t)$ attains its limits at $t\to\pm\infty$ sufficiently rapidly. In this case there are asymptotic states $|n\rangle_\pm$ at $t\to\pm\infty$, and there is a transition probability W_{mn} between states $|m\rangle_-$ and $|n\rangle_+$. Suppose, for simplicity, that the limiting Hamiltonians (H_+ and H_-) coincide. Then

$$W_{mn} = |\langle m|T(g_0)|n\rangle|^2, \quad g_0 = \begin{pmatrix} \alpha & \beta \\ \bar{\beta} & \bar{\alpha} \end{pmatrix}, \quad |\alpha|^2 - |\beta|^2 = 1. \tag{18.2.10}$$

Using the expression for $\langle m|T|n\rangle$ from [171], we get finally

$$W_{mn} = \frac{n_<!}{n_>!} \sqrt{1-\varrho}\,|P^{\frac{1}{2}|m-n|}_{\frac{1}{2}(m+n)}(\sqrt{1-\varrho})|^2, \quad \varrho = \frac{|\beta|^2}{|\alpha|^2}. \tag{18.2.11}$$

B) Next consider the case of periodical time dependence

$$\omega(t+T) = \omega(t).$$

The Schrödinger equation has solutions with definite quasi-energies:

$$|\psi_\varepsilon(t+T)\rangle = \exp\left(-\mathrm{i}\,\frac{\varepsilon T}{\hbar}\right)|\psi_\varepsilon(t)\rangle. \tag{18.2.12}$$

To find the quasi-energy spectrum first consider the evolution operator of the system

$$U(t, t_0)|\psi(t_0) = |\psi(t)\rangle,$$

and use it to construct the unitary operator

$$S = U(t_0 + T, t_0). \tag{18.2.13}$$

Write the latter as an exponential

$$S = \exp\left(-\frac{\mathrm{i}}{\hbar}\,T\tilde{H}\right), \tag{18.2.14}$$

where $\tilde{H}$ is an Hermitian operator. The spectrum of this operator is exactly the spectrum of quasienergies.

The operator S is a finite transformation of the $SU(1,1)$ group, and the operator $\tilde{H}$ belongs to a representation of the corresponding Lie algebra, so

$$\tilde{H} = \hbar(\Omega_0 K_0 - \Omega_1 K_1 - \Omega_2 K_2). \tag{18.2.15}$$

A unitary operator U transforms $\tilde{H} \to U\tilde{H}U^+$ to a standard form. Different physical situations arise, depending on the vector $\Omega = (\Omega_0, \Omega_1, \Omega_2)$. Each case corresponds to a certain orbit of the adjoint representation of the $SU(1,1)$ group.

1a) Let

$$\Omega^2 = \Omega_0^2 - \Omega_1^2 - \Omega_2^2 > 0, \qquad \Omega_0 > 0. \tag{18.2.16}$$

The unitary transformation $\tilde{H} \to \tilde{H}' = U\tilde{H}U^+$ leads to the form

$$\tilde{H}' = \hbar\Omega K_0 . \tag{18.2.17}$$

The quasi-energy spectrum is discrete; it is bounded from below and has the form

$$\varepsilon_n = \hbar\Omega(k+n). \tag{18.2.18}$$

The ground state of the Hamiltonian is a coherent state relevant to the representation T^k of the discrete series of the $SU(1,1)$ group.

1b) Let

$$\Omega^2 = \Omega_0^2 - \Omega_1^2 - \Omega_2^2 > 0, \qquad \Omega_0 < 0. \tag{18.2.19}$$

Now $\tilde{H}' = -\hbar\Omega K_0$. The quasi-energy spectrum is discrete and bounded from above

$$\varepsilon_n = -\hbar\Omega(k+n)$$

2) Let

$$\Omega_0^2 - \Omega_1^2 - \Omega_2^2 = -\lambda^2 < 0.$$

The operator $\tilde{H}$ can be reduced to

$$\tilde{H}' = -\hbar\lambda K_1 . \tag{18.2.20}$$

The quasi-energy spectrum is continuous and occupies the whole axis: $-\infty < \varepsilon < +\infty$. The classical motion is unstable here.

3a) If $\Omega^2 = 0$, $\Omega_0 > 0$, then $\tilde{H}' = \hbar\Omega_0(K_0 - K_1)$, and the spectrum is continuous and bounded from below, $0 < \varepsilon < \infty$.

3b) Finally, if $\Omega^2 = 0$, $\Omega_0 < 0$ then

$$\tilde{H}' = -\hbar\Omega_0(K_0 - K_1). \tag{18.2.21}$$

Here the spectrum is also continuous and bounded from above, $-\infty < \varepsilon < 0$. For a classical oscillator, Cases 3a and 3b correspond to the boundaries of the instability domain [175].

18.3 Quantum Singular Oscillator

18.3.1 The Stationary Case

The singular oscillator is a system described by the Hamiltonian

$$H = H_0 + V,$$

$$H_0 = -\frac{1}{2}\frac{d^2}{dx^2} + \frac{x^2}{2} = a^+ a + \frac{1}{2}, \qquad V = g^2 x^{-2}, \tag{18.3.1}$$

(using the units $\hbar = m = \omega = 1$) where

$$a = \frac{1}{\sqrt{2}}\left(x + \frac{d}{dx}\right), \qquad a^+ = \frac{1}{\sqrt{2}}\left(x - \frac{d}{dx}\right)$$

are the usual annihilation and creation operators.

First we show that the operator H generates the algebra $SU(1,1)$. One can easily verify that the operators

$$B_2^+ = (a^+)^2 - g^2/x^2, \qquad B_2 = a^2 - g^2/x^2 \qquad \text{satisfy} \tag{18.3.2}$$

$$[H, B_2^+] = 2B_2^+, \qquad [H, B_2] = -2B_2, \tag{18.3.3}$$

$$[B_2, B_2^+] = 4H. \tag{18.3.4}$$

Thus, thc operators H, B_2^+ and B_2 are generators of the $SU(1,1)$ Lie algebra. It follows from (18.3.3) that if $\psi_E(x)$ is an eigenfunction of the Hamiltonian

$$H\psi_E = E\psi_E,$$

then $B_2^+\psi_E$ and $B_2\psi_E$ are also eigenfunctions of H with the eigenvalues $(E+2)$ and $(E-2)$, respectively. Thus, B_2^+ (B_2) is a raising (lowering) operator. As for harmonical oscillators, one can find the spectrum and eigenfunctions of the Hamiltonian algebraically without solving the Schrödinger equation.

Actually, the wave function ψ_0 of the ground state, i.e., the state with the lowest energy, satisfies

$$B_2\psi_0 = 0, \qquad \text{so that} \tag{18.3.5}$$

$$\psi_0 = c_0 x^\alpha \exp\left(-\frac{x^2}{2}\right), \qquad \alpha = \frac{1}{2} + \left(\frac{1}{4} + 2g^2\right)^{1/2}. \tag{18.3.6}$$

The equation

$$H\psi_0 = E_0\psi_0 \tag{18.3.7}$$

shows that

$$E_0=\alpha+\tfrac{1}{2}. \tag{18.3.8}$$

For an arbitrary state the wave function ψ_n is obtained from ψ_0 by the n-fold application of the raising operator B_2^+

$$\psi_n(x)=c_n(B_2^+)^n\psi_0(x), \qquad E_n=2n+E_0. \tag{18.3.9}$$

Such algebraic generation of the wave functions and energy spectrum is important since it can be applied to the system of N interacting particles where it is impossible to find all solutions of the Schrödinger equation in the conventional way.

Note the relationship of the operators H, B_2^+ and B_2 to the standard generators K_0, K_1 and K_2 of the $SU(1,1)$ algebra:

$$K_0=\frac{H}{2}, \quad K_+=K_1+\mathrm{i}K_2=-\frac{B_2^+}{2}, \quad K_-=K_1-\mathrm{i}K_2=-\frac{B_2}{2}. \tag{18.3.10}$$

The Casimir operator

$$C_2=K_1^2+K_2^2-K_0^2=\frac{1}{4}\left(\frac{B_2B_2^+ +B_2^+B_2}{2}-H^2\right), \tag{18.3.11}$$

i.e., the operator commuting with all the generators of the $SU(1,1)$ Lie algebra, is obtained by substituting the explicit expressions for B_2, B_2^+ and H,

$$C_2=k(1-k)=\tfrac{3}{16}-\tfrac{1}{4}\,\alpha(\alpha-1).$$

Hence

$$k=\tfrac{1}{2}\,(\tfrac{1}{2}+\alpha). \tag{18.3.12}$$

This means that the whole set of the wave functions transforms by an irreducible representation of discrete series of the group $SU(1,1)$ $T_k^+(g)$; where $k=\frac{1}{2}(\alpha+\frac{1}{2})$.

Using the standard method, namely, applying the lhs and rhs of (18.3.4) to the wave function ψ_n, gives

$$B_2^+\psi_n=-2\sqrt{(n+1)(n+\alpha+\tfrac{1}{2})}\,\psi_{n+1} \tag{18.3.13}$$

$$B_2\psi_n=-2\sqrt{n(n+\alpha+\tfrac{1}{2})}\,\psi_{n-1}. \tag{18.3.13'}$$

The phase factor (-1) is due to a special choice of the phase factor in the wave functions $\psi_n(x)$.

Applying the operator $(B_2^+)^n$ to the wave function of the ground state ψ_0 one has

$$\psi_n(x)=\frac{(-1)^n}{2^n}\frac{\sqrt{\Gamma(\alpha+\frac{1}{2})}}{\sqrt{n!\,\Gamma(n+\alpha+\frac{1}{2})}}(B_2^+)^n\psi_0(x). \tag{18.3.14}$$

The wave function $\psi_n(x)$ contains the factor $x^\alpha \exp(-x^2/2)$. Separating the latter we introduce new operators

$$\begin{aligned} A_2^+ &= x^{-\alpha}\exp\left(\frac{x^2}{2}\right)B_2^+\left[x^\alpha \exp\left(-\frac{x^2}{2}\right)\right] \\ &= \frac{1}{2}\left[\frac{d^2}{dx^2}+2\left(\frac{\alpha}{x}-2x\right)\frac{d}{dx}+4\left(x^2-\alpha-\frac{1}{2}\right)\right] \end{aligned} \tag{18.3.15}$$

$$A_2 = x^{-\alpha}\exp\left(\frac{x^2}{2}\right)B_2\left[x^\alpha \exp\left(-\frac{x^2}{2}\right)\right]=\frac{1}{2}\left(\frac{d^2}{dx^2}+2\frac{\alpha}{x}\frac{d}{dx}\right) \tag{18.3.16}$$

$$\begin{aligned} \tilde{H} &= x^{-\alpha}\exp\left(\frac{x^2}{2}\right)H\left[x^\alpha \exp\left(-\frac{x^2}{2}\right)\right] \\ &= \frac{1}{2}\left[-\frac{d^2}{dx^2}-2\left(\frac{\alpha}{x}-x\right)\frac{d}{dx}+2\alpha+1\right]. \end{aligned} \tag{18.3.17}$$

Evidently, the operators A_2, A_2^+ and H are also generators of $SU(1,1)$, and the commutation relations are the same as in (18.3.3, 4).

The equation $H\psi_n = E_n\psi_n$ is reduced to the usual equation for the Laguerre polynomials (defined as in [176])

$$\left[\frac{d^2}{dx^2}+2\left(\frac{\alpha}{x}-x\right)\frac{d}{dx}+4n\right]L_n^{\alpha-\frac{1}{2}}(x^2)=0. \tag{18.3.18}$$

From (18.3.13, 13′) we obtain the recursive relations for the Laguerre polynomials

$$A_2^+ L_n^{\alpha-\frac{1}{2}}(x^2) = -2(n+1)\,L_{n+1}^{\alpha-\frac{1}{2}}(x^2) \tag{18.3.19}$$

$$A_2 L_n^{\alpha-\frac{1}{2}}(x^2) = -2\left(n+\alpha-\tfrac{1}{2}\right)L_{n-1}^{\alpha-\frac{1}{2}}(x^2). \tag{18.3.20}$$

Hence we get an expression for the normalized eigenfunctions of the Hamiltonian

$$\psi_n(x)=\sqrt{\frac{2(n!)}{\Gamma(n+\alpha+\frac{1}{2})}}\,L_n^{\alpha-\frac{1}{2}}(x^2)x^\alpha \exp\left(-\frac{x^2}{2}\right). \tag{18.3.21}$$

Note that as $g\to 0$, $\alpha\to 1$ and the wave functions are reduced to the oscillator wave functions. However, only the oscillator functions vanishing at $r=0$ are obtained in this case. So, for $g\to 0$, $E_n\to[2n+(3/2)]$ and the energy spectrum is different from the oscillator energy spectrum on the whole axis, being the energy spectrum in the Hilbert space $L^2(0, \infty)$ with the boundary condition $\psi(0)=0$, i.e., for the oscillator problem on the halfaxis. At $g \neq 0$, the spectrum is that of the oscillator problem on the halfaxis, shifted by $\Delta E = \alpha - 1$.

The wave functions $\psi_n(x)$ form a basis $\{|n\rangle\}$ in the Hilbert space $\mathscr{H}$ of the square-integrable functions on the halfaxis $0 < x < \infty$. The CS system $\{|\zeta\rangle\}$, $|\zeta| < 1$ in this space is (cf. Chap. 5)

$$|\zeta\rangle = (1-|\zeta|^2)^k \sum_n \sqrt{\frac{\Gamma(n+2k)}{\Gamma(n+1)\Gamma(2k)}}\, \zeta^n |n\rangle. \tag{18.3.22}$$

The CS $|\zeta\rangle$ can be defined as a state annihilated by the operator

$$\tilde{K}_- = \exp(\zeta K_+) K_- \exp(-\zeta K_+) = K_- - 2\zeta K_0 + \zeta^2 K_+ \tag{18.3.23}$$

$$\tilde{K}_- |\zeta\rangle = 0. \tag{18.3.24}$$

Hence

$$\langle x|\zeta\rangle = \psi_\zeta(x) = \frac{\sqrt{2}}{\Gamma(2k)} \frac{(1-|\zeta|^2)^k}{(1-\zeta)^{2k}}\, x^\alpha \exp\left(-\frac{1}{2}\frac{1+\zeta}{1-\zeta}\, x^2\right),$$

$$2k = \alpha + \tfrac{1}{2}. \tag{18.3.25}$$

Comparing this expression with that in (18.3.22) gives the generating function for the Laguerre polynomials

$$(1-\zeta)^{-(\alpha+1)} \exp\left(\frac{x\zeta}{\zeta - 1}\right) = \sum_{n=0}^{\infty} L_n^\alpha(x)\zeta^n. \tag{18.3.26}$$

18.3.2 The Nonstationary Case

Suppose the quantum singular oscillator has a variable frequency. The Schrödinger equation for the system is

$$\mathrm{i}\frac{d}{dt}|\psi(t)\rangle = H(t)|\psi(t), \quad \text{where} \tag{18.3.27}$$

$$H(t) = \frac{1}{2}[p^2 + \omega^2(t)x^2] + \frac{g^2}{x^2}, \quad \omega(0) \equiv \omega_0 = 1. \tag{18.3.28}$$

Equation (18.3.27), just as in Sect. 18.2, can be rewritten as

$$\mathrm{i}\frac{d}{dt}|\psi(t)\rangle = (\boldsymbol{\Omega K})|\psi(t)\rangle, \quad \text{where} \tag{18.3.29}$$

$$(\boldsymbol{\Omega K}) = \Omega_0 K_0 - \Omega_1 K_1 - \Omega_2 K_2 \, ,$$

$$\Omega_{0,1} = \omega_0 \left[\left(\frac{\omega(t)}{\omega_0} \right)^2 \pm 1 \right], \qquad \Omega_2 = 0, \tag{18.3.30}$$

and the operators K_0, K_1 and K_2 are given in (18.3.10, 2).

Thus, all the results obtained in Sect. 18.2 are still valid for this case. In particular, the Schrödinger equation has the following solutions

$$|\psi(t)\rangle = \exp\left[-\mathrm{i}\varphi(t)\right]|\zeta(t)\rangle, \qquad |\zeta| < 1, \tag{18.3.31}$$

where $\zeta(t)$ and $\varphi(t)$ satisfy the classical equations (18.2.8, 9).

If the frequency tends to its limits rapidly enough, as $t \to \pm\infty$, then the transition probability from state $|n, \omega_-\rangle$ to state $|m, \omega_+\rangle$ is

$$W_{mn} = |\langle m, \omega_+ | \hat{S} | n, \omega_- \rangle|^2, \qquad \text{where} \tag{18.3.32}$$

$$\hat{S} = T^k(g_0), \quad g_0 = \begin{pmatrix} \cosh \frac{\tau}{2} & \sinh \frac{\tau}{2} \\ \sinh \frac{\tau}{2} & \cosh \frac{\tau}{2} \end{pmatrix}, \quad \tanh^2 \frac{\tau}{2} = \varrho. \tag{18.3.33}$$

Here ϱ is the reflection coefficient for the potential barrier [116]

$$k^2(x) = 2\left[E - U(x)\right] = \omega^2(x).$$

Using the explicit formula [83] for the matrix elements of the representation $T^k(g)$, we obtain the final expression:

$$W_{mn} = \frac{m!}{[(m-n)!]^2 n!} \frac{\Gamma(\frac{1}{2} - \alpha - n)}{\Gamma(\frac{1}{2} - \alpha - m)} \left(\sinh \frac{\tau}{2} \right)^{2(m-n)} \left(\cosh \frac{\tau}{2} \right)^{2(m+n+\alpha)+1}$$
$$\times \left| F\left(m + \alpha + \frac{1}{2}, \, m+1, \, m+1-n; \, -\sinh^2 \frac{\tau}{2} \right) \right|^2 . \tag{18.3.34}$$

Here $F(a, b; c; x)$ is the hypergeometric function.

For a periodical time dependence of $\omega(t)$, the quasi-energy spectrum has the same structure as given in Sect. 18.2.

18.3.3 The Case of N Interacting Particles

Consider the system of N interacting particles [177] described by the Hamiltonian

$$\begin{aligned} H &= H_0 + V \\ H_0 &= -\frac{1}{2}\sum_{j=1}^{N}\frac{\partial^2}{\partial x_j^2} + \frac{1}{2N}\sum_{j<k}(x_j - x_k)^2, \\ V &= g^2\sum_{j<k}(x_j - x_k)^{-2}. \end{aligned} \tag{18.3.35}$$

Since we are dealing with translation-invariant solutions of the Schrödinger equation $H\psi = E\psi$, it is suitable to introduce translation-invariant variables

$$\xi_j = x_j - X, \quad X = \frac{1}{N}\sum_{j=1}^{N} x_j, \quad j = 1, 2, \ldots, N. \tag{18.3.36}$$

The variables ξ_j are not independent,

$$\sum_{j=1}^{N}\xi_j = 0. \tag{18.3.37}$$

Let us also introduce translation-invariant differential operators, called formal ξ_j derivatives

$$\frac{\partial}{\partial\xi_j} = \frac{\partial}{\partial x_j} - \frac{\partial}{\partial X}, \quad \frac{\partial}{\partial X} = \frac{1}{N}\sum_{k=1}^{N}\frac{\partial}{\partial x_k}, \quad \sum_j \frac{\partial}{\partial\xi_j} = 0. \tag{18.3.38}$$

From (18.3.36, 38) we find

$$\frac{\partial}{\partial\xi_j}\xi_k = \delta_{jk} - \frac{1}{N}. \tag{18.3.39}$$

One readily gets

$$r^2 = \frac{1}{N}\sum_{j<k}(x_j - x_k)^2 = \sum_{j=1}^{N}\xi_j^2. \tag{18.3.40}$$

Taking account of this identity, the Hamiltonian (18.3.35) is

$$H = -\frac{N}{2}\left(\frac{\partial}{\partial X}\right)^2 - \frac{1}{2}\sum_{j=1}^{N}\frac{\partial^2}{\partial\xi_j^2} + \frac{1}{2}\sum_{j=1}^{N}\xi_j^2 + \frac{g}{2}\sum_{j\neq k}(\xi_j - \xi_k)^{-2}. \tag{18.3.41}$$

The first term in (18.3.41) is inessential for translation-invariant states so we drop it. Let us introduce translation-invariant operators b_j^+ and b_k proportional to creation and annihilation operators

$$b_j^+=\left(\xi_j-\frac{\partial}{\partial \xi_j}\right),\quad b_j=\left(\xi_j+\frac{\partial}{\partial \xi_j}\right),\quad \sum_{j=1}^N b_j^+=\sum_{j=1}^N b_j=0. \tag{18.3.42}$$

Here

$$[b_j, b_k]=[b_j^+, b_k^+]=0,\quad [b_j, b_k^+]=2\left(\delta_{ik}-\frac{1}{N}\right). \tag{18.3.43}$$

The Hamiltonian (18.3.35) can be expressed in terms of b_j and b_k^+

$$H=H_0+V$$
$$H_0=\frac{1}{2}\left(\sum_{j=1}^N b_j^+ b_j+N-1\right),\quad V=\frac{g^2}{2}\sum_{j\neq k}(\xi_j-\xi_k)^2. \tag{18.3.44}$$

In analogy to the one-particle case, we introduce the operators B_2 and B_2^+

$$B_2^+=\frac{1}{2}\sum_{j=1}^N (b_j^+)^2-V,\quad B_2=\frac{1}{2}\sum_{j=1}^N b_j^2-V. \tag{18.3.45}$$

It can be easily verified that the commutation relations are the same,

$$[H, B_2^+]=2B_2^+,\quad [H, B_2]=-2B_2,\quad [B_2, B_2^+]=4H. \tag{18.3.46}$$

Hence B_2 and B_2^+ are the lowering and raising operators, respectively.

Let us show that (18.3.46) hold when the potential $V(x)$ is an arbitrary homogeneous function of degree (-2). The third relation in (18.3.46) stems directly from (18.3.45). To provc the second relation it is suitable to make a transformation:

$$\tilde{H}=\exp\left(\frac{r^2}{2}\right)H\exp\left(-\frac{r^2}{2}\right),\quad \tilde{B}_2=\exp\left(\frac{r^2}{2}\right)B_2\exp\left(-\frac{r^2}{2}\right)$$

under which $b_j\to\frac{\partial}{\partial\xi_j}$, $b_2^+\to 2\xi_j-\frac{\partial}{\partial\xi_j}$, while the first relation can be proved with

$$\tilde{\tilde{H}}=\exp\left(-\frac{r^2}{2}\right)H\exp\left(\frac{r^2}{2}\right),\quad \tilde{\tilde{B}}_2^+=\exp\left(-\frac{r^2}{2}\right)B_2^+\exp\left(\frac{r^2}{2}\right)$$

where $b_j^+\to-\frac{\partial}{\partial\xi_j}$, $b_k\to 2\xi_k+\frac{\partial}{\partial\xi_k}$.

Equation (18.3.46) shows that the $SU(1,1)$ Lie algebra arises even in this more general case. The Casimir operator C_2, as given in (18.3.11), is again the identity times a number,

$$C_2=k(1-k),\quad k=\frac{E_0}{2}, \tag{18.3.47}$$

where E_0 is the ground-state energy. Hence we obtain the set of normalized wave functions

$$\psi_n = \frac{(-1)^n}{2^n} \frac{\sqrt{\Gamma(2k)}}{\sqrt{n!\,\Gamma(n+2k)}} (B_2^+)^n \psi_0 , \tag{18.3.48}$$

which are transformed according to the irreducible representation T_k^+ of the group $SU(1,1)$. The ground-state wave function is $\psi_0 = Z \exp(-r^2/2)$, so $B_2\psi_0 = 0$ and we get an equation for Z:

$$\frac{1}{2} \sum_{j=1}^{N} \frac{\partial^2 Z}{\partial \xi_j^2} - VZ = 0. \tag{18.3.49}$$

Clearly, Z is a homogeneous function of ξ_j. Let its homogeneity degree be α', then $H\psi_0 = E\psi_0$ leads to

$$E_0 = \frac{N-1}{2} + \alpha', \qquad k = \frac{E_0}{2} = \frac{N-1}{4} + \frac{\alpha'}{2}. \tag{18.3.50}$$

A specific feature of this set of wave functions is that all functions are

$$\psi_n(\xi) = \varphi_n(r)\psi_0(\xi) = Z \exp\left(-\frac{r^2}{2}\right)\varphi_n(r). \tag{18.3.51}$$

To verify this fact, we employ new operators A_2, A_2^+ and $\tilde{H}$ which act on functions $\varphi_n(r)$ (cf. (18.3.15–17).

$$A_2^+ = Z^{-1} \exp\left(\frac{r^2}{2}\right) B_2^+ Z \exp\left(-\frac{r^2}{2}\right) = \frac{1}{2} \sum_j \frac{\partial^2}{\partial \xi_j^2} - \sum_j \frac{\partial}{\partial \xi_j} \frac{1}{Z} \frac{\partial Z}{\partial \xi_j} \tag{18.3.52}$$

$$A_2 = Z^{-1} \exp\left(\frac{r^2}{2}\right) B_2 Z \exp\left(-\frac{r^2}{2}\right) = \frac{1}{2} \sum_j \frac{\partial^2}{\partial \xi_j^2} + \sum_j \frac{1}{Z} \frac{\partial Z}{\partial \xi_j} \frac{\partial}{\partial \xi_j} \tag{18.3.53}$$

$$\tilde{H} = Z^{-1} \exp\left(\frac{r^2}{2}\right) H \exp\left(-\frac{r^2}{2}\right)$$

$$= -\frac{1}{2} \sum_j \frac{\partial^2}{\partial \xi_j^2} + \sum_j \left(\xi_j - \frac{1}{2} \frac{\partial Z}{\partial \xi_j}\right) \frac{\partial}{\partial \xi_j} + E_0 . \tag{18.3.54}$$

The action of these operators on the functions depending on r only is equivalent to that of the corresponding operators in (18.3.15–17) with the replacement $\alpha \to \alpha' + (N-2)/2$, $x \to r$. Thus, the normalized wave functions are

$$\psi_n(\xi) = \mathrm{N}_0 \sqrt{\frac{n!\,\Gamma\left(\frac{N-1}{2} + \alpha'\right)}{\Gamma\left(n + \frac{N-1}{2} + \alpha'\right)}}\, Z \exp\left(-\frac{r^2}{2}\right) L_n^{\alpha' + \frac{N-3}{2}}(r^2). \tag{18.3.55}$$

If V is an arbitrary homogeneous function of degree (-2), the function Z and its homogeneity degree are not known. For the potential (18.3.35), $Z=D^{\alpha}$, $D=\prod_{j<k}(\xi_j-\xi_k)$, $\alpha'=\frac{N(N-1)}{2}\alpha$, and the functions (18.3.55) are reduced to those obtained in [177]. The operators A_2^+ and A_2, acting on $\varphi_n(r^2)$, give recursive relations (18.3.19, 20) for the Laguerre polynomials.

We have obtained the simplest series of wave functions for the many-body problem. Note that the arbitrary wave function has the form

$$\psi_{n\mu}(\xi)=Z\exp\left(-\frac{r^2}{2}\right)\varphi_{n\mu}(r)P_{\mu}(\xi), \tag{18.3.56}$$

where $P_{\mu}(\xi)$ is a homogeneous function of degree μ.

The equations for $\varphi_{n\mu}(r)$ and $P_{\mu}(\xi)$ can be easily obtained from the Schrödinger equation

$$H\psi_{n\mu}=E_{n\mu}\psi_{n\mu}.$$

Namely,

$$-\frac{1}{2}\frac{d^2\varphi_{n\mu}}{dr^2}-\frac{\frac{N-2}{2}+\alpha'+\mu}{r}-r\frac{d\varphi_{n\mu}}{dr}+(E_0+\mu)\varphi_{n\mu}$$

$$=E_{n\mu}\varphi_{n\mu}, \quad \text{hence} \tag{18.3.57}$$

$$\varphi_{n\mu}(r)=N_{\mu}\sqrt{\frac{n!\,\Gamma\left(\frac{N-1}{2}+\alpha'+\mu\right)}{\Gamma\left(n+\frac{N-1}{2}+\alpha'+\mu\right)}}\,L_n^{\frac{1}{2}(N-3)+\alpha'+\mu}(r^2), \tag{18.3.58}$$

$$\mathrm{E}_{n\mu}=2n+\mu_0+E_0. \tag{18.3.59}$$

The equation for $P_{\mu}(\xi)$ is

$$\left[\frac{1}{2}\sum_{j=1}^{N}\frac{\partial^2}{\partial\xi_j^2}+\frac{1}{Z}\sum_{j=1}^{N}\frac{\partial Z}{\partial\xi_j}\frac{\partial}{\partial\xi_j}\right]P_{\mu}(\xi)=0. \tag{18.3.60}$$

When operators A_2^+ and A_2, (18.3.52, 53), are applied to the function $\varphi_{n\mu}(r)P_{\mu}(\xi)$, they produce recursive relations for $\varphi_{n\mu}(r)$. There are no simple recursive relations for $P_{\mu}(\xi)$.

Thus, every new solution of (18.3.60) determines a new series of solutions

$$\psi_{n\mu}(\xi)=Z\exp\left(-\frac{r^2}{2}\right)\varphi_{n\mu}(r)P_{\mu}(\xi).$$

For potential (18.3.44), $Z=D^{\alpha}$ and (18.3.60) is reduced to the equation obtained in [178]

$$\sum_j \frac{\partial^2 P_\mu}{\partial \xi_j^2}+2\alpha \sum_{j<k} \frac{1}{\xi_j-\xi_k}\left(\frac{\partial}{\partial \xi_j}-\frac{\partial}{\partial \xi_k}\right)P_\mu=0. \tag{18.3.61}$$

It was shown in [178] that (18.3.61) has polynomial solutions, i.e., $\mu=m$ is an integer, and $P_m(\xi)$ is completely symmetrical under coordinate permutation. As seen from (18.3.59), the equidistant spectrum is a result of the fact that the solution is a polynomial. The symmetry of $P_m(\xi)$ indicates that the number of the solution $g(m)$ of degree m is equal to the number of completely symmetrical harmonical polynomials of degree m, which, in turn, is given by the number of solutions of

$$m=3n_3+\ldots+Nn_N\,,$$

where $n_3,\ldots,n_N$ are nonnegative integers.

Hence we obtain an expression for the generating function of $g(m)$,

$$G(t)=\sum_{m=0}^{\infty} g(m)t^m=\frac{1}{(1-t^3)\ldots(1-t^N)}. \tag{18.3.62}$$

Now one can easily show that the degeneracy $f(s)$ for an energy $E_s=E_0+s$ is equal to the number of solutions of the equation (for nonnegative integers)

$$s=2n_2+3n_3+\ldots+Nn_N\,.$$

The generating function for $f(s)$ has the form

$$F(t)=\sum_{m=0}^{\infty} f(m)t^m=\frac{1}{(1-t^2)\ldots(1-t^N)}. \tag{18.3.63}$$

The explicit expressions for $f(s)$ and $g(s)$ at small N are given in [177].

The N-particle problem reduces now to finding all solutions of (18.3.61). The solution of this problem is discussed in [177].

To conclude this section, note that using the invariance of the system of N interacting particles with respect to the $SU(1,1)$ group makes it possible to solve also the problem with a variable frequency, where Hamiltonian H_0 in (18.3.35) is

$$H_0=-\frac{1}{2}\sum_{j=1}^{N}\frac{\partial^2}{\partial x_j^2}+\frac{\omega^2(t)}{2N}\sum_{j<k}(x_j-x_k)^2. \tag{18.3.64}$$

As can be easily verified, the results of Sect. 18.2 can be extended to this case as well.

18.4 Oscillator with Variable Frequency Acted on by an External Force

Let us consider an oscillator with variable frequency $\omega(t)$ acted on by a force $f(t)$. The Schrödinger equation is

$$i\frac{d}{dt}|\psi(t)\rangle = \left[-\frac{1}{2}\frac{\partial^2}{\partial x^2}+\frac{1}{2}\omega^2(t)x^2 - f(t)x\right]\psi(x,t). \tag{18.4.1}$$

The time dependence of $\omega(t)$ and $f(t)$ are arbitrary with the natural boundary conditions,

$$f(t)\to 0 \quad \text{at} \quad t\to\pm\infty \tag{18.4.2}$$

$$\omega(t)\to\begin{cases}\omega_- & \text{at} \quad t\to-\infty\\ \omega_+ & \text{at} \quad t\to+\infty\end{cases}; \tag{18.4.2$'$}$$

the limits $\omega_\pm$ can differ.

Here the Hamiltonian $H(t)$ in (18.4.1) is a linear combination of the operators K_0, K_1, K_2 given in (18.2.5) and of the operators p, x and $\hat{I}$. These six operators form a basis for a six-parametrical Lie algebra, which is associated with the so-called inhomogeneous symplectic group I Sp(2, ℝ). Thus, just as in the preceding case, a solution of (18.4.1) exists which is CS for the group I Sp(2, ℝ),

$$|\psi(t)\rangle = \exp[-i\varphi(t)]|\zeta(t),\alpha(t)\rangle, \tag{18.4.3}$$

where CS $|\zeta,\alpha\rangle$ is given by

$$|\zeta,\alpha\rangle = N\exp\left[\tfrac{1}{2}\zeta a^{+2}+\alpha a^+\right]|0\rangle, \quad |\zeta|<1, \quad \text{and} \tag{18.4.4}$$
$$N=(1-|\zeta|^2)^{-1/2}\exp\left\{-\tfrac{1}{2}|\alpha|^2\right\}$$

is the normalization factor.

Let us find the transition probability W_{mn} from the state $|n,\omega_-\rangle$ at $t\to-\infty$ into the state $|m,\omega_+\rangle$ at $t\to+\infty$

$$W_{mn} = |\langle m,\omega_+|U(\infty,-\infty)|n,\omega_-\rangle|^2, \tag{18.4.5}$$

where $U(t_2,t_1)$ is the evolution operator. The generating function for W_{mn} can be easily obtained using CS:

$$H(z_1,z_2)=\sum\frac{z_1^m}{\sqrt{m!}}\langle m,\omega_+|S|n,\omega_-\rangle\frac{z_2^n}{\sqrt{n!}}. \tag{18.4.6}$$

Omitting details, the final result [173] is

$$H(z_1,z_2)=(1-\varrho)^{1/4}\exp\left\{-\frac{v}{2}(1-\sqrt{\varrho}\cos 2\varphi)\right.$$
$$+\tfrac{1}{2}[\sqrt{\varrho}(z_2^2-z_1^2)+2\sqrt{1-\varrho}\,z_1z_2]$$
$$\left.+\sqrt{v}[\sqrt{1-\varrho}\,\mathrm{e}^{-\mathrm{i}\varphi}z_1-(\mathrm{e}^{\mathrm{i}\varphi}-\sqrt{\varrho}\,\mathrm{e}^{-\mathrm{i}\varphi})z_2]\right\}. \tag{18.4.7}$$

The parameters ϱ, v and φ involved here are determined as follows.

Let $\xi(t)$ be the solution of the classical equation of motion for the oscillator with variable frequency

$$\ddot{\xi}+\omega^2(t)\xi=0, \tag{18.4.8}$$

where the initial condition is

$$\xi(t)\sim \mathrm{e}^{\mathrm{i}\omega_- t} \quad \text{at} \quad t\to-\infty. \tag{18.4.9}$$

The parameter ϱ is determined by the $t\to+\infty$ asymptotics of $\xi(t)$

$$\varrho=\left|\frac{c_2}{c_1}\right|^2, \quad \xi(t)\sim c_1\mathrm{e}^{\mathrm{i}\omega_+ t}-c_2\mathrm{e}^{-\mathrm{i}\omega_+ t}, \quad t\to+\infty. \tag{18.4.10}$$

Here $0\le\varrho<1$ and ϱ is determined completely by the form of $\omega(t)$ and is independent of the external force.

The parameter $v=|d|^2$ represents the excitation induced by the external force and is given by

$$d=\frac{\mathrm{i}}{\sqrt{2\omega_-}}\int_{-\infty}^{\infty}\xi(t')f(t')dt'. \tag{18.4.11}$$

The phase shift φ has the following form

$$\varphi=\frac{\delta_1+\delta_2}{2}-\beta, \tag{18.4.12}$$

where δ_1, δ_2 and β are the phases of complex numbers c_1, c_2 and d, respectively.

Calculation of the probabilities W_{mn} is reduced now to expanding the function $H(z_1,z_2)$ in Taylor series; the coefficients can be expressed in terms of the Hermite polynomials of two variables. According to [176], these polynomials are

$$\exp\left[\sum_{i,j=1}^{2}a_{ij}\left(y_iz_j-\frac{1}{2}z_iz_j\right)\right]=\sum_{n_1,n_2}\frac{z_1^{n_1}z_2^{n_2}}{\sqrt{n_1!n_2!}}H_{n_1n_2}(y_1,y_2). \tag{18.4.13}$$

In our case

$$[a_{ij}]=\begin{pmatrix} \sqrt{\varrho} & -\sqrt{1-\varrho} \\ -\sqrt{1-\varrho} & -\sqrt{\varrho} \end{pmatrix}, \qquad \begin{aligned} y_1 &= \sqrt{v(1-\varrho)}\,e^{i\varphi} \\ y_2 &= -\sqrt{v}\,(e^{-i\varphi}-\sqrt{\varrho}\,e^{i\varphi}) \end{aligned} \tag{18.4.14}$$

(note that $a^2=I$, $a^{-1}=a$). Finally, the expression for the transition probability is

$$W_{mn}=\frac{\sqrt{1-\varrho}}{m!n!}\,|H_{mn}(y_1,y_2)|^2 \exp\left[-v(1-\sqrt{\varrho}\cos 2\varphi)\right]. \tag{18.4.15}$$

For the simplest values of m and n, W_{mn} are given here explicitly:

$$\begin{aligned} &W_{00}=\sqrt{1-\varrho}\exp\left[-v(1-\sqrt{\varrho}\cos 2\varphi)\right], \\ &W_{10}=v(1-\varrho)\,W_{00}, \quad W_{01}=v(1-2\sqrt{\varrho}\cos 2\varphi+v^2\varrho)\,W_{00}, \\ &W_{11}=(1-\varrho)\left[(1-v)^2+2v(1-v)\sqrt{\varrho}\cos 2\varphi+v^2\varrho\right]W_{00}. \end{aligned} \tag{18.4.16}$$

If one of the variables z_1, z_2 is set at zero in (18.4.13), the generating function is reduced to that for standard Hermite polynomials. Hence, if at $t\to-\infty$ the oscillator was not excited ($n=0$), the parameters W_{m0} are expressed in terms of the Hermite polynomials,

$$W_{m0}=\frac{\sqrt{1-\varrho}}{2^m m!}\,\varrho^{m/2}\left|H_m\left(\frac{\sqrt{2v(1-\varrho)}\,e^{-i\varphi}}{\sqrt[4]{\varrho}}\right)\right|^2 \exp\left[-v(1-\sqrt{\varrho}\cos 2\varphi)\right]. \tag{18.4.17}$$

A similar formula holds for W_{0n}:

$$W_{0n}=\frac{\sqrt{1-\varrho}}{2^n n!}\,\varrho^{n/2}\left|H_n\left(\frac{\sqrt{2v}\,(\sqrt{\varrho}\,e^{-i\varphi}-e^{i\varphi})}{\sqrt[4]{\varrho}}\right)\right|^2 \times \exp\left[-v(1-\sqrt{\varrho}\cos 2\varphi)\right]. \tag{18.4.18}$$

It is seen from (18.4.17, 18) that $W_{0n}\neq W_{n0}$, in general, while for vanishing external force $v=0$, and the probabilities W_{mn} are symmetrical under $m\leftrightarrow n$.

General principles of quantum mechanics state that the probabilities W_{mn} are symmetrical only if $\omega(-t)=\omega(t)$. As is seen from (18.2.11), in fact, the symmetry $W_{mn}=W_{nm}$ arises for an arbitrary time dependence of $\omega(t)$. The reason for this additional symmetry can be understood by relating ϱ to the reflection coefficient from a one-dimensional barrier. To this end, we construct a linear combination of $\xi(t)$ and $\bar{\xi}(t)$ with the following properties:

$$\xi_1(t)=\xi(t)+R\bar{\xi}(t)=\begin{cases} e^{i\omega_- t}+Re^{-i\omega_- t} & \text{at} \quad t\to-\infty \\ De^{i\omega_+ t} & \text{at} \quad t\to+\infty \end{cases} \tag{18.4.19}$$

Hence it is seen that $\xi_1(t)$ coincides with the wave function of the one-dimensional Schrödinger equation if t is replaced by x, and $\omega(t)$ by $k(x)$, while coefficients R and D stand for the amplitudes of reflected and penetrating waves. Comparing (18.4.19 and 3.10), then

$$c_1 = \frac{D}{1-|R|^2}, \quad c_2 = \frac{R\bar{D}}{1-|R|^2}, \quad \varrho = |R|^2. \tag{18.4.20}$$

Thus, the solution (18.4.19) corresponds to the incident wave falling to the barrier from the left. The time inversion $t \to -t$ corresponds to replacing it by the wave falling from the right. Denoting the reflection coefficients for the two waves by ϱ and ϱ',

$$W_{mn}(\varrho) = W_{nm}(\varrho').$$

Since $\varrho = \varrho'$, [116], the probabilities W_{mn} have an additional $m \leftrightarrow n$ symmetry. Clearly, from (18.4.15), W_{mn} are symmetrical also at $\varrho = 0$. If $v \neq 0$, and

$$\omega(-t) = \omega(t), \quad f(-t) = f(t), \tag{18.4.21}$$

such a symmetry exists if

$$\cos 2\varphi = \sqrt{\varrho}, \tag{18.4.22}$$

as shown in [174]. Now the variables ϱ and φ are no longer independent, and the probabilities W_{mn} are determined by two parameters ϱ and v.

19. Particles in External Electromagnetic Fields

Problems of particles in homogeneous variable electromagnetic fields are quite appropriate for the CS method. Spin precession in a variable magnetic field and the creation of particle pairs (bosons or fermions) in a homogeneous electric field are considered as examples. The quantum problems are reduced exactly to their classical analogues, as well as the oscillator problems.

19.1 Spin Motion in a Variable Magnetic Field

Consider a neutral particle with spin j and magnetic moment μ in a variable magnetic field $\boldsymbol{H}(t)$. The time evolution of states of such a system is determined by the Schrödinger equation

$$\mathrm{i}\frac{d}{dt}|\psi(t)\rangle = -\boldsymbol{A}(t)\boldsymbol{J}|\psi(t)\rangle = \mathrm{i}(aJ_+ - \bar{a}J_- - \mathrm{i}bJ_0)|\psi(t)\rangle, \tag{19.1.1}$$

where

$$\begin{aligned} &\boldsymbol{\mu} = \frac{\mu}{j}\boldsymbol{J}, \quad \boldsymbol{J} = (J_1, J_2, J_3), \quad \boldsymbol{A} = \frac{\mu}{j}\boldsymbol{H}, \\ &a = -\tfrac{1}{2}(A_1 - \mathrm{i}A_2), \quad b = -A_3, \end{aligned} \tag{19.1.2}$$

and J_j is the operator of infinitesimal rotation around the axis x_j, $J_\pm = J_1 \pm \mathrm{i}J_2$. For the vector $\boldsymbol{H}(t)$, assume only that it tends sufficiently rapidly to some limits at $t \to \pm\infty$ so that at $t \to \pm\infty$ the asymptotic states $|\psi^\pm\rangle$ exist.

It has long been known [179, 116] that the problem for a particle with an arbitrary spin can be reduced to that for spin-1/2 particles. Using the spin CS makes it possible to obtain the solution in the simplest way.

Substituting into (19.1.1) the wave function

$$|\psi(t)\rangle = \exp[-\mathrm{i}\varphi(t)]|\zeta(t)\rangle, \quad \text{we get} \tag{19.1.3}$$

$$\mathrm{i}\frac{d}{dt}|\zeta(t)\rangle = [H(t) - \dot{\varphi}]|\zeta(t)\rangle. \tag{19.1.4}$$

On the other hand, the explicit formula for CS shows that

$$\zeta J_+|\zeta\rangle=(j+J_0)|\zeta\rangle, \tag{19.1.5}$$

$$J_-|\zeta\rangle=\zeta(j-J_0)|\zeta\rangle, \quad \text{and} \tag{19.1.6}$$

$$\frac{d}{dt}|\zeta(t)\rangle=\left[\frac{-j}{1+|\zeta|^2}\frac{d}{dt}(1+|\zeta|^2)\right]|\zeta(t)\rangle+\left(\frac{\dot{\zeta}}{\zeta}\right)(J_0+j)|\zeta(t)\rangle. \tag{19.1.7}$$

Hence we find equations for $\zeta(t)$ and $\varphi(t)$:

$$\dot{\zeta}=a+\bar{a}\zeta^2-\mathrm{i}b\zeta, \tag{19.1.8}$$

$$\mathrm{i}\frac{1}{j}\dot{\varphi}=-\mathrm{i}\frac{\dot{\zeta}}{\zeta}+\mathrm{i}\left[\frac{1}{1+|\zeta|^2}\frac{d}{dt}(1+|\zeta|^2)\right]+\frac{a}{\zeta}+\bar{a}\zeta. \tag{19.1.9}$$

Note that an equality results from (19.1.8),

$$\frac{d}{dt}(1+|\zeta|^2)=(a\bar{\zeta}+\bar{a}\zeta)(1+|\zeta|^2). \tag{19.1.10}$$

It can be used to get

$$\mathrm{i}\dot{\varphi}=j(-\bar{\zeta}a+\zeta\bar{a}-\mathrm{i}b). \tag{19.1.11}$$

Thus, the quantum problem is reduced to a simpler problem, i.e., to solve (19.1.8, 11). Mapping the ζ plane to the unit sphere, we get from (19.1.8) the following equation for the unit vectors:

$$\boldsymbol{n}=-[\boldsymbol{a}(t),\boldsymbol{n}]. \tag{19.1.12}$$

Thus, the complex ζ plane (or the sphere S^2) takes the place of the pasc plane for the classical dynamical system. However, unlike the conventional classical dynamical systems, the invariant metrics on the plane ζ are not Euclidean.

If $a(t)\to 0$, $b(t)\to$const, as t goes to infinity, $|\zeta(t)|^2\to\varrho=$const because of (19.1.8). Hence one gets at once an expression for the transition probability from the initial state $|0\rangle=|j,-j\rangle$ into a final state $|m\rangle=|j,-j+m\rangle$

$$W_m=\frac{(2j)!}{m!(2j-m)!}\frac{\varrho^m}{(1+\varrho)^{2j}}. \tag{19.1.13}$$

The general formula for the transition probabilities is

$$W_{mn}=|\mathscr{D}^j_{\mu\nu}(\theta)|^2, \tag{19.1.14}$$

where $\mu=m-j$, $\nu=n-j$, $\varrho=\tan^2\frac{\theta}{2}$ and $\mathscr{D}^j_{\mu\nu}(\theta)$ are the known representation matrix elements.

As an example, consider the case of an external magnetic field

$$H(t)=H_0+H_1(t)=H_0e_3+H_1(e_1\cos\omega t+e_2\sin\omega t), \tag{19.1.15}$$

which is the sum of a constant magnetic field directed along the z axis and the field $H_1(t)$, rotating in the (x, y) plane with an angular velocity ω. Equation (19.1.8) can be easily solved, giving

$$\zeta(t)=\frac{\mathrm{i}\omega_\perp\sin\Omega t\exp[\mathrm{i}(\omega_\parallel-\omega)t]}{2\Omega\cos\Omega t-\mathrm{i}(\omega-\omega_\parallel)\sin\Omega t}\quad\text{where} \tag{19.1.16}$$

$$\Omega=\frac{1}{2}\sqrt{(\omega-\omega_\parallel)^2+\omega_\perp^2},\qquad\omega_\parallel=\left(\frac{\mu_0}{j}\right)H_0,\qquad\omega_\perp=\left(\frac{\mu_0}{j}\right)H_1. \tag{19.1.17}$$

In the simplest case $j=1/2$ we get an expression for the "spin-flip" probability

$$W_{\frac{1}{2},-\frac{1}{2}}(t)=\frac{\omega_\perp^2\sin^2\Omega t}{(\omega-\omega_\parallel)^2+\omega_\perp^2}. \tag{19.1.18}$$

Note that the spin-flip probability depends on ω resonantly and acquires the maximum value ($W=1$) only at $\omega=\omega_\parallel$.

19.2 Boson Pair Production in a Variable Homogeneous External Field

An interesting quantum effect in classical external field is pair production of particles and antiparticles from a vacuum. This phenomenon has been studied in a great number of theoretical works, references may be found in [180].

This section conside the group-theoretical aspects of the problem of boson pair production in a variable homogeneous electric field. This problem has been considered in [181]. In the simplest case of a scalar particle, the problem reduces to that of a quantum oscillator with variable frequency [181, 182]. The problem of a quantum oscillator with variable frequency has been already considered in Sect. 18.2.

19.2.1 Dynamical Symmetry for Scalar Particles

Let us first consider a scalar charged particle with mass m in a homogeneous electric field $E(t)$. It is described by the wave function $\varphi(r, t)$ which satisfies

$$\ddot{\varphi}+[p-A(t)]^2\varphi+m^2\varphi=0 \tag{19.2.1}$$

(hereafter we use the units $\hbar=c=e=1$), where $p=-\mathrm{i}\nabla$ is the momentum operator, $A(t)=-\int E(t')dt'$ is the vector potential and the point stands for the time derivative. It is seen immediately that a solution of the type $\varphi(\boldsymbol{r}, t)=\exp(\mathrm{i}\boldsymbol{pr})\varphi_0(t)$ exists and $\varphi_0(t)$ satisfies the equation for the one-dimensional classical oscillator with variable frequency

$$\ddot{\varphi}_0(t)+\omega^2(t)\varphi_0(t)=0. \tag{19.2.2}$$

Here $\omega^2(t)=m^2+\boldsymbol{\pi}^2(t)$, $\boldsymbol{\pi}(t)=\boldsymbol{p}+\int\limits_{-\infty}^{t}\boldsymbol{E}(t')dt'$ is the classical momentum of the particle at a time moment t, $\boldsymbol{p}$ is the particle momentum at $t\to-\infty$ [assuming that $\boldsymbol{E}(t)$ falls fast at $t\to\pm\infty$, so that $\boldsymbol{\pi}(t)$ and $\omega(t)$ have definite limits, $\boldsymbol{\pi}_\pm$ and $\omega_\pm$, respectively, as $t\to\pm\infty$].

After the second quantization, the function $\varphi(\boldsymbol{r}, t)$ becomes the Heisenberg operator

$$\hat{\varphi}(\boldsymbol{r}, t)=\frac{1}{(2\pi)^{3/2}}\int\frac{d\boldsymbol{p}}{\sqrt{2\omega_-(\bar{\boldsymbol{p}})}}[\hat{a}_{\boldsymbol{p}}(t)\,\mathrm{e}^{\mathrm{i}\boldsymbol{pr}}+\hat{b}^+_{-\boldsymbol{p}}(t)\mathrm{e}^{-\mathrm{i}\boldsymbol{pr}}] \tag{19.2.3}$$

which satisfies (19.2.1) while the operator $\hat{a}_{\boldsymbol{p}}(t)$ and $\hat{b}_{\boldsymbol{p}}(t)$ satisfy (19.2.2). At an arbitrary time, the operators $\hat{a}_{\boldsymbol{p}}(t)$ and $\hat{b}^+_{-\boldsymbol{p}}(t)$ are related to the operators $\hat{a}_{\boldsymbol{p}}$ and $\hat{b}^+_{-\boldsymbol{p}}$ corresponding to the free motion at $t\to-\infty$, by means of the unitary transformation

$$\begin{aligned}\hat{a}_{\boldsymbol{p}}(t)&=S^+(t)\hat{a}_{\boldsymbol{p}}S(t)=u'_{\boldsymbol{p}}(t)\hat{a}_{\boldsymbol{p}}+v'_{\boldsymbol{p}}(t)\hat{b}^+_{-\boldsymbol{p}}\\ \hat{b}^+_{-\boldsymbol{p}}(t)&=S^+(t)\hat{b}^+_{-\boldsymbol{p}}S(t)=\bar{v}'_{\boldsymbol{p}}(t)\hat{a}_{\boldsymbol{p}}+\bar{u}'_{\boldsymbol{p}}(t)\hat{b}^+_{-\boldsymbol{p}}.\end{aligned} \tag{19.2.4}$$

The functions $u'_{\boldsymbol{p}}(t)$ and $v'_{\boldsymbol{p}}(t)$ satisfy (19.2.2) and the boundary conditions: at $t\to-\infty$, $u'_{\boldsymbol{p}}(t)\sim\exp(-\mathrm{i}\omega_-t)$, $v'_{\boldsymbol{p}}(t)\to 0$. The transformation (19.2.4) is specified by the matrix

$$g'-\begin{pmatrix}u'_{\boldsymbol{p}} & v'_{\boldsymbol{p}}\\ \bar{v}'_{\boldsymbol{p}} & \bar{u}'_{\boldsymbol{p}}\end{pmatrix}.$$

Note that as the transformation $S(t)$ is unitary, the transformation (19.2.4) is canonical, i.e., it preserves the commutation relations. Hence it follows that $|u_{\boldsymbol{p}}|^2-|v_{\boldsymbol{p}}|^2=1$ and therefore the set of all transformations is the group $G=SU(1,1)$. Here, as seen from (19.2.4), each $g(t)$ corresponds to two operators $S(t)$ and $-S(t)$, and therefore operators $S(t)$ belong to a representation of a group $\tilde{G}$ which is the double covering of the group G.

One should bear in mind that if $\omega_+\neq\omega_-$ the system of stationary states for $t\to+\infty$ does not coincide with that for $t\to-\infty$. However, transition from states corresponding to frequency ω_- to states corresponding to frequency ω_+ is given by an operator R of the same type as S. The matrix element of transition from an initial state ψ_- $(t\to-\infty, \omega=\omega_-)$ to a final state ψ_+ $(t\to+\infty, \omega=\omega_+)$, is given by a matrix element of the operator $T=RS$. Therefore, without loss of generality, we assume in the following that $\omega_-=\omega_+$. Thus, the operator T

performs a transformation of the type (19.2.4) in which u' and v' should be substituted for

$$u=\exp[\mathrm{i}(\alpha_1+\alpha_2)/2]\cosh\frac{\tau}{2} \quad \text{and} \quad v=\exp[\mathrm{i}(\alpha_1-\alpha_2)/2]\sinh\frac{\tau}{2}$$

and the transition probability W depends on a single parameter $\tau=\lim_{t\to+\infty}\tau(t)$

$$W=|\langle\psi_+|T(g_0)|\psi_-\rangle|^2, \qquad g_0=\begin{pmatrix}\cosh\frac{\tau}{2} & \sinh\frac{\tau}{2}\\ \sinh\frac{\tau}{2} & \cosh\frac{\tau}{2}\end{pmatrix}. \tag{19.2.5}$$

Equality (19.2.4) also describes the distribution of pairs of charged scalar particles created in a gravitational field of homogeneous expanding universe. Since the theory of this process may be found elsewhere [183–185], we present here only an expression for the effective oscillator frequency

$$\omega^2(t)=m^2+k^2(t)-\frac{1}{4}\frac{\ddot{g}}{g}+\frac{3}{16}\left(\frac{\dot{g}}{g}\right)^2. \tag{19.2.6}$$

Here $-k^2(t)$ is the eigenvalue of the Laplace operator

$$\Delta=g^{\alpha\beta}\partial_\alpha\partial_\beta, \qquad \partial_\alpha=\frac{\partial}{\partial x^\alpha};$$

$g^{\alpha\beta}$ is the spatial part of the metric tensor in a synchronous coordinate system ($g^{00}=1$, $g^{0\alpha}=0$), $g=\det|g_{\alpha\beta}|$.

To study the group structure of the problem, consider infinitesimal operators of representation $T(g)$. They are

$$K_+=a^+b^+, \qquad K_-=ab, \qquad K_0=\tfrac{1}{2}(a^+a+b^+b+1) \tag{19.2.7}$$

and satisfy the standard commutation relations of the $SU(1,1)$ Lie algebra

$$[K_0,K_\pm]=\pm K_\pm, \qquad [K_-,K_+]=2K_0 \tag{19.2.8}$$

(in the following, for simplicity the subscripts of a_p and b^+_{-p} are omitted).

It can be easily checked that the operator

$$C_2=\frac{K_+K_-+K_-K_+}{2}-K_0^2 \tag{19.2.9}$$

is an invariant operator (Casimir's operator), commuting with the operators K_+, K_- and K_0. Substituting in (19.2.9) the explicit expressions for K_+, K_- and

K_0 results in

$$C_2 = \tfrac{1}{4} - \tfrac{1}{4}(a^+a - b^+b)^2. \tag{19.2.10}$$

Thus, for states

$$|m,n\rangle = \frac{(a^+)^m(b^+)^n}{\sqrt{m!n!}}|0,0\rangle,$$

for which $(m-n)$ is fixed,

$$C_2 = \mathrm{const} = k(1-k)$$

and, as follows from *Bargmann's* work [83], the states $\{|n+n_0, n\rangle\}$ are a basis for irreducible representation T^k of discrete series of group $SU(1,1)$.

The matrix elements of these representations are known [83], and in the simplest case $n_0=0$, $k=1/2$ where thc initial state is a vacuum

$$W_n = |\langle n,n|T^{1/2}|0,0\rangle|^2 = (1-\varrho)\varrho^n, \qquad \varrho = \tanh^2\frac{\tau}{2} = \frac{|v|^2}{|u|^2} \tag{19.2.11}$$

The parameter ϱ has simple physical meaning, namely, it coincides with the reflection coefficient from a potential barrier of the form $\omega^2(x)$, as was argued in Sect. 18.4.

Consider the states appearing during the system evolution. Let the state be a vacuum at the initial moment of time, i.e., $a|\psi(0)\rangle = b|\psi(0)\rangle = 0$. Then at an arbitrary moment of time $|\psi(t)\rangle = S(t)|\psi(0)\rangle$, and this state satisfies

$$(S(t)aS^+(t))|\psi(t)\rangle = 0, \qquad (S(t)bS^+(t))|\psi(t)\rangle - 0.$$

Thus, it is now appropriate to consider the transformation inverse to (19.2.4). It is easy to verify that it has the form

$$\begin{aligned}\tilde{a} &= SaS^+ = \bar{u}a - vb^+\\ \tilde{b} &= SbS^+ = -va^+ + \bar{u}b.\end{aligned} \tag{19.2.12}$$

Because of conditions $\tilde{a}|\psi(t)\rangle = 0$, $\tilde{b}|\psi(t)\rangle = 0$,

$$|\psi(t)\rangle = \exp[-\mathrm{i}\varphi(t)]|\zeta(t)\rangle, \qquad \text{where} \tag{19.2.13}$$

$$|\zeta\rangle = \sqrt{1-|\zeta|^2}\exp(\zeta b^+a^+)|0,0\rangle, \qquad \zeta = \frac{v}{\bar{u}}, \qquad |\zeta| < 1. \tag{19.2.14}$$

It can be easily seen that the states $|\zeta\rangle$ are generalized coherent states for the representation $T^{1/2}$ of the $SU(1,1)$ group.

19.2.2 The Multidimensional Case: Coherent States

Turning to the pair production of bosons with spin s, note that arguments like those of the preceding section lead to a canonical transformation for $N=2s+1$ degrees of freedom

$$\begin{aligned}\tilde{a}_i &= T^+ a_i T = A_{ij} a_j + B_{ij} b_j^+ \\ \tilde{b}_i^+ &= T^+ b_i^+ T = C_{ij} a_j + D_{ij} b_j^+ .\end{aligned} \tag{19.2.15}$$

The time dependences of the coefficients A_{ij}, B_{ij}, C_{ij} and D_{ij} are determined by equations for the Heisenberg field operators $\varphi_\mu(\boldsymbol{r}, t)$ and can be essentially different for different spins.

It can be easily seen that to make the transformation canonical, i.e., leaving the commutation relations intact, one should impose the following conditions:

$$AA^+ - BB^+ = \mathrm{I}, \qquad DD^+ - CC^+ = \mathrm{I}, \qquad AC^+ = BD^+ . \tag{19.2.16}$$

These conditions can be rewritten as $MEM^+ = E$, where

$$M = \begin{pmatrix} A & B \\ C & D \end{pmatrix}, \qquad E = \begin{pmatrix} \mathrm{I} & 0 \\ 0 & -\mathrm{I} \end{pmatrix}, \tag{19.2.17}$$

so the transformations (19.2.15) are elements of the group $G = SU(N, N)$. It follows from (19.2.17) that the inverse transformation matrix is

$$M^{-1} = EM^+E, \qquad M^{-1} = \begin{pmatrix} A^+ & -C^+ \\ -B^+ & D^+ \end{pmatrix} \tag{19.2.18}$$

and the conditions

$$A^+A - C^+C = \mathrm{I}, \qquad D^+D - B^+B = \mathrm{I}, \qquad A^+B = C^+D \tag{19.2.19}$$

must also be fulfilled.

The operators $T(g)$ represent the group $\tilde{G}$ which covers group G doubly. It is convenient to unite the infinitesimal operators of this representation in a matrix

$$N = \begin{pmatrix} a_i^+ a_j & a_i^+ b_j^+ \\ b_i a_j & b_i b_j^+ \end{pmatrix}. \tag{19.2.20}$$

It can be seen immediately that the operator

$$C_2 = \mathrm{tr}\,(NENE) \tag{19.2.21}$$

is an invariant operator (Casimir's operator), i.e., it commutes with all the infinitesimal operators. The eigenvalues of C_2 can be found easily with the commutation relations. The result is

$$C_2 = (a_i^+ a_i - b_i^+ b_i)^2 - (a_i^+ a_i - b_i^+ b_i) - (N^2 - N). \tag{19.2.22}$$

(summation in repeated indices is implied). Thus, for states $\{|m,n\rangle\}$, $|m,n\rangle = |m_1, \ldots, m_N, n_1, \ldots, n_N\rangle$ such that $m-n=\sum m_i - \sum n_j = \text{const}$, the operator C_2 is identity times a constant,

$$C_2 = [(m-n)^2 - (m-n) - (N^2 - N)]\hat{1}. \tag{19.2.23}$$

It can be shown that higher-order Casimir operators C_p are also constant in this case. Hence it follows that the set of states $\{|m,n\rangle\}$ with $m-n=\text{const}$ forms the basis of an irreducible representation of the $SU(N,N)$ group. The $G=SU(N,N)$ group contains the compact Cartan subgroup $K=SU(N)\otimes SU(N)\otimes U(1)$ and thus, according to *Harish-Chandra* [186], has a discrete series of representations. Specifically, the representation we deal with is of this type.

Let us consider in more detail the case where the initial state is the vacuum $|\psi(0)\rangle = |0,0\rangle$, $a_i|0,0\rangle = 0$, $b_j|0,0\rangle = 0$. For any time t the state $|\psi(t)\rangle$ is given by

$$|\psi(t)\rangle = T(g(t))|\psi(0)\rangle = T(g)|0,0\rangle \tag{19.2.24}$$

and satisfies

$$\tilde{a}_i|\psi(t)\rangle = 0, \quad \tilde{b}_j|\psi(t)\rangle = 0, \quad \text{where} \tag{19.2.25}$$

$$\tilde{a}_i = T a_i T^+, \quad \tilde{b}_j = T b_j T^+. \tag{19.2.26}$$

Calculating the operators $\tilde{a}_i$ and $\tilde{b}_j$ gives

$$\begin{aligned} \tilde{a}_i &= A^+_{ij} a_j - C^+_{ij} b^+_j \\ \tilde{b}_i &= -(B^{\mathrm{T}})_{ij} a^+_j + (D^{\mathrm{T}})_{ij} b_j . \end{aligned} \tag{19.2.27}$$

Equations (19.2.25) have the solution

$$|\psi(t)\rangle = \exp[-\mathrm{i}\varphi(t)]|\zeta(t)\rangle, \quad \langle\zeta|\zeta\rangle = 1, \quad \text{where} \tag{19.2.28}$$

$$|\zeta\rangle = N\exp(\zeta_{ij} \mathrm{a}^+_i \mathrm{b}^+_j)|0,0\rangle, \quad \zeta = (A^+)^{-1} C^+ = BD^{-1}, \tag{19.2.29}$$

$$N = [\det(1 - \zeta^+\zeta)]^{1/2}. \tag{19.2.30}$$

Matrix elements of the operator $T(g)$ are obtained from (19.2.29),

$$\langle m,n|T|0,0\rangle = N \sum_{\{n_{ij}\}} \prod_{i,j} \frac{\sqrt{m_i! n_j!}}{n_{ij}!} (\zeta_{ij})^{n_{ij}}. \tag{19.2.31}$$

The sum is over all integers n_{ij} satisfying the conditions

$$\sum_j n_{ij} = m_i = \text{const}, \quad \sum_i n_{ij} = n_j = \text{const}. \tag{19.2.32}$$

As in the preceding case, the states $\{|\zeta\rangle\}$ are generalized coherent states. Such a state is represented by an element of the coset space G/H, where H is the stationary subgroup of the vector $|\psi_0\rangle$. In our case $|\psi_0\rangle=|0,0\rangle$, $G=SU(N,N)$, $H=SU(N)\otimes SU(N)\otimes U(1)$, and the coset space G/H is realized by complex $N\times N$ matrices ζ under the condition that the matrix $1-\zeta^+\zeta$ is positively definite.

The set $\{|\zeta\rangle\}$ has all properties specific to CS. This is an overcomplete system (completeness of some subsystems of such systems have been considered previously [73]), and its states are nonorthogonal to each other. That is, the scalar products are

$$\langle\zeta|\zeta'\rangle=[\det(1-\zeta^+\zeta)]^{1/2}[\det(1-\zeta'^+\zeta')]^{1/2} \times[\det(1-\zeta^+\zeta')]^{-1}. \tag{19.2.33}$$

The following relations also hold:

$$a_k^+a_l|\zeta\rangle=\zeta_{lj}a_k^+b_j^+|\zeta\rangle \tag{19.2.34}$$

$$b_k^+b_l^+|\zeta\rangle=\zeta_{il}a_i^+b_k^+|\zeta\rangle \tag{19.2.35}$$

$$a_kb_l|\zeta\rangle=\zeta_{kl}|\zeta\rangle+\zeta_{il}\zeta_{kj}a_i^+b_j^+|\zeta\rangle. \tag{19.2.36}$$

The action of the operator $T(g)$ on the CS $|\zeta\rangle$ is given by

$$T(g)|\zeta\rangle=\exp(\mathrm{i}\varphi)|\zeta'\rangle, \quad \text{where} \tag{19.2.37}$$

$$\zeta'=(A\zeta-B)(-C\zeta+D)^{-1} \tag{19.2.38}$$

$$\varphi=2k\arg[\det(D-C\zeta)]. \tag{19.2.39}$$

The matrix element $\langle\eta|T|\xi\rangle$ is

$$\langle\eta|\mathrm{T}|\xi\rangle=[\det(1-\eta\eta^+)]^{1/2}[\det(1-\xi\xi^+)]^{1/2} \times[\det(D-C\xi+\eta^+B-\eta^+A\xi)]^{-1}. \tag{19.2.40}$$

Hence we get a generating function for the matrix elements in the canonical basis,

$$\sum\langle m,n|T(g)|m',n'\rangle\prod\bar{C}^{ij}_{mn}C^{i'j'}_{mn}\xi^{n'_{i'j'}}_{i'j'}\eta^{n_{ij}}_{ij} =[\det(D-C\xi+\eta^+B-\eta^+A\xi)]^{-1}. \tag{19.2.41}$$

Here C^{ij}_{mn} are coefficients in the expansion of CS $|\zeta\rangle$,

$$|\zeta\rangle=N\sum_{\{n_{ij}\}}\prod_{ij}C^{ij}_{mn}\zeta^{n_{ij}}_{ij}|m,n\rangle, \quad \sum_j n_{ij}=m_i, \quad \sum_i n_{ij}=n_j. \tag{19.2.42}$$

Now we are in a position to calculate the transition probabilities

$$W_n=\sum_{\{m,n\}}|\langle m,n|T|0,0\rangle|^2, \quad \sum m_i=\sum n_j=n. \tag{19.2.43}$$

Let us consider the operator

$$S = T^+ R T, \tag{19.2.44}$$

where T is the canonical transformation of (19.2.15), and R is defined by

$$R^+ a_i R = e^{-i\theta} a_i, \qquad R^+ b_i^+ R = e^{i\theta} b_i^+. \tag{19.2.45}$$

The operator S is also a canonical transformation:

$$\begin{aligned} S^+ a_i S &= \tilde{A}_{ij} a_j + \tilde{B}_{ij} b_j^+ \\ S^+ b_i^+ S &= \tilde{C}_{ij} a_j + \tilde{D}_{ij} b_j^+ . \end{aligned} \tag{19.2.46}$$

The calculations give

$$\begin{aligned} &\tilde{A} = e^{-i\theta} A^+ A - e^{i\theta} C^+ C, \qquad \tilde{B} = e^{-i\theta} A^+ B - e^{i\theta} C^+ D, \\ &\tilde{C}^+ = -e^{-i\theta} B^+ A + e^{i\theta} D^+ C, \qquad \tilde{D} = -e^{-i\theta} B^+ B + e^{i\theta} D^+ D \end{aligned} \tag{19.2.47}$$

and the vacuum-to-vacuum matrix element

$$\langle 0,0|S|0,0\rangle = \langle 0,0|T^+ R T|0,0\rangle \tag{19.2.48}$$

is obtained taking account of (19.2.45),

$$\langle m,n|R|m',n'\rangle = \prod \delta_{m_i m_i'} \delta_{n_i n_i'} e^{-2in\theta}. \tag{19.2.49}$$

The result is

$$\langle 0|S|0\rangle = \sum_{n=0}^{\infty} |\langle m,n|T|0,0\rangle|^2 e^{-2in\theta} = \sum W_n e^{-2in\theta}. \tag{19.2.50}$$

On the other hand,

$$\langle 0|S|0\rangle = N = |\det \tilde{D}|^{-2} \quad \text{and} \tag{19.2.51}$$

$$\begin{aligned} |\det \tilde{D}| &= |\det (D^+ D - e^{-2i\theta} B^+ B)| \\ &= |\det D|^2 |\det (\mathrm{I} - e^{-2i\theta} \zeta^+ \zeta)|. \end{aligned} \tag{19.2.52}$$

Finally, the generating function for the transition probabilities is

$$\mathscr{G}(t) = \sum W_n t^n = \frac{\det (\mathrm{I} - \zeta^+ \zeta)}{\det (\mathrm{I} - t\zeta^+ \zeta)}. \tag{19.2.53}$$

This expression can be rewritten as

$$\mathscr{G}(t) = W_0 \exp\left(-\sum_{k=1}^{\infty} \frac{S_k}{k} t^k \right), \qquad S_k = \operatorname{tr} \{(\zeta^+ \zeta)^k\}. \tag{19.2.54}$$

Note that $\zeta^+\zeta$ is a Hermitian nonnegative definite matrix, so it can be reduced to diagonal form. Let ϱ_i, $0 \leq \varrho_i < 1$, be the eigenvalues of this matrix. It is seen that the probability W_n of n-pair production is determined by N numbers ϱ_i, and the generating function for the probabilities W_n is the product of N generating functions for the one-dimensional case.

19.2.3 The Multidimensional Case: Nonstationary Problem

Let $H(t)$ be a time-dependent Hamiltonian, linear in the infinitesimal operator of representation $T(g)$,

$$H = A_{ij} a_i^+ b_j^+ + \bar{A}_{ij} a_i b_j + C_{ij} a_i^+ a_j + D_{ij} b_i^+ b_j,$$
$$C^+ = C, \qquad D^+ = D. \tag{19.2.55}$$

Solutions of the nonstationary Schrödinger equation with the Hamiltonian (19.2.55) have the form $|\psi(t)\rangle = \exp[-\mathrm{i}\varphi(t)]|\zeta(t)\rangle$. To calculate the time derivative of $|\zeta\rangle$ we use

$$\frac{d}{dt}(\det A) = \det A \operatorname{tr}(\dot{A}A^{-1}). \tag{19.2.56}$$

Hence one gets a useful relation

$$\frac{d}{dt}|\zeta\rangle = \tfrac{1}{2}\operatorname{tr}\left[(1-\zeta^+\zeta)^{-1}\frac{d}{dt}(1-\zeta^+\zeta)\right]|\zeta\rangle + \dot{\zeta}_{kl} a_k^+ b_l^+ |\zeta\rangle. \tag{19.2.57}$$

With (19.2.34–36), we obtain equations for $\zeta(t)$ and $\varphi(t)$:

$$\mathrm{i}\dot{\zeta} = (A + \zeta A^+ \zeta + C\zeta + \zeta D^{\mathrm{T}}), \tag{19.2.58}$$

$$\dot{\varphi} = A_{kl}\zeta_{kl} - \frac{\mathrm{i}}{2}\operatorname{tr}\left\{(1-\zeta^+\zeta)^{-1}\frac{d}{dt}(1-\zeta^+\zeta)\right\}. \tag{19.2.59}$$

In view of (19.2.58)

$$\mathrm{i}\frac{d}{dt}(1-\zeta^+\zeta) = (1-\zeta^+\zeta)A^+\zeta - \zeta^+ A(1-\zeta^+\zeta)$$
$$+ (1-\zeta^+\zeta)D^{\mathrm{T}} - D^{\mathrm{T}}(1-\zeta^+\zeta) \tag{19.2.60}$$

so that (19.2.59) is reduced to

$$\dot{\varphi} = \tfrac{1}{2}\operatorname{tr}\{A^+\zeta + \zeta^+ A\}. \tag{19.2.61}$$

Equations (19.2.58, 61) determine the evolution of ζ and φ completely.

In the simplest case, A and $\bar{A}\to 0$, C and $D\to$const, as t goes to infinity. Then, because of (19.2.59),

$$\zeta(t)=\exp(-\mathrm{i}Ct)\zeta^{(0)}\exp(-\mathrm{i}D^{\mathrm{T}}t). \tag{19.2.62}$$

So all eigenvalues of the matrix $\zeta^+\zeta$ have definite limits, $\varrho_i\to\varrho_i^{(0)}$, which determine the pair production probability according to (19.2.53).

Thus, the pair production probabilities are functions of $\varrho_1,\ldots,\varrho_N$, which are obtained from the solution of (19.2.58). The time dependence of the coefficients A, C and D is determined by the original equation for the field operators as in the one-dimensional case.

It is noteworthy that (19.2.58) can be simplified in some cases. For instance, if the homogeneous electric field has variable magnitude but fixed direction (say, along the z axis) the particle spin projection on the z axis is conserved. Therefore, the canonical transformation (19.2.15) couples only a_j and b_j^+. Thus, the problem is essentially one-dimensional. However, the time dependence of the transformation coefficients can be more complicated than in the one-dimensional case.

19.3 Fermion Pair Production in a Variable Homogeneous External Field

This section considers group-theoretical aspects of fermion pair production in variable homogeneous external fields. We show that the $SU(2N)$ group, where $N=2s+1$, s being the particle spin, is the dynamical symmetry group of the problem. At an arbitrary moment of time the system states are CS associated with a representation of the $SU(2N)$ group. The pair production probability is given by the modulus squared of the representation matrix element. For a spin-1/2 particle the presence of dynamical symmetry enables the problem to be reduced to the more simple one, of spin motion in a variable magnetic field [182, 187–189]. In this section the results of [187, 190] are given in some detail. For other problems related to calculating the pair production probability, which are beyond the presented exposition, see [180, 189].

19.3.1 Dynamical Symmetry for Spin-1/2 Particles

Let us consider a charged spin-1/2 particle with mass m in a homogeneous electric field $\boldsymbol{E}(t)$. It is described by bispinor $\psi(\boldsymbol{r},t)$ which satisfies the Dirac equation

$$\mathrm{i}\dot{\psi}=H\psi,\qquad H=[\hat{\boldsymbol{p}}-\boldsymbol{A}(t)]\boldsymbol{\alpha}+m\beta, \tag{19.3.1}$$

(using the units $\hbar = c = e = 1$) where $\boldsymbol{A}(t) = -\int^{t} \boldsymbol{E}(t')dt'$ is the vector potential, $\boldsymbol{\alpha}$, β are the standard Dirac matrices, $\hat{\boldsymbol{p}} = -\mathrm{i}\nabla$ is the momentum operator and the point stands for the time derivative. It can be easily seen that the solution of (19.3.1) is of the form

$$\psi(\boldsymbol{r}, t) = \exp(\mathrm{i}\boldsymbol{p}\boldsymbol{r})\psi_0(t).$$

Hence the equation for $\psi_0(t)$ is

$$\mathrm{i}\dot{\psi}_0 = [\boldsymbol{\pi}(t)\boldsymbol{\alpha} + m\beta]\psi_0. \tag{19.3.2}$$

Here $\boldsymbol{\pi}(t) = \boldsymbol{p} - \boldsymbol{A}(t) = \boldsymbol{p} + \int^{t} \boldsymbol{E}(t')dt'$ is the classical particle momentum at a time moment t, $\boldsymbol{p}$ is the particle momentum at $t \to -\infty$ [we assume that $\boldsymbol{E}(t)$ falls sufficiently rapidly as $t \to \pm\infty$, so that $\boldsymbol{\pi}(t)$ has definite limits $\boldsymbol{\pi}_{\pm}$]. In a number of special cases (19.3.2) can be reduced to a more simple form, though this cannot be done in general.

1a) $\dot{\boldsymbol{\pi}} = \boldsymbol{E}(t) = E(t)\boldsymbol{n}$, where the unit vector $\boldsymbol{n}$ is invariable. Then $\boldsymbol{\pi} = \boldsymbol{p}_{\perp} + (\boldsymbol{\pi}\boldsymbol{n})\boldsymbol{n}$, $\boldsymbol{p}_{\perp}\boldsymbol{n} = 0$, and (19.3.2) becomes

$$\mathrm{i}\dot{\psi}_0 = (\Omega_1 J_1 + \Omega_3 J_3)\psi_0, \quad \text{where} \tag{19.3.3}$$

$$\Omega_1 = 2\sqrt{p_{\perp}^2 + m^2}, \qquad \Omega_3 = 2\boldsymbol{\pi}\boldsymbol{n}, \tag{19.3.4}$$

$$J_1 = \Omega_1^{-1}(\boldsymbol{p}_{\perp}\boldsymbol{\alpha} + m\beta), \qquad J_3 = \tfrac{1}{2}\,\boldsymbol{n}\boldsymbol{\alpha}. \tag{19.3.5}$$

With an extra operator $J_2 = \mathrm{i}\Omega_1^{-1}(\boldsymbol{p}_{\perp}\boldsymbol{\alpha} + m\beta)(\boldsymbol{n}\boldsymbol{\alpha})$, the set J_k satisfies the standard commutation relations of the $SU(2)$ Lie algebra: $[J_k, J_l] = \mathrm{i}\varepsilon_{klm} J_m$. Consequently, (19.3.3) is similar to the equation describing the motion of a spin-1/2 object in a variable planar magnetic field,

$$\boldsymbol{H}(t) = \boldsymbol{\Omega}(t) = (\Omega_1(t), 0, \Omega_3(t)).$$

In particular, with the exact solution of the Dirac equation given in [191], one can get a new exact solution of the spin precession problem for a magnetic field of the type

$$\boldsymbol{H}(t) = (c, 0, a + b\tanh\gamma t).$$

Note that besides J_k the Dirac matrices can be used to construct operators K_l generating another $SU(2)$ Lie algebra and commuting with the operators J_k. For $\boldsymbol{p}_{\perp} = \text{const}$, one can work with another reference frame in which $\boldsymbol{p}_{\perp} = 0$, $\boldsymbol{\pi}(t) = (0, 0, \pi(t))$, since the operators

$$K_1 = -\frac{\mathrm{i}}{2}\,\alpha_2\alpha_3\beta, \qquad K_2 = \frac{\mathrm{i}}{2}\,\alpha_1\alpha_3\beta, \qquad K_3 = -\frac{\mathrm{i}}{2}\,\alpha_1\alpha_2 \tag{19.3.6}$$

satisfy the same conditions. The operators K_1, K_2 and K_3 commute with the Hamiltonian, so they are conserved quantities. Therefore a particular solution of (19.3.2) with $\boldsymbol{\pi}(t)=(0, 0, \pi(t))$ exists, which is an eigenfunction of the operator $K_3=-\frac{\mathrm{i}}{2}\alpha_1\alpha_2$. This fact has a simple physical explanation: the conservation of K_3 is associated with invariance of the Hamiltonian H with respect to rotations around the third axis.

b) Let $\boldsymbol{p}_\perp=0$, $\boldsymbol{\pi}(t)=(0, 0, \pi(t))$ and the particle has an anomalous magnetic moment μ. Here the Dirac equation is

$$\mathrm{i}\dot{\psi}_0=[\pi(t)\alpha_3+m\beta+\mathrm{i}\mu\dot{\pi}(t)\,\beta\alpha_3]\psi_0, \tag{19.3.7}$$

or, equivalently,

$$\mathrm{i}\dot{\psi}_0=(\boldsymbol{\Omega J})\psi_0, \quad \text{where} \tag{19.3.8}$$

$$\boldsymbol{\Omega}=2(m, \mu\dot{\pi}, \pi), \qquad \boldsymbol{J}=\tfrac{1}{2}(\beta\Lambda\,\mathrm{i}\beta\alpha_3, \alpha_3). \tag{19.3.9}$$

Thus, in this case $SU(2)$ is also a symmetry group.

c) For homogeneous expanding Universe the Dirac equation can be reduced to the same form [184].

2a) Suppose the electric field vector $\boldsymbol{E}(t)=\dot{\boldsymbol{\pi}}(t)$ lies in a fixed plane, e.g., the plane (x_1, x_2).

Consider the operators M_i, N_j

$$\begin{aligned} &M_1+N_1=\frac{\mathrm{i}}{2\varepsilon}(p_3\alpha_3+m\beta)\alpha_2, \qquad M_2+N_2=\frac{\mathrm{i}}{2\varepsilon}(p_3\alpha_3+m\beta)\alpha_1 \\ &M_3+N_3=-\frac{\mathrm{i}}{2}\alpha_1\alpha_2, \qquad \varepsilon=\sqrt{p_3^2+m^2}, \end{aligned} \tag{19.3.10}$$

$$\begin{aligned} &M_1-N_1=\tfrac{1}{2}\alpha_1, \qquad M_2-N_2=\tfrac{1}{2}\alpha_2, \\ &M_3-N_3=\frac{1}{2\varepsilon}(p_3\alpha_3+m\beta). \end{aligned} \tag{19.3.11}$$

It can be easily verified that $[M_i, N_j]=0$, and the operators M_i and N_j satisfy the standard commutation relations for the $SU(2)$ Lie algebra. Now the Dirac equation (19.3.2) is as follows:

$$\mathrm{i}\dot{\psi}_0=\boldsymbol{\Omega}(\boldsymbol{M}-\boldsymbol{N})\psi_0, \quad \text{where} \tag{19.3.12}$$

$$\boldsymbol{\Omega}=2(\pi_1(t), \pi_2(t), \varepsilon). \tag{19.3.13}$$

Consequently, the dynamical symmetry group is $SU(2)\times SU(2)$. Now $\psi_0(t)=S(t)\psi_0(0)$, $S(t)=S_1(t)S_2(t)$, and the operators $S_1(t)$ and $S_2(t)$ satisfy

$$\mathrm{i}\dot{S}_1 = (\boldsymbol{\Omega M})S_1, \quad \mathrm{i}\dot{S}_2 = -(\boldsymbol{\Omega N})S_2 \tag{19.3.14}$$

with the initial conditions $S_1(0) = \hat{\mathrm{I}}$, $S_2(0) = \hat{\mathrm{I}}$.

Again the problem is reduced to the spin motion in a variable magnetic field $\boldsymbol{H}(t) = \pm\boldsymbol{\Omega}(t)$. In particular, if the electric field intensity is a uniformly rotating vector in the plane x_1, x_2 [$\pi_1(t) = \pi\cos\omega t$, $\pi_2(t) = \pi\sin\omega t$, $\pi_3 = 0$], the pair production probability is given by the formula for spin-flip probability obtained by *Rabi* [192]. Note that $K_3 = -i\alpha_1\alpha_2\beta/2$ is an operator commuting with the Hamiltonian, since the problem has the symmetry plane (x_1, x_2).

b) Let the particle have an anomalous magnetic moment $\mu, p_3 = 0$, and the vector $\boldsymbol{E}(t)$ lies in the plane (x_1, x_2). The Dirac equation is

$$\mathrm{i}\dot{\psi}_0 = (\boldsymbol{\Omega}_1\boldsymbol{M} - \boldsymbol{\Omega}_2\boldsymbol{N})\psi_0, \tag{19.3.15}$$

where the operators $\boldsymbol{M}$ and $\boldsymbol{N}$ are given by (19.3.10, 11) with $p_3 = 0$, and

$$\begin{aligned}\boldsymbol{\Omega}_1 &= 2(\pi_1 + \mu\dot{\pi}_2, \pi_2 - \mu\dot{\pi}_1, m)\\ \boldsymbol{\Omega}_2 &= 2(\pi_1 - \mu\dot{\pi}_2, \pi_2 + \mu\dot{\pi}_1, m).\end{aligned} \tag{19.3.16}$$

As in the preceding case, $K_3 = -\mathrm{i}\alpha_1\alpha_2\beta/2$ is conserved.

c) Let the particle have an anomalous magnetic moment μ, and the electric field is $\boldsymbol{E}(t) = E(t)\boldsymbol{n}$, where the unit vector $\boldsymbol{n}$ is time independent and directed, for instance, along the axis x_3. The Dirac equation is reduced to

$$\mathrm{i}\dot{\psi}_0 = (\boldsymbol{\Omega}_1\boldsymbol{M} - \boldsymbol{\Omega}_2\boldsymbol{N})\psi_0, \quad \text{where} \tag{19.3.17}$$

$$\boldsymbol{M} - \boldsymbol{N} = \tfrac{1}{2}((\boldsymbol{n}_\perp\boldsymbol{\alpha}), \alpha_3, \beta), \quad \boldsymbol{M} + \boldsymbol{N} = \tfrac{1}{2}(-\mathrm{i}\alpha_3\beta, \mathrm{i}\beta(\boldsymbol{n}_\perp\boldsymbol{\alpha}), -\mathrm{i}(\boldsymbol{n}_\perp\boldsymbol{\alpha})\alpha_3) \tag{19.3.18}$$

$$\boldsymbol{\Omega}_1 = 2(p_\perp + \mu\dot{\pi}_3, \pi_3, m), \quad \boldsymbol{\Omega}_2 = 2(p_\perp - \mu\dot{\pi}_3, \pi_3, m). \tag{19.3.19}$$

Now the conserved operator is $K_3 = \mathrm{i}(\boldsymbol{\alpha n}_\perp)\alpha_3\beta/2$.

3. In the general case of arbitrary $\boldsymbol{E}(t)$ there is no additional integral of motion, and $SU(4)$ is the dynamical symmetry group.

19.3.2 Heisenberg Representation

Up to now $\psi(t)$ was not considered as a quantized field. After the second quantization, the bispinor $\psi(\boldsymbol{r}, t)$ becomes a Heisenberg operator

$$\hat{\psi}(\boldsymbol{r}, t) = \frac{1}{(2\pi)^{3/2}} \sum_{\sigma=\pm1} \int d^3p \sqrt{\frac{m}{\varepsilon_p}}\, [\hat{a}_{\boldsymbol{p}\sigma}(t)u_{\boldsymbol{p}\sigma}\mathrm{e}^{\mathrm{i}\boldsymbol{pr}} + \hat{b}^+_{\boldsymbol{p}\sigma}(t)v_{\boldsymbol{p}\sigma}\mathrm{e}^{-\mathrm{i}\boldsymbol{pr}}] \tag{19.3.20}$$

Here $\varepsilon_p=\sqrt{p^2+m^2}$, $u_{p\sigma}$ and $v_{p\sigma}$ are bispinors which are eigenvectors of the operator K_3: $2K_3u_{p\sigma}=\sigma u_{p\sigma}$, $2K_3v_{p\sigma}=\sigma v_{p\sigma}$; $\overset{+}{a}_{p\sigma}$, $a_{p\sigma}(\overset{+}{b}_{p\sigma}, b_{p\sigma})$ are the creation-annihilation operators of the particle (antiparticle) with momentum $\boldsymbol{p}$ and the K_3 eigenvalue $(\sigma/2)$. The operator $\hat{\psi}(\boldsymbol{r},t)$ satisfies $\mathrm{i}\dot{\hat{\psi}}=[\hat{\mathscr{H}}(t),\hat{\psi}]$, which is equivalent to (18.2.1), and the Hamiltonian operator $\hat{\mathscr{H}}(t)$ is obtained in a standard way from the nonquantized operator (t). The time evolution of the Heisenberg operator $\hat{\psi}(\boldsymbol{r},t)$, as well as $\hat{a}_{p\sigma}$ and $\hat{b}_{p\sigma}$, is given by a unitary transformation

$$\hat{a}_{p\sigma}(t)=\hat{S}^+(t)\hat{a}_{p\sigma}\hat{S}(t), \qquad \hat{b}^+_{p\sigma}(t)=\hat{S}^+(t)\hat{b}^+_{p\sigma}\hat{S}(t),$$

which conserves the commutation relations. As the equation for $\hat{\psi}(\boldsymbol{r},t)$ is linear, the transformation of the operators $\hat{a}_{p\sigma}$ and $\hat{b}^+_{p\sigma}$ is also linear. Besides, in cases 1 and 2 σ is conserved, and the subscripts $\boldsymbol{p}\sigma$ in $a_{p\sigma}$ and $\overset{+}{b}_{p\sigma}$ can be omitted, so

$$\hat{a}(t)=A'(t)\hat{a}+B'(t)\hat{b}^+, \qquad \hat{b}^+(t)=-\bar{B}'(t)\hat{a}+\bar{A}'(t)\hat{b}^+. \tag{19.3.21}$$

As the transformation (19.3.21) is canonical, $|A'|^2+|B'|^2=1$, the transformations of (19.3.21) belong to $SU(2)$, and operators $\hat{S}(t)$ realize a representation of this group.

One should bear in mind that two systems of states stationary at $t\to\pm\infty$ do not coincide in general. However, the transition from one system to the other is given by an operator $\hat{R}$ of the same type as $\hat{S}$. The matrix element for the considered transition from an initial state $\psi_-(t\to-\infty)$ to a final state $\psi_+(t\to+\infty)$ coincides with the matrix element of the operator $T=RS$. The operator T determines the transformation of the type (19.3.21) in which one should replace A' and B' for

$$A=\exp[\mathrm{i}(\varphi_1+\varphi_2)/2]\cos(\theta/2)$$

and

$$B=\exp[\mathrm{i}(\varphi_1-\varphi_2/2]\sin(\theta/2).$$

In this case the transition probability W depends only on a parameter $\theta=\lim\limits_{t\to+\infty}\theta(t)$

$$W=|\langle\psi_+|T(g_0)|\psi_-\rangle|^2, \qquad g_0=\begin{pmatrix}\cos\dfrac{\theta}{2} & \sin\dfrac{\theta}{2}\\ -\sin\dfrac{\theta}{2} & \cos\dfrac{\theta}{2}\end{pmatrix}. \tag{19.3.22}$$

Equation (19.3.22) also describes the charged pair distribution of spinor particles created in the gravitational field of homogeneous expanding Universe.

Let us consider the infinitesimal operators of the representation $T(g)$,

$$J_+ = \hat{a}^+ \hat{b}^+, \quad J_- = \hat{b}\hat{a}, \quad J_0 = \tfrac{1}{2}(\hat{a}^+\hat{a} + \hat{b}^+\hat{b} - 1). \tag{19.3.23}$$

They satisfy the standard commutation relations for the $SU(2)$ Lie algebra

$$[J_0, J_\pm] = \pm J_\pm, \quad [J_-, J_+] = -2J_0. \tag{19.3.24}$$

The operator

$$C_2 = \frac{J_+ J_- + J_- J_+}{2} + J_0^2 \tag{19.3.25}$$

is an invariant (Casimir's) operator. With the explicit expressions for J_+, J_- and J_0,

$$C_2 = \tfrac{3}{4}[1 - (\hat{a}^+\hat{a} - \hat{b}^+\hat{b})^2]. \tag{19.3.26}$$

Thus, for states $|0,0\rangle$ and $|1,1\rangle = b^+a^+|0,0\rangle$, $C_2 = 3/4$ and for states $|0,1\rangle$ and $|1,0\rangle$, $C_2 = 0$. Consequently, states $|0,0\rangle$ and $|1,1\rangle$ realize the spinor representation of the $SU(2)$ group, and the pair creation probability is

$$W = |\langle 1,1|T(g)|0,0\rangle|^2 = \sin^2(\theta/2),$$
$$g = \begin{pmatrix} A & B \\ -\bar{B} & \bar{A} \end{pmatrix}, \quad |B|^2 = \sin^2(\theta/2). \tag{19.3.27}$$

In other words, the pair creation probability is equal to the spin-flip probability induced by magnetic-field variation.

Let us consider the evolution of the system states. Initially the vacuum was $|\psi(0)\rangle$, where $\hat{a}|\psi(0)\rangle = \hat{b}|\psi(0)\rangle = 0$. At a moment t, $|\psi(t)\rangle = S(t)|\psi(0)\rangle$ and the state vector satisfies

$$S(t)\hat{a}S^+(t)|\psi(t)\rangle = 0, \quad S(t)\hat{b}S^+(t)|\psi(t)\rangle = 0. \tag{19.3.28}$$

We are led to a transformation inverse to that of (19.3.21) and it can be easily verified that it has the same form,

$$\tilde{\hat{a}} = S\hat{a}S^+ = \bar{A}\hat{a} - B\hat{b}^+$$
$$\tilde{\hat{b}} = S\hat{b}S^+ = B\hat{a}^+ + \bar{A}\hat{b}. \tag{19.3.29}$$

In view of the conditions $\tilde{\hat{a}}|\psi(t)\rangle = 0$, $\tilde{\hat{b}}|\psi(t)\rangle = 0$ we conclude that

$$|\psi(t)\rangle = \exp[-i\varphi(t)]|\zeta(t)\rangle, \quad \text{where} \tag{19.3.30}$$

$$|\zeta\rangle = (1 + |\zeta|^2)^{-1/2}\exp(\zeta\hat{a}^+\hat{b}^+)|0,0\rangle, \quad \zeta = B/\bar{A}. \tag{19.3.31}$$

The states $|\zeta\rangle$ are generalized coherent states.

19.3.3 The Multidimensional Case: Coherent States

If the fermion spin is s, arguments analogous to those in Sect. 19.2.2, lead to canonical transformations for $N=2s+1$ degrees of freedom,

$$\tilde{\hat{a}}_i = T^+\hat{a}_i T = A_{ij}\hat{a}_j + B_{ij}\hat{b}_j^+ ,$$
$$\tilde{\hat{b}}_i^+ = T^+\hat{b}_i^+ T = C_{ij}a_j + D_{ij}\hat{b}_j^+ . \tag{19.3.32}$$

It is easy to see that this transformation is canonical, i.e., conserves the commutation relations, if the following conditions hold

$$AA^+ + BB^+ = \mathrm{I}, \quad DD^+ + CC^+ = \mathrm{I}, \quad AC^+ = -BD^+ \tag{19.3.33}$$

or, equivalently,

$$MM^+ = \mathrm{I}, \quad \text{where} \quad M = \begin{pmatrix} A & B \\ C & D \end{pmatrix}. \tag{19.3.34}$$

Clearly, transformations (19.3.32) belong to the group $G = SU(2N)$, and the operators $T(g)$ realize a representation of this group.

Besides, it is seen that the inverse transformation matrix is

$$M^{-1} = M^+, \quad M^{-1} = \begin{pmatrix} A^+ & C^+ \\ B^+ & D^+ \end{pmatrix} \tag{19.3.35}$$

and the following conditions must be fulfilled

$$A^+A + C^+C = \mathrm{I}, \quad D^+D + B^+B = \mathrm{I}, \quad A^+B = -C^+D. \tag{19.3.36}$$

It is suitable to combine the infinitesimal operators of the representation $T(g)$ in a single matrix

$$\hat{N} = \begin{pmatrix} \hat{a}_i^+\hat{a}_j & \hat{a}_i^+\hat{b}_j^+ \\ \hat{b}_i\hat{a}_j & \hat{b}_i\hat{b}_j^+ \end{pmatrix}. \tag{19.3.37}$$

A direct test shows that

$$\hat{C}_2 = \mathrm{tr}\{\hat{N}\hat{N}\} \tag{19.3.38}$$

is an invariant operator (Casimir's operator), as it commutes with all infinitesimal operators. With the help of the canonical commutation relations, the expression for $\hat{C}_2$ can be written as

$$\hat{C}_2 = -(\hat{a}_i^+\hat{a}_i - \hat{b}_j^+\hat{b}_j)^2 + (\hat{a}_i^+\hat{a}_i - \hat{b}_j^+\hat{b}_j) + (N^2 - N)\hat{\mathrm{I}}. \tag{19.3.39}$$

Thus, for the states

$$\{|\boldsymbol{m}, \boldsymbol{n}\rangle\}, \quad |\boldsymbol{m}, \boldsymbol{n}\rangle = |m_1, \ldots, m_N, n_1, \ldots, n_N\rangle,$$

$$m_i = 0, 1, \quad n_j = 0, 1, \quad m - n = \sum m_i - \sum n_j = \text{const},$$

the operator C_2 is the identity times a constant,

$$\hat{C}_2 = [-(m-n)^2 + (m-n) + N^2 - N]\hat{1}. \tag{19.3.40}$$

It can be shown that the higher order Casimir operators $\hat{C}_p$ $(p = 3, \ldots 2N+1)$ are also reduced to constants. Thus, the set of states $\{|\boldsymbol{m}, \boldsymbol{n}\rangle\}$ with $(m - n = \text{const})$ forms a basis for an irreducible representation of the $SU(2N)$ group, namely, the representation given by the Young diagram $[1^N]$.

Note that the isotropy subgroup of the state vector $|\psi_0\rangle$ is $G_0 = SU(N) \times SU(N) \times U(1)$. Consequently, according to [15] the representation in view can be realized in the space of functions analytical on the coset space $M = G/G_0$.

Let us analyze the simplest case where the initial state is the vacuum: $|\psi(0)\rangle = |\mathbf{0}, \mathbf{0}\rangle$, $a_i|\mathbf{0}, \mathbf{0}\rangle = b_i|\mathbf{0}, \mathbf{0}\rangle = 0$. For $t > 0$ the state $|\psi(t)\rangle = T(g(t))|\mathbf{0}, \mathbf{0}\rangle$ satisfies

$$\tilde{\hat{a}}_i|\psi(t)\rangle = 0, \quad \tilde{\hat{b}}_j|\psi(t)\rangle = 0. \tag{19.3.41}$$

Here

$$\tilde{\hat{a}}_i = T\hat{a}_i T^+, \quad \tilde{\hat{b}}_j = T\hat{b}_j T^+. \tag{19.3.42}$$

Thus, we arrive at a transformation inverse to that in (19.3.32),

$$\tilde{\hat{a}}_i = A_{ij}^+ \hat{a}_j + C_{ij}^+ \hat{b}_j^+, \quad \tilde{\hat{b}}_i = B_{ij}' a_j^+ + D_{ij}' \hat{b}_j. \tag{19.3.43}$$

It can be easily verified that the solution of (19.3.41) is

$$|\psi(t)\rangle = \exp[-\mathrm{i}\varphi(t)]|\zeta(t)\rangle, \quad \langle\zeta(t)|\zeta(t)\rangle = 1, \tag{19.3.44}$$

where

$$|\zeta\rangle = \mathcal{N} \exp(\zeta_{ij}\hat{a}_i^+ \hat{b}_j^+)|\mathbf{0}\,\mathbf{0}\rangle \tag{19.3.45}$$

$$\zeta = -(A^+)^{-1} C^+ = BD^{-1}, \quad \mathcal{N} = [\det(1 + \zeta^+ \zeta)]^{-1/2}. \tag{19.3.46}$$

Expanding the exponent in (19.3.45) into the Taylor series yields an expression for the matrix elements of the operator $T(g)$

$$\langle \boldsymbol{m}, \boldsymbol{n}|T(g)|\mathbf{0},\mathbf{0}\rangle = \mathcal{N} \sum_{P_i, P_j} (-1)^{\frac{n(n-1)}{2}} \frac{1}{n!} \varepsilon_{P_i} \varepsilon_{P_j} \zeta_{i_1' j_1'} \ldots \zeta_{i_n' j_n'} \tag{19.3.47}$$

where $\varepsilon_P = \pm 1$ is the parity of permutation P, and

$$|\boldsymbol{m}, \boldsymbol{n}\rangle = |0, \ldots, 1_{i_1}, 0, \ldots, 1_{i_n}, \ldots, 0, \ldots, 1_{j_1}, \ldots, 1_{j_n}, \ldots\rangle.$$

The summation here is performed over all permutations of indices $i_1, \ldots, i_n$, $j_1, \ldots, j_n$.

$$P_i(i_1, \ldots, i_n) = (i'_1, \ldots, i'_n) \tag{19.3.48}$$
$$P_j(j_1, \ldots, j_n) = (j'_1, \ldots, j'_n).$$

The state $|\zeta\rangle$ is a generalized coherent state. It is given by the point ζ in the coset space G/G_0 where $G = SU(2N)$, G_0 is a isotropy subgroup of $|\psi_0\rangle$. In the present case,

$$|\psi_0\rangle = |\mathbf{0}, \mathbf{0}\rangle, \qquad G = SU(2N), \qquad G_0 = SU(N) \times SU(N) \times U(1)$$

and the coset space consists of the complex $N \times N$ matrices.

The system $\{|\zeta\rangle\}$, just as the usual CS system, is overcomplete and the states are nonorthogonal:

$$\langle\zeta_2|\zeta_1\rangle = [\det(1 + \zeta_2^+ \zeta_2)]^{-1/2} [\det(1 + \zeta_1^+ \zeta_1)]^{-1/2} \det(1 + \zeta_2^+ \zeta_1). \tag{19.3.49}$$

The following relations for the infinitesimal operators of the representation $T(g)$ are useful:

$$\hat{a}_k^+ \hat{a}_l |\zeta\rangle = \zeta_{lj} \hat{a}_k^+ \hat{b}_j^+ |\zeta\rangle, \tag{19.3.50}$$

$$\hat{b}_k^+ \hat{b}_l |\zeta\rangle = \zeta_{il} \hat{a}_i^+ \hat{b}_k^+ |\zeta\rangle, \tag{19.3.51}$$

$$\hat{a}_k \hat{b}_l |\zeta\rangle - - \zeta_{kl} |\zeta\rangle + \zeta_{il} \zeta_{kj} \hat{a}_i^+ \hat{b}_j^+ |\zeta\rangle. \tag{19.3.52}$$

The action of the operator $T(g)$ on the coherent states $|\zeta\rangle$ is given by

$$T(g)|\zeta\rangle = \exp(\mathrm{i}\varphi)|\zeta_g\rangle, \tag{19.3.53}$$

where

$$\zeta_g = (A\zeta + B)(C\zeta + D)^{-1} \tag{19.3.54}$$

$$\varphi = \arg\det(C\zeta + D). \tag{19.3.55}$$

It is now easy to get an explicit expression for the representation matrix element,

$$\langle\eta|T(g)|\xi\rangle = [\det(1 + \eta^+ \eta)]^{-1/2} [\det(1 + \xi^+ \xi)]^{-1/2} \times \det(D + C\xi + \eta^+ B + \eta^+ A\xi). \tag{19.3.56}$$

Hence one gets the generating function for the matrix elements of the operator $T(g)$

$$\sum \langle \boldsymbol{m}, \boldsymbol{n}|T(g)|\boldsymbol{m}', \boldsymbol{n}'\rangle \bar{C}_{\boldsymbol{mn}}(\eta) C_{\boldsymbol{m}'\boldsymbol{n}'}(\xi) = \det(D + C\xi + \eta^+ B + \eta^+ A\xi). \quad (19.3.57)$$

Here $C_{\boldsymbol{mn}}(\xi)$ are the expansion coefficients of the coherent state $|\xi\rangle$ in the canonical basis,

$$|\xi\rangle = \det(1 + \xi^+ \xi)]^{-1/2} \sum C_{\boldsymbol{m},\boldsymbol{n}}(\xi)|\boldsymbol{m}, \boldsymbol{n}\rangle. \quad (19.3.58)$$

Now we are in a position to calculate the transition probabilities,

$$W_n = \sum_{\{\boldsymbol{m},\boldsymbol{n}\}} |\langle \boldsymbol{m}, \boldsymbol{n}|T(g)|\boldsymbol{0}, \boldsymbol{0}\rangle|^2, \quad \sum m_i = \sum n_i = n. \quad (19.3.59)$$

Consider $S = T^+ RT$, where the operator T corresponds to the canonical transformation (19.3.32) and the operator R determines another transformation

$$R^+ \hat{a}_i R = \mathrm{e}^{-\mathrm{i}\theta} \hat{a}_i, \qquad R^+ \hat{b}_i^+ R = \mathrm{e}^{\mathrm{i}\theta} \hat{b}_i^+. \quad (19.3.60)$$

The operator S corresponds to a canonical transformation,

$$\begin{aligned} S^+ \hat{a}_i S &= \tilde{A}_{ij} \hat{a}_j + \tilde{B}_{ij} \hat{b}_j^+, \\ S^+ \hat{b}_i^+ S &= \tilde{C}_{ij} \hat{a}_j + \tilde{D}_{ij} \hat{b}_j^+. \end{aligned} \quad (19.3.61)$$

After necessary computations

$$\begin{aligned} \tilde{A} &= \mathrm{e}^{-\mathrm{i}\theta} A^+ A + \mathrm{e}^{\mathrm{i}\theta} C^+ C, \qquad \tilde{B} = \mathrm{e}^{-\mathrm{i}\theta} A^+ B + \mathrm{e}^{\mathrm{i}\theta} C^+ D \\ \tilde{C} &= \mathrm{e}^{-\mathrm{i}\theta} B^+ A + \mathrm{e}^{\mathrm{i}\theta} D^+ C, \qquad \tilde{D} = \mathrm{e}^{-\mathrm{i}\theta} B^+ B + \mathrm{e}^{\mathrm{i}\theta} D^+ D. \end{aligned} \quad (19.3.62)$$

Let us find the matrix element

$$\langle \boldsymbol{0}, \boldsymbol{0}|S|\boldsymbol{0}, \boldsymbol{0}\rangle = \langle \boldsymbol{0}, \boldsymbol{0}|T^+ RT|\boldsymbol{0}, \boldsymbol{0}\rangle.$$

Because of (19.3.60)

$$\langle \boldsymbol{m}, \boldsymbol{n}|R|\boldsymbol{m}', \boldsymbol{n}'\rangle = \mathrm{e}^{-\mathrm{i}n\theta} \prod_i \delta_{m_i m_i'} \delta_{n_i n_i'} \quad (19.3.63)$$

so that

$$\langle \boldsymbol{0}|S|\boldsymbol{0}\rangle = \sum |\langle \boldsymbol{m}, \boldsymbol{n}|T|\boldsymbol{0}, \boldsymbol{0}\rangle|^2 \mathrm{e}^{-2\mathrm{i}n\theta} = \sum W_n \mathrm{e}^{-2\mathrm{i}n\theta}. \quad (19.3.64)$$

On the other hand

$$\langle \boldsymbol{0}|S|\boldsymbol{0}\rangle = \tilde{N} = |\det \tilde{D}|, \quad (19.3.65)$$

$$
\begin{aligned}
|\det \tilde{D}| &= |\det (D^+ D + e^{-2i\theta} B^+ B)| \\
&= |\det D|^2 |\det (I + e^{-2i\theta} \zeta^+ \zeta)|.
\end{aligned}
\tag{19.3.66}
$$

Finally, the generating function for the transition probabilities is

$$
\mathscr{F}(t) = \sum W_n t^n = \frac{\det (I + t\zeta^+ \zeta)}{\det (I + \zeta^+ \zeta)}.
\tag{19.3.67}
$$

20. Generating Function for Clebsch-Gordan Coefficients of the $SU(2)$ Group

This chapter shows that using coherent states enables the generating function for the Clebsch-Gordan coefficients of the $SU(2)$ group to be obtained easily [193]. The method considered can also be used in a number of other cases.

Recall from Chap. 4 that polarizational states of a particle with spin j can be described by polynomials $f(z)$ belonging to the space $\mathscr{F}_j$, i.e., polynomials satisfying the condition

$$\|f\|_j^2=\int |f(z)|^2 d\mu_j(z)<\infty, \qquad z=x+\mathrm{i}y,$$

$$d\mu_j(z)=\frac{(2j+1)}{\pi}(1+|z|^2)^{-(2+2j)}dx\,dy. \tag{20.1}$$

The standard basis in space $\mathscr{F}_j$ is the set of functions

$$f_{j\mu}(z)=C_{j\mu}z^{j+\mu}, \qquad C_{j\mu}=\sqrt{\frac{(2j)!}{(j-\mu)!(j+\mu)!}}, \qquad -j\leq\mu\leq j. \tag{20.2}$$

Likewise, the functions

$$f_{j_1\mu_1,j_2\mu_2}(z_1,z_2)\equiv f_{j_1\mu_1}(z_1)f_{j_2\mu_2}(z_2), \quad -j_1\leq\mu_1\leq j_1, \quad -j_2\leq\mu_2\leq j_2 \tag{20.3}$$

form the basis of space

$$\mathscr{F}_{j_1j_2}=\mathscr{F}_{j_1}\otimes\mathscr{F}_{j_2}.$$

Here $\mathscr{F}_{j_1j_2}$ is the space of polynomials $f(z_1,z_2)$ of two complex variables, satisfying the condition

$$\begin{aligned}&\|f\|_{j_1j_2}^2=\int |f(z_1,z_2)|^2 d\mu_{j_1j_2}(z_1,z_2)<\infty\\ &d\mu_{j_1j_2}(z_1,z_2)=d\mu_{j_1}(z_1)\,d\mu_{j_2}(z_2).\end{aligned} \tag{20.4}$$

Here, as for space $\mathscr{F}_j$, a vector in space $\mathscr{F}_{j_1j_2}$ describes a spin state of a two-particle system with spins j_1 and j_2.

Note the well-known fact that the space $\mathscr{F}_{j_1j_2}$ is decomposed as a direct sum,

$$\mathscr{F}_{j_1j_2}=\bigoplus_j \mathscr{F}^j, \qquad |j_1-j_2|\leq j\leq j_1+j_2, \tag{20.5}$$

where space $\mathscr{F}^j$ consists of those vectors of $\mathscr{F}_{j_1,j_2}$ which transform according to representation T^j of the $SU(2)$ group. Note also that there is a vector with the lowest weight, $f_j \equiv f_{j,-j}$, in space $\mathscr{F}^j$, which satisfies the conditions

$$\boldsymbol{J}^2 f_j = j(j+1) f_j, \qquad J_3 f_j = -j f_j, \tag{20.6}$$

or, equivalently,

$$J_- f_j = (\partial_1 + \partial_2) f_j = 0, \quad J_3 f_j = (z_1\partial_1 + z_2\partial_2) f_j = -jf, \quad \partial_j = \partial/\partial z_j. \tag{20.7}$$

Here

$$J_j = J_j^{(1)} + J_j^{(2)}, \qquad J_\pm = J_1 \pm \mathrm{i} J_2 .$$

Equations (20.7) can be easily solved, giving

$$f_j = C_j (z_1 - z_2)^k, \qquad k = j_1 + j_2 - j, \qquad 0 \le k \le 2 j_{\min}. \tag{20.8}$$

The action of operators $\exp(\zeta J_+)$ on this function gives the system of coherent states

$$\begin{aligned} f_\zeta^j(z_1, z_2) &\equiv f_{j_1 j_2 j}(z_1, z_2, \zeta) = \exp(\zeta J_+) f_j(z_1, z_2) \\ &= D(z_1 - z_2)^{j_1 + j_2 - j}(1 + \zeta z_1)^{j_1 - j_2 + j}(1 + \zeta z_2)^{j_2 - j_1 + j}, \end{aligned} \tag{20.9}$$

where

$$D = \left(\frac{(2j+1)!\,(2j_1)!\,(2j_2)!}{(j_1 + j_2 - j)!\,(j_1 - j_2 + j)!\,(j_2 - j_1 + j)!\,(j_1 + j_2 + j + 1)!} \right)^{1/2}. \tag{20.10}$$

This expression for $f_{j_1 j_2 j}(z_1, z_2, \zeta)$ was obtained in [193]; it is the generating function for the Clebsch-Gordan coefficients of the $SU(2)$ group:

$$\begin{aligned} & f_{j_1 j_2 j}(z_1, z_2, z) \\ &= \sum_{\mu_1} \sum_{\mu_2} \sum_{\mu} \left(C_{j-\mu}^{2j} C_{j_1-\mu_1}^{2j_1} C_{j_2-\mu_2}^{2j_2} \right)^{1/2} z_1^{j_1-\mu_1} z_2^{j_2-\mu_2} z^{j-\mu} \langle j_1 \mu_1 ; j_2 \mu_2 | j \mu \rangle, \end{aligned} \tag{20.11}$$

where

$$\langle j_1 \mu_1 ; j_2 \mu_2 | j \mu \rangle$$

is the Clebsch-Gordan coefficient.

Let us look at some examples taken from [193].

$$f_{\frac{1}{2},\frac{1}{2},0} = \frac{1}{\sqrt{2}}(z_2 - z_1), \qquad f_{\frac{1}{2},\frac{1}{2},1} = (1 + z z_1)(1 + z z_2),$$

$$f_{1,\frac{1}{2},\frac{1}{2}}=\sqrt{\tfrac{2}{3}}\,(z_2-z_1)(1+zz_1), \quad f_{1,\frac{1}{2},\frac{3}{2}}=(1+zz_1)^2(1+zz_2),$$

$$f_{\frac{3}{2},\frac{1}{2},1}=\sqrt{\tfrac{3}{4}}\,(z_2-z_1)(1+zz_1)^2, \quad f_{\frac{3}{2},\frac{1}{2},2}=(1+zz_1)^3(1+zz_2), \tag{20.12}$$

$$f_{j,\frac{1}{2},j-\frac{1}{2}}=\sqrt{\frac{2j}{2j+1}}\,(z_2-z_1)(1+zz_1)^{2j-1}, \quad f_{j,\frac{1}{2},j+\frac{1}{2}}=(1+zz_1)^{2j}(1+zz_2),$$

$$f_{1,1,0}=\frac{1}{\sqrt{3}}\,(z_2-z_1)^2, \quad f_{1,1,1}=(z_2-z_1)(1+zz_1)(1+zz_2),$$

$$f_{1,1,2}=(1+zz_1)^2(1+zz_2)^2.$$

21. Coherent States and the Quasiclassical Limit

This chapter gives an exact mathematical formulation and proof that in representations of a compact Lie group G specified by large quantum numbers the generators of the corresponding Lie algebra $\mathscr{G}$ can be replaced by c-numbers [194]. This is an extension of the fact that if the angular momentum is large ($j \to \infty$), the angular momentum operators J_x, J_y, J_z can be replaced by the following c-numbers;

$$J_x \to j \sin\theta \cos\varphi, \qquad J_y \to j \sin\theta \sin\varphi, \qquad J_z \to j \cos\theta.$$

Let us start with the group $SU(2)$. In this case generators of the Lie algebra are reduced to c-numbers as j goes to infinity. In other words, the "classical spin vector" appears, which is represented by a point on the two-dimensional sphere S^2. In this connection, *Fuller* and *Lenard* [195] considered a sequence of spherical harmonic representations of $SO(n)$ and showed that the limiting space is the Grassmann manifold $G(n,2)$, i.e., the manifold of all oriented two-dimensional planes in the real n-dimensional space $\mathbb{R}^n$.

Simon [194] presented a method of finding such spaces in the most general case. His idea was to use the coherent states. In the classical limit we get the orbits of the coadjoint representation of the Lie group, while different orbits are involved for different representations. For instance, for the vector representation of $SO(4)$ the $S^2 \times S^2$ space applies, while for the spinor representation of the same group the $S^2 \cup S^2$ space applies.

Let us now formulate and prove the Simon theorem. We suppose here that the reader is familiar with the main concepts of the theory of compact Lie groups [70].

Let G be a compact simple Lie group and $T_1 \equiv T^\lambda$ be a fixed representation of group G characterized by a highest weight λ. Denote by T_N and $\mathscr{H}^N$ a representation of G and the space of this representation characterized by the highest weight $N\lambda$. Let $\mathscr{H}_\alpha^N$ for fixed N be a copy of space $\mathscr{H}^N$ labeled by the point $\alpha \in \Lambda$, where Λ is the finite subset of lattice $\mathbb{Z}^\nu$. Let $\mathscr{H}_\Lambda^N = \bigotimes_{\alpha\in\Lambda} \mathscr{H}_\alpha^N$.

Define an operator $S_\alpha(X)$ in space $\mathscr{H}_\Lambda^N$ ($\alpha \in \Lambda$, $X \in \mathscr{G}$, $\mathscr{G}$ is the Lie algebra of group G) as the tensor product of a generator X of the Lie algebra in the representation $T_N = T^{N\lambda}$ corresponding to the point α. All the other points of the set Λ correspond to the unit operator in the tensor product.

Let $X_1, \ldots, X_m$ be a fixed basis in the Lie algebra $\mathscr{G}$, and H be a multiaffine function of $|\Lambda| m$ vectors $S_{\alpha,i}$ ($\alpha \in \Lambda$, $i = 1, \ldots, m$, signifying that the function is a sum of monomials, and the degree of the variable $S_{\alpha,i}$ in every monomial is zero or one). The operator $H(S_\alpha(X_i))$ is determined unambiguously, since each monomial contains only commuting operators.

Now one can evaluate the quantum partition function of the operator H, as given by

$$Z_Q^N(\beta) = d_N^{-|\Lambda|} \operatorname{tr}\{\exp[-H(\beta S_\alpha(X_i)/N)]\}, \tag{21.1}$$

where d_N is the dimensionality of space $\mathscr{H}^N$.

Let us find the corresponding classical partition function. Let O^λ be an orbit of the coadjoint representation of group G passing through point $\lambda \in \mathscr{G}^*$ and $d\mu$ be a G-invariant measure on the orbit such that $\mu(B \subset O^\lambda)$ is the group measure for the set $\{x \in G | \mathrm{Ad}^*(x)\lambda \in B\}$. Let O_α be a copy of O^λ corresponding to the point $\alpha \in \Lambda$ and $O^{|\Lambda|} = \bigotimes_\alpha O_\alpha$. Then the classical partition function is

$$Z_{\mathrm{cl}}(\beta) = \int_{O^{|\Lambda|}} \exp[-H(\beta l_\alpha(X_i))] \prod_\alpha d\mu(l_\alpha). \tag{21.2}$$

Simon [194] proved the following theorem, which is a generalization of the *Lieb* theorem [82].

Theorem. Quantities $Z_Q^N(\beta)$ and $Z_{\mathrm{cl}}(\beta)$ given in (21.1, 2) satisfy the following inequalities:

$$Z_{\mathrm{cl}}(\beta) \le Z_Q^N(\beta) \le Z_{\mathrm{cl}}(\beta(1 + aN^{-1})), \tag{21.3}$$

where $a = 2(\lambda, \delta)/(\lambda, \lambda)$, λ is the highest weight of the fundamental representation of the group G, δ is the halfsum of positive roots of the Lie algebra $\mathscr{G}$, (λ, δ) stands for the Killing-Cartan inner product of vectors λ and δ in the root space.

Remark. The lower bound in (21.3) is valid also when λ is not a fundamental weight; perhaps this is also valid for the upper bound.

Proof. Let $|\varphi\rangle$ be a vector in space $\mathscr{H}$ which corresponds to a maximal weight λ and $P(\lambda) = |\varphi\rangle\langle\varphi|$ be a projection operator on state $|\varphi\rangle$. Note that $\langle\varphi|\hat{X}|\varphi\rangle = N\lambda(X)$. Moreover, since $N\lambda$ is the maximal weight, any unit vector $|\chi\rangle$ in $\mathscr{H}^N$ such that $\langle\chi|\hat{X}|\chi\rangle = N\lambda(X)$, $X \in \mathscr{H}$ ($\mathscr{H}$ is the Cartan subalgebra of the Lie algebra $\mathscr{G}$) must have the form $|\chi\rangle = \exp(\mathrm{i}\gamma)|\varphi\rangle$, because the weight space is one-dimensional. Further,

$$\langle T(g)\varphi|\hat{X}|T(g)\varphi\rangle = N(\mathrm{Ad}^*(g)\lambda)(X), \quad \text{so} \tag{21.4}$$

$$T(g)P(\lambda)T(g)^{-1} = P(\lambda) \tag{21.5}$$

if and only if $\mathrm{Ad}^*(g)\lambda = \lambda$.

Let $l \in O^\lambda$ and $g \in G$, where $\mathrm{Ad}^*(g)\lambda = l$. Introduce an operator,

$$P(l) = T(g)P(\lambda)T(g)^{-1}. \tag{21.6}$$

It is easy to see that $P(l)$ does not depend on the choice of g, provided that $\mathrm{Ad}^*(g)\lambda = l$. Note that

$$\mathrm{tr}\,[\hat{X}_N P(l)] = Nl(x), \tag{21.7}$$

as

$$\begin{aligned}\mathrm{tr}\,[\hat{X}_N P(l)] &= \mathrm{tr}\,\{\hat{X}_N T_N(g)P(\lambda)T_N(g)^{-1}\}\\ &= \mathrm{tr}\,\{[\mathrm{Ad}\,(g^{-1})X_N]P(\lambda)\} = N\lambda(\mathrm{Ad}\,(g^{-1})X)\\ &= N(\mathrm{Ad}^*\,(g)\lambda)\,(X) = Nl(X).\end{aligned} \tag{21.8}$$

Moreover, if $d\mu(g)$ is the Haar measure on G, and $\mu = d_N\tilde{\mu}$, then

$$\int_{O^\lambda} P(l)\,d\mu(l) = \hat{\mathrm{I}}, \qquad \text{since} \tag{21.9}$$

$$\int_{O^\lambda} P(l)\,d\mu(l) = d_N \int T(g)P(\lambda)T(g)^{-1}\,d\mu(g) = c\hat{\mathrm{I}}, \tag{21.10}$$

as a consequence of Schur's lemma. Calculating the trace of the operators in (21.10) yields (21.9). Let $l = \{l_\alpha\} \in O^{|\lambda|}$, $P(\{l_\alpha\}) = \bigotimes_\alpha P(l_\alpha)$; as a consequence of (21.9)

$$\int_{O^{|\lambda|}} P(l) \bigotimes_\alpha d\mu(l_\alpha) = \hat{\mathrm{I}}, \tag{21.11}$$

so that $\{P(l)\}$ is the set of projectors to the coherent states. According to (21.7), the symbol of operator $H(\beta S_\alpha(X_i))/N$ is exactly $H(\beta l_\alpha(X_i))$, so that the lower bound in (21.3) is a consequence of inequality (1.6.11). It is possible to prove as well [194] that the contravariant symbol of the same operator is $H((1 + aN^{-1})l_\alpha(X_i))$. Thus, the upper bound (21.3) is also a consequence of inequality (1.6.11).

Remark. Note that the *Lieb* theorem [82] is a special case of this result for $G = SO(3)$.

Example 1. Let us consider the spherical harmonic representations of the $SO(2n)$ group. These representations correspond to orbits passing through the point

$$\begin{pmatrix} \begin{matrix} 0 & 1 \\ -1 & 0 \end{matrix} & 0 \\ 0 & 0 \end{pmatrix} \tag{21.12}$$

and the orbits are the Grassman manifolds $G(2n, 2)$ of 2-planes in the $(2n)$-space. Then $(\lambda, \delta)=(n-1)(\lambda, \lambda)$ and the theorem gives the result of [195] for the $SO(2n)$ group. A similar calculation leads to the result of [195] for the $SO(2n+1)$ group.

Remark 2. Let us consider the spinor representation for the $SO(2n)$ group. Then $\lambda=(\omega_1+\ldots+\omega_n)/2$, where $\omega_1, \ldots, \omega_n$ are fundamental weight vectors for the Lie algebra of $SO(2n)$, and the orbit passes through the point

$$\frac{1}{2}\begin{pmatrix} \sigma & 0 & & & 0 \\ 0 & \sigma & & & \\ & & \cdot & & \\ & & & \cdot & \\ & & & & \cdot \\ 0 & & & & \sigma \end{pmatrix} \tag{21.13}$$

where $\sigma=\begin{pmatrix} 0 & 1 \\ -1 & 0 \end{pmatrix}$. The orbit is

$$O=\{\tfrac{1}{2}M \mid M^t=-M, M^tM=MM^t\}, \tag{21.14}$$

i.e., $\tilde{O}=\{2K \mid K\in O\}$ is the set of antisymmetric orthogonal matrices. A direct calculation indicates that

$$(\delta, \lambda)/(\lambda, \lambda)=n-1.$$

Thus, if Z_{cl} is defined in terms of $\tilde{O}$, then

$$Z_{\text{cl}}\left(\frac{\beta}{2}\right)\leq Z_Q^{2N}(\beta)\leq Z_{\text{cl}}\left(\frac{\beta}{2}\left(1+\frac{n-1}{N}\right)\right). \tag{21.15}$$

22. $1/N$ Expansion for Gross-Neveu Models

This chapter considers a class of quantum field theoretical models – the Gross-Neveu models – following [196]. These are models of an N-component fermion field in two-dimensional space-time with $(\bar{\psi}\psi)^2$ interaction; a number of exact solutions can be found for equations of motion in these models. Following *Berezin* [94], we show that the parameter $1/N$ is analogous to Planck's constant, so obtaining some classical Hamiltonian systems as $N \to \infty$. The phase space is, however, nonlinear and the Poisson brackets are not canonical. The symmetry group for these models is $SO(2n)$. It is remarkable that the constructions in this chapter are mainly algebraic and therefore the results obtained slightly depend on particular features of the models.

22.1 Description of the Model

The original *Gross-Neveu* model [197] is given by the Lagrangian

$$\mathscr{L} = \mathrm{i}\bar{\psi}_k \gamma_a \partial_a \psi_k + \frac{g^2}{2}(\bar{\psi}_k \psi_k)^2, \qquad a=0,1;\, k=1,\dots,N. \tag{22.1.1}$$

Here and in the following, summation over repeated indices is implied; the field ψ_k is a two-component spinor. It is suitable to use the Majorana representation of matrices

$$\gamma_a\colon \quad \gamma_0 = \sigma_2, \qquad \gamma_1 = \mathrm{i}\sigma_1, \tag{22.1.2}$$

where σ_1 and σ_2 are the usual Pauli matrices. We use in the following

$$\psi_k = \frac{1}{\sqrt{2}}(\chi^{1,k} + \mathrm{i}\chi^{2,k})$$

and a $2N$-component field

$$\chi^\mu = (\chi^{1,1},\dots,\chi^{1N},\, \chi^{2,1},\dots,\chi^{2N}), \qquad \mu = 1,2,\dots,2N \tag{22.1.3}$$

Here each component is a spinor: $\chi^\mu \equiv \{\chi^\mu_\alpha\}$, $\alpha = 1,2$.

The fields χ^μ_α are operators, and they satisfy equal-time canonical anti-commutation relations

$$\{\chi^\mu_\alpha(x),\ \chi^\nu_\beta(y)\}_+ = \hbar\delta_{\mu\nu}\delta_{\alpha\beta}\delta(x-y). \tag{22.1.4}$$

Correspondingly, the Hamiltonian has the form

$$H=\tfrac{1}{2}\int dx\,dy\,\delta(x-y)\,[H_{1,y}(x,y)+H_{2,x}(x,y)-g^2\Omega^2(x,y)], \tag{22.1.5}$$

where the operator densities H_1, H_2 and Ω are given by

$$H_1(x,y)=\Phi_{11}(x,y), \qquad H_2(x,y)=\Phi_{22}(x,y)$$

$$\Omega(x,y)=\Phi_{12}(x,y)=-\Phi_{21}(y,x), \tag{22.1.6}$$

$$\Phi_{\alpha\beta}(x,y)\equiv\frac{1}{2\mathrm{i}}\,[\chi^\mu_\alpha(x),\ \chi^\mu_\beta(y)],$$

bracket stands for the usual commutator. The functions H_1 and H_2 are antisymmetric when the variables x and y are interchanged

$$\begin{aligned} H_1(x,y)&=-H_1(y,x),\\ H_2(x,y)&=-H_2(y,x). \end{aligned} \tag{22.1.7}$$

Note that in the classical limit ($\hbar\to 0$) the Clifford algebra (22.1.4) is reduced to the Grassmann algebra, while (22.1.6, 7) are not changed.

Because of (22.1.6, 7), the model is invariant with respect to global transformations belonging to the $SO(2N)$ group. On the other hand, the bilocal operators $\Phi_{\alpha\beta}(x,y)$ in (22.1.6) belong to the infinite-dimensional orthogonal algebra A. Hence it follows, in particular, that the Heisenberg equations for the operators $\Phi_{\alpha\beta}(x,y)$ form a closed system,

$$\begin{aligned} &H_{1,t}(x,y)+H_{1,x}(x,y)+H_{1,y}(x,y)\\ &\quad=\frac{g^2}{2}\,[\Omega(x,y)\Omega(y,y)+\Omega(y,y)\Omega(x,y)\\ &\quad-\Omega(x,x)\Omega(y,x)-\Omega(y,x)\Omega(x,x)] \end{aligned} \tag{22.1.8}$$

$$\begin{aligned} &H_{2,t}(x,y)-H_{2,x}(x,y)-H_{2,y}(x,y)\\ &\quad=\frac{g^2}{2}\,[\Omega(y,x)\Omega(y,y)+\Omega(y,y)\Omega(y,x)\\ &\quad-\Omega(x,y)\Omega(x,x)-\Omega(x,x)\Omega(x,y)] \end{aligned} \tag{22.1.9}$$

$$\Omega_t(x,y)+\Omega_x(x,y)-\Omega_y(x,y)$$
$$=\frac{g^2}{2}\,[\Omega(x,x)H_2(x,y)+H_2(x,y)\Omega(x,x)$$
$$-\Omega(y,y)H_1(x,y)-H_1(x,y)\Omega(y,y)] \tag{22.1.10}$$

It is suitable now to replace the continuous variables x and y by discrete ones n and m: $n, m=1,\ldots,\Lambda$; in the final results the continuous variables are easily recovered. The anticommutation relations (22.1.4) become

$$\{\chi_\alpha^\mu(n),\ \chi_\beta^\nu(m)\}_+=\hbar\delta_{\mu\nu}\delta_{\alpha\beta}\delta_{mn} \tag{22.1.11}$$

and the algebra generated by bilocal fields $\Phi_{\alpha\beta}(m,n)$ is the Lie algebra of $SO(2\Lambda)$.

The next step is to introduce the Fock representation

$$a_n^\mu=\frac{1}{\sqrt{2}}\,[\chi_1^\mu(n)-\mathrm{i}\chi_2^\mu(n)],\qquad (a_n^\mu)^+=\frac{1}{\sqrt{2}}\,[\chi_1^\mu(n)+\mathrm{i}\chi_2^\mu(n)],$$
$$\{a_n^\mu,\ a_m^\nu\}=0,\qquad \{a_n^\mu,\ (a_m^\nu)^+\}=\hbar\delta_{\mu\nu}\delta_{mn},\qquad a_n^\mu|0\rangle=0. \tag{22.1.12}$$

Respectively, we substitute $\Phi_{\alpha\beta}(m,n)$ for another set of $SO(2N)$-invariant operators

$$A_{mn}=\tfrac{1}{4}\,[(a_m^\mu)^+a_n^\mu-a_n^\mu(a_m^\mu)^+],\qquad B_{mn}=\tfrac{1}{2}\,a_m^\mu a_n^\mu,$$
$$B_{mn}^+=\tfrac{1}{2}\,(a_n^\mu)^+(a_m^\mu)^+\quad\text{or}\quad B_{mn}^*=-B_{nm}^+=\tfrac{1}{2}\,(a_m^\mu)^+(a_n^\mu)^+. \tag{22.1.13}$$

The operators are skew-symmetrical in the subscripts, $B_{nm}=-B_{mn}$, $B_{nm}^+=-B_{mn}^+$, and they generate the Lie algebra of $SO(2\Lambda)$:

$$[A_{nm},\ A_{kl}]=\frac{\hbar}{2}\,(A_{nl}\delta_{mk}-A_{km}\delta_{nl}),$$
$$[A_{nm},\ B_{kl}]=\frac{\hbar}{2}\,(\delta_{nl}B_{km}+\delta_{nk}B_{ml}), \tag{22.1.14}$$
$$[B_{mn},\ B_{kl}^*]=\frac{\hbar}{2}\,(\delta_{nk}A_{lm}-\delta_{mk}A_{ln}+\delta_{ml}A_{kn}-\delta_{nl}A_{lm}).$$

Using the matrix notations

$$A=[A_{mn}],\qquad B=[B_{mn}],$$
$$B^*=[B_{mn}^*]=-[B_{nm}^+],$$

one has

$$A^+ = A, \quad B^t = -B, \quad (B^+)^t = -B^+, \quad \text{and} \tag{22.1.15}$$

$$\Omega = [\Omega(m,n)], \quad H_1 = [H_1(m,n)], \quad H_2 = [H_2(m,n)],$$

$$H_1^t = -H_1, \quad H_2^t = -H_2. \tag{22.1.16}$$

The two sets of the $SO(2N)$-invariant operators are related by

$$\begin{aligned} A &= \tfrac{1}{2}[\Omega + \Omega^t + \mathrm{i}(H_1 + H_2)], \\ B &= \tfrac{1}{4}[\Omega - \Omega^t + \mathrm{i}(H_1 - H_2)], \\ B^* &= \tfrac{1}{4}[\Omega^t - \Omega + \mathrm{i}(H_1 - H_2)]. \end{aligned} \tag{22.1.17}$$

Now the Fock space $\mathscr{F}$ defined by (22.1.12) can be decomposed over irreducible representations of the Lie algebra $\mathscr{G} = SO(2\Lambda)$. For example, the subspace $\mathscr{F}_0$ spanned by arbitrary linear superpositions of states

$$B_{mn}^+ B_{kl}^+ \ldots |0\rangle, \tag{22.1.18}$$

invariant with respect to the global transformations $SO(2N)$, form an irreducible representation space for the Lie algebra $SO(2\Lambda)$. Actually, a space containing states which transform according to an irreducible representation of $SO(2N)$ is also an irreducible representation space for $SO(2\Lambda)$.

Let us consider, for example, the quadratic Casimir operator of the Lie algebra $SO(2\Lambda)$ and show that this operator can be expressed via the Casimir operators of the Lie algebra $SO(2N)$. Evidently, for $SO(2\Lambda)$

$$C_2 = 4\,\mathrm{tr}\,[A^2 + \tfrac{1}{2}(BB^* + B^*B)]. \tag{22.1.19}$$

On the other hand, the quadratic Casimir operator for $SO(2N)$ is

$$Q^2 = \tfrac{1}{2} Q^{\mu\nu} Q^{\nu\mu}, \quad \text{where} \tag{22.1.20}$$

$$Q^{\mu\nu} = \frac{1}{2\mathrm{i}} \sum_{n,\alpha} [\chi_\alpha^\mu(n), \chi_\alpha^\nu(n)]. \tag{22.1.21}$$

A direct computation shows that

$$C_2 = \hbar^2 N\Lambda(\Lambda + N - 1) + Q^2. \tag{22.1.22}$$

A similar result is valid also for higher Casimir operators $C_4, C_6, C_8, \ldots$, they can be expressed in terms of

$$Q^{\mu\nu} Q^{\nu\mu}, \quad Q^{\mu\nu} Q^{\nu\lambda} Q^{\lambda\varrho} Q^{\varrho\mu}, \ldots .$$

All states of type (22.1.18) are annihilated by Q^2 (singlet states); the eigenvalue of C_2 which is specific for irreducible subspace $\mathscr{F}_0$ is reduced to

$$C_2(\mathscr{F}_0)=\hbar^2 N\Lambda(\Lambda+N-1). \tag{22.1.23}$$

Following [94], we shall now show that the classical limit occurs in subspace $\mathscr{F}_0$, as N goes to infinity. To this end, we replace the discrete basis (22.1.18) in $\mathscr{F}_0$ by a system of generalized coherent states

$$|z\rangle=\exp\left(\sum_{m,n} z_{m,n}B^+_{m,n}\right)|0\rangle \tag{22.1.24}$$

($z_{m,n}=-z_{n,m}$ are complex numbers), whose properties have been considered in Chap. 9.

Let $H=H(A_N, (B_N)^+, B_N)$ be an operator containing only

$$A^N_{m,n}=\frac{1}{N}\sum_{\mu=1}^{N}(a^\mu_m)^+a^\mu_n\,,\qquad B^N_{m,n}=\frac{1}{N}\sum_{\mu=1}^{N}a^\mu_m a^\mu_n\,,$$

$$(B^N_{m,n})^+=\frac{1}{N}\sum_{\mu=1}^{N}(a^\mu_n)^+(a^\mu_m)^+\,.$$

Such operators will be called admissible. The subspace $\mathscr{F}_0$ is invariant under the action of an arbitrary admissible operator. Take an admissible operator $\hat{H}_N$ and evaluate its covariant symbol

$$H_N(z,\bar{z})=\frac{\langle z|\hat{H}_N|z\rangle}{\langle z|z\rangle}\,. \tag{22.1.25}$$

Evidently, the symbols of the operators A_N, A_N^+, B_N are independent of N. Furthermore, it can be easily seen that

$$H_N(z,\bar{z})=H(z,\bar{z})+\frac{1}{N}\tilde{H}_N(z,\bar{z}), \tag{22.1.26}$$

where $\tilde{H}_N(z,\bar{z})$ has a definite limit as $N\to\infty$. In space $\mathscr{F}_0$ the Heisenberg equations are

$$\frac{\hbar}{\mathrm{i}N}\frac{d\hat{A}_N}{dt}=[\hat{H}_N,\hat{A}_N],\qquad \frac{\hbar}{\mathrm{i}N}\frac{d\hat{B}_N}{dt}=[\hat{H}_N,\hat{B}_N]. \tag{22.1.27}$$

The corresponding equations for the symbols of the operators are obtained using the results of Chap. 16,

$$\hbar\frac{dA}{dt}=\{H,A\}_{\mathrm{P.B.}}+\frac{1}{N}Q_N\,,\qquad \hbar\frac{dB}{dt}=\{H,B\}_{\mathrm{P.B.}}+\frac{1}{N}R_N\,, \tag{22.1.28}$$

where the Poisson brackets have a noncanonical form (due to the nonlinearity of the phase space in classical limit)

$$\{f,h\}_{\mathrm{P.B.}} = \frac{2}{\mathrm{i}}\,\mathrm{tr}\left[\frac{\partial f}{\partial \bar{z}}\,(\mathrm{I}-\bar{z}z)\,\frac{\partial h}{\partial z}\,(\mathrm{I}-z\bar{z}) - \frac{\partial h}{\partial \bar{z}}\,(\mathrm{I}-\bar{z}z)\,\frac{\partial f}{\partial z}\,(\mathrm{I}-z\bar{z})\right]. \qquad (22.1.29)$$

In the limit $N \to \infty$, (22.1.28) become the classical equations of mechanics on the manifold $\mathscr{M}_F$ (Chap. 9) with the Hamiltonian $\hbar^{-1} H(z,\bar{z})$.

Thus, the limit $N \to \infty$ is equivalent to the usual quasiclassical limit. The role of Planck's constant in such a limiting process is played by the quantity $\kappa = N^{-1}$. The resulting equations of motion are

$$\begin{aligned} \frac{dA}{dt} &= -2\mathrm{i}\left(\frac{\partial H}{\partial \bar{A}}\,B + \frac{\partial H}{\partial B}\,A + B\,\frac{\partial H}{\partial \bar{A}} + A\,\frac{\partial H}{\partial \bar{B}}\right) \\ \frac{dB}{dt} &= -2\mathrm{i}\left(\frac{\partial H}{\partial B}\,B + \frac{\partial H}{\partial \bar{A}}\,\bar{A} - B\,\frac{\partial H}{\partial B} - A\,\frac{\partial H}{\partial A}\right) \end{aligned} \qquad (22.1.30)$$

and coincide completely with the classical equations of motion.

22.2 Dimensionality of Space $\tilde{\mathscr{H}}_N = \mathscr{F}_0$ in the Fermion Case

The Gross-Neveu model was introduced originally in a space $\mathscr{H}_N$ which has a larger dimensionality than that of $\tilde{\mathscr{H}}_N$. The question is, which information is lost when passing from $\mathscr{H}_N$ to $\tilde{\mathscr{H}}_N$. As suggested in [94], the quantity we are concerned with is the ratio of dimensionalities of spaces $\tilde{\mathscr{H}}_N$ and $\mathscr{H}_N$ for fermion systems with a finite number of degrees of freedom $(n+1)$. Then

$$\dim \mathscr{H}_1 = 2^{n+1}, \qquad \dim \mathscr{H}_N = 2^{N(n+1)}. \qquad (22.2.1)$$

According to a general relation given in [94]

$$\dim \tilde{\mathscr{H}}_N = \frac{C_n(N)}{C_n(\infty)}\,; \qquad (22.2.2)$$

$C_n(N)$ are defined Sect. 9.3. Using (9.3.14) we get

$$\dim \tilde{\mathscr{H}}_N = \frac{\Gamma(N+n+1)\ldots\Gamma(N+2n)\,\Gamma(1)\,\Gamma(3)\ldots\Gamma(2n-1)}{\Gamma(N+1)\ldots\Gamma(N+2n-1)\,\Gamma(n+1)\,\Gamma(n+2)\ldots\Gamma(2n)}. \qquad (22.2.3)$$

It is convenient to transform this expression to

$$\dim \tilde{\mathscr{H}}_N = \prod_1^N T_k, \quad T_k = \frac{(n+k)(n+k+1)\dots(2n+k-1)}{k(k+2)\dots(k+2n-2)}. \tag{22.2.4}$$

Clearly,

$$T_{2s} = 2^{-n}\frac{(2n+2s-1)!(s-1)!}{(n+2s-1)!(s+n-1)!}, \quad T_{2s+1} = 2^n\frac{(n+s)!(2s)!}{(n+2s)!\,s!} \tag{22.2.5}$$

and applying the Stirling formula yields

$$\ln(\dim \tilde{\mathscr{H}}_N) = nN\ln 2 - c\ln 2n + O(1). \tag{22.2.6}$$

Thus, the leading term in the $N \to \infty$ asymptotics of $\dim \tilde{\mathscr{H}}_N$ is just $\dim \mathscr{H}_N$. Hence it is plausible that for large N no significant loss of information occurs if $\mathscr{H}_N$ is substituted for $\tilde{\mathscr{H}}_N$.

22.3 Quasiclassical Limit

Turning back to the Gross-Neveu model, evaluate the symbols of operators,

$$A(z,\bar{z}) = \frac{\langle z|\hat{A}|z\rangle}{\langle z|z\rangle}, \quad B(z,\bar{z}) = \frac{\langle z|\hat{B}|z\rangle}{\langle z|z\rangle}, \quad B^+(z,\bar{z}) = \frac{\langle z|\hat{B}^+|z\rangle}{\langle z|z\rangle}. \tag{22.3.1}$$

Here $|z\rangle$ is the CS of (22.1.24).

As shown above, the quantum dynamics tends to classical dynamics as $N \to \infty$, so the operators can be replaced by their symbols. According to Chap. 9, the matrices A, B and B^+ can be written as

$$A = \hbar\left[-\frac{N}{2}\mathrm{I} + \xi^+\xi\right], \quad B = \hbar\,[N\mathrm{I} - (\xi^+\xi)']^{1/2}\,\xi,$$
$$B^+ = \hbar\xi^+[N\mathrm{I} - (\xi^+\xi)']^{1/2}. \tag{22.3.2}$$

Thus the effective Hamiltonian for the singlet states acquires the form

$$H = H_0(\xi, \xi^+, \hbar, N).$$

In the limit $N \to \infty$ new variables are appropriate,

$$\xi = \sqrt{N}\,\zeta, \quad \xi^+ = \sqrt{N}\,\zeta^+, \tag{22.3.3}$$

where ζ and ζ^+ can be considered as c numbers and, following [196], the limit is

$$\begin{aligned} &A=\hbar N[-\tfrac{1}{2}\mathrm{I}+\zeta^+\zeta], \qquad B=\hbar N[\mathrm{I}-\zeta\zeta^+]^{1/2}\zeta, \\ &B^+=\hbar N\zeta^+[\mathrm{I}-\zeta\zeta^+]^{1/2}. \end{aligned} \tag{22.3.4}$$

The corresponding Heisenberg equations (22.1.27) become the usual Hamiltonian equations

$$\mathrm{i}\zeta_t=-\frac{\partial H_0}{\partial\zeta^+}, \qquad \mathrm{i}\zeta_t^+=\frac{\partial H_0}{\partial\zeta}. \tag{22.3.5}$$

We have obtained a c-number Hamiltonian system associated with the singlet sector of the Gross-Neveu model. Note that Planck's constant $\hbar$ and the parameter N are involved as a single combination $\hbar N$.

Restoring the continuous variables x, y instead of discrete m, n, one gets

$$\begin{aligned} &H_{1,t}(x,y)+H_{1,x}(x,y)+H_{1,y}(x,y) \\ &\qquad =g^2[\Omega(y,y)\Omega(x,y)-\Omega(x,x)\Omega(y,x)] \\ &H_{2,t}(x,y)-H_{2,x}(x,y)-H_{2,y}(x,y) \\ &\qquad =g^2[\Omega(y,y)\Omega(y,x)-\Omega(x,x)\Omega(x,y)] \\ &\Omega_t(x,y)+\Omega_x(x,y)-\Omega_y(x,y) \\ &\qquad =g^2[\Omega(x,x)H_2(x,y)-\Omega(y,y)H_1(x,y)]. \end{aligned} \tag{22.3.6}$$

However, since one deals with the subspace $\mathscr{F}_0$, there are some additional restrictions consequent to the fact that $\Lambda(2\Lambda-1)$ independent real generators of the Lie algebra $SO(2\Lambda)$ are parametrized by the $\Lambda(\Lambda-1)$ real elements of the antisymmetric matrix ζ. So there are $\Lambda(2\Lambda-1)-\Lambda(\Lambda-1)=\Lambda^2$ real constraints contained in the identities

$$A^2+B^+B=\left(\frac{\hbar N}{2}\right)^2\mathrm{I}, \qquad BA=A^tB. \tag{22.3.7}$$

These conditions have already been discussed by *Berezin* [94].

The respective conditions for the bilocal quantities are

$$\begin{aligned} &\int dz\,[\Omega(x,z)\Omega(y,z)-H_1(x,z)H_1(z,y)]=(\hbar N)^2\delta(x-y), \\ &\int dz\,[\Omega(z,x)\Omega(z,y)-H_2(x,z)H_2(z,y)]=(\hbar N)^2\delta(x-y), \\ &\int dz\,[H_1(x,z)\Omega(z,y)+\Omega(x,z)H_2(z,y)]=0, \\ &H_1(x,y)=-H_1(y,x), \qquad H_2(x,y)=-H_2(y,x). \end{aligned} \tag{22.3.8}$$

Equations (22.3.6, 8) are still too complicated, and their general solution is not immediately obvious. So it is useful to indicate a relation between the system in view and that considered by *Neveu* and *Papanicolau* [198], and a more general system studied by *Zakharov* and *Mikhailov* [199].

The system (22.3.8) is related to the symplectic group $\mathrm{Sp}(2n, \mathbb{R})$. Let quantities u^a and v^b be transformed according to the $(2n)$-dimensional representation of $\mathrm{Sp}(2n, \mathbb{R})$. The system investigated in [198] is described by

$$u_t^a + u_x^a = g^2 (u^b v_b) v^a, \qquad v_t^a - v_x^a = -g^2 (u^b v_b) u^a, \tag{22.3.9}$$

where the matrix

$$[\varepsilon_{ab}] = [\varepsilon^{ab}] = \begin{pmatrix} 0 & \mathrm{I}_N \\ -\mathrm{I}_N & 0 \end{pmatrix} \tag{22.3.10}$$

lifts and lowers the indices. A simple calculation shows that the bilocal fields

$$\begin{aligned} &h_1(x,y,t) = u^a(x,t) u_a(y,t), \qquad h_2(x,y,t) = v^a(x,t) v_a(y,t), \\ &\omega(x,y,t) = u^a(x,t) v_a(y,t) \end{aligned} \tag{22.3.11}$$

satisfy (22.3.6). Therefore, we can use the results of [198] to obtain solutions of (22.3.8). One should bear in mind, however, that not all solutions of (22.3.9) satisfy (22.3.6) with the additional constraints (22.3.8). The simplest example is the plane-wave solution for the $\mathrm{Sp}(2, \mathbb{R})$ group. Then

$$\begin{aligned} &u^a = \frac{1}{g}\sqrt{\omega_k + k}\begin{pmatrix} \cos\theta \\ \sin\theta \end{pmatrix}, \qquad u_a = \frac{1}{g}\sqrt{\omega_k + k}\begin{pmatrix} \sin\theta \\ -\cos\theta \end{pmatrix}, \\ &v^a = \frac{1}{g}\sqrt{\omega_k - k}\begin{pmatrix} -\sin\theta \\ \cos\theta \end{pmatrix}, \qquad v_a = \frac{1}{g}\sqrt{\omega_k - k}\begin{pmatrix} \cos\theta \\ \sin\theta \end{pmatrix}, \\ &\theta = \omega_k t - kx, \qquad \omega_k = \sqrt{k^2 + (\mu g^2)^2}. \end{aligned} \tag{22.3.12}$$

Here

$$\begin{aligned} &h_1(x,y;k) = \frac{\omega_k + k}{g^2} \sin[k(x-y)], \\ &h_2(x,y;k) = \frac{\omega_k - k}{g^2} \sin[k(x-y)], \\ &\omega(x,y;k) = \mu \cos[k(x-y)]. \end{aligned} \tag{22.3.13}$$

These h_1, h_2 and ω satisfy (22.3.6) identically, but not the constraints (22.3.8), which hold for the following superposition of fields:

$$H_1(x,y)=\hbar N g^2 \int \frac{dk}{2\pi\omega_k}\, h_1(x,y;k)=\hbar N \int \frac{k\,dk}{2\pi\omega_k} \sin\,[k(x-y)],$$

$$H_2(x,y)=-\hbar N \int \frac{k\,dk}{2\pi\omega_k} \sin\,[k(x-y)], \tag{22.3.14}$$

$$\Omega(x,y)=\hbar N \mu g^2 \int \frac{dk}{2\pi\omega_k} \cos\,[k(x-y)].$$

Moreover, (22.3.6) are fulfilled if the parameter μ satisfies an integral equation

$$\mu=\hbar N \mu g^2 \int\limits_{-\infty}^{\infty} \frac{dk}{2\pi} \frac{1}{\sqrt{k^2+(\mu g^2)^2}}. \tag{22.3.15}$$

Besides the trivial solution ($\mu=0$), (22.3.15) has a well-known solution of the superconductor type with $\mu \neq 0$.

It is not difficult to get now the ground-state energy setting the bilocal fields (22.3.14) into the energy functional (22.1.5). The calculation employing the energy-gap given by (22.3.15) leads to

$$\varepsilon_0=-L(2\hbar N) \int\limits_0^{\infty} \frac{dk}{2\pi} \frac{k^2+m^2/2}{\sqrt{k^2+m^2}}, \qquad m\equiv \mu g^2, \tag{22.3.16}$$

where L is the system volume.

As expected, the ground-state energy depends only on a single dimensional parameter m which takes the place of the dimensionless coupling constant. For the trivial solution $m=0$, (22.3.16) is reduced to the standard sum over the Dirac sea of negative energy states for massless fermion excitations. The integral in (22.3.16) diverges as $k\to\infty$. The divergence can be eliminated, as usual, via the standard regularization. The excited states can be considered along the same lines.

Equation (22.3.9) have soliton-type solutions. We will not present here explicit formulae for the single-soliton solution since it can be found in [196]. The multisoliton formulae can be obtained using the technique described in [199].

23. Relaxation to Thermodynamic Equilibrium

This chapter gives two examples of application of the CS method to problems in non-equilibrium statistical physics. It describes the evolution toward thermodynamic equilibrium for quantum systems with equidistant energy spectra (the quantum oscillator and a spinning particle in a magnetic field) set in thermostat.

23.1 Relaxation of Quantum Oscillator to Thermodynamic Equilibrium

We shall follow here [200, 201]. The evolution of a quantum oscillator is described by a kinetic equation for the density matrix $\hat{\varrho}$. This is the simplest model revealing statistical properties of a coherent light beam propagating in a weakly absorbing medium [202–209]. The problem in view is also important because it is a rare example of a problem in nonequilibrium quantum statistical mechanics which admits the exact solution.

23.1.1 Kinetic Equation

The state of a quantum oscillator posed in a thermostat under temperature T is determined by a density matrix satisfying the following equation (for a derivation of this equation see [200, 204, 210]):

$$\dot{\varrho} = -\tfrac{1}{2}\gamma[(\nu+1)(a^+a\varrho - 2a\varrho a^+ + \varrho a^+ a) + \nu(aa^+\varrho - 2a^+\varrho a + \varrho aa^+)], \quad \dot{\varrho} = \frac{d\varrho}{dt}. \tag{23.1.1}$$

Here a and a^+ are the annihilation and creation operators for the oscillator, γ is the oscillator damping factor in thermodynamic equilibrium, and

$$\nu = \xi/(1-\xi) = [\exp(\hbar\omega/kT) - 1]^{-1}, \quad \xi = \exp(-\hbar\omega/kT). \tag{23.1.2}$$

The kinetic equation (23.1.1) is contained, in essence, in the famous paper by *Landau* [211] (for $\nu=0$). Actually, Eq. (31) in his work can be rewritten as

$$\dot{\varrho}=\frac{2\mathrm{i}}{3\hbar c}(\boldsymbol{D}\varrho\dddot{\boldsymbol{D}}^{+}-\dddot{\boldsymbol{D}}^{-}\varrho\boldsymbol{D}+\boldsymbol{D}\dddot{\boldsymbol{D}}^{-}\varrho-\varrho\dddot{\boldsymbol{D}}^{+}\boldsymbol{D}), \tag{23.1.3}$$

where $\boldsymbol{D}$ is the dipole moment operator, $\boldsymbol{D}^{\pm}$ its positive- and negative-frequency parts. Retaining in (23.1.3) only the resonance terms and taking into account the equalities which hold for a one-dimensional oscillator,

$$\boldsymbol{D}^{+}=\mathrm{e}(\hbar/2m\omega)^{1/2}\mathrm{e}^{\mathrm{i}\omega t}\boldsymbol{a}^{+},\qquad \boldsymbol{D}^{-}=\mathrm{e}(\hbar/2m\omega)^{1/2}\mathrm{e}^{-\mathrm{i}\omega t}\boldsymbol{a},$$

$$\dddot{\boldsymbol{D}}^{\pm}=\mp\mathrm{i}\omega^{3}\boldsymbol{D}^{\pm}, \tag{23.1.4}$$

we see that (23.1.3) does, in fact, coincide with (23.1.1) for $\nu=0$, $\gamma=2(e^2\omega^2/mc^3)/3$. This equation can be obtained also from the general theory of relaxation of quantum systems [210]. A simple derivation, whose advantage is an obvious transition to the classical limit, has been presented in [200]. It is noteworthy that (23.1.1) can describe not only damping but also linear excitation of the oscillator (in a medium with negative temperature). To this end, one should change the sign of γ and take $\nu<-1$. Note also that (23.1.1) is a unique equation which satisfies the following general requirements:

1) linearity in ϱ;
2) hermiticity ($\varrho^{+}(t)=\varrho(t)$);
3) conservation of the norm ($\mathrm{tr}\,\varrho(t)=1$);
4) the one-photon absorption approximation, i.e., the restriction to a pair of operators a and a^{+} in (23.1.1).

Because of the correspondence principle, the constant γ is just the damping factor for the classical oscillator.

Following [200, 201], let us consider a few methods for solving the kinetic equation (23.1.1).

23.1.2 Characteristic Functions and Quasiprobability Distributions

The density operator ϱ may be given not only in terms of its matrix elements ϱ_{mn}, but also by the characteristic functions $\chi_k(\eta)$, which are determined as follows (Sect. 1.7);

$$\chi_N(\eta)=\mathrm{tr}\,(\varrho\mathrm{e}^{\eta a^{+}}\mathrm{e}^{-\bar{\eta}a}),$$

$$\chi_O(\eta)=\mathrm{tr}\,(\varrho\mathrm{e}^{(\eta a^{+}-\bar{\eta}a)}),$$

$$\chi_A(\eta)=\mathrm{tr}\,(\varrho\mathrm{e}^{-\bar{\eta}a}\mathrm{e}^{\eta a^{+}}). \tag{23.1.5}$$

Here

$$\chi_k(\eta)=\exp(-\sigma_k|\eta|^2)\chi_N(\eta),\qquad \sigma_k=\begin{cases}0 & \text{for}\quad k=N\\ \frac{1}{2} & \text{for}\quad k=O\\ 1 & \text{for}\quad k=A.\end{cases} \tag{23.1.6}$$

In terms of functions $\chi_k(\eta)$, (23.1.5) looks like

$$\dot{\chi}_k = -\gamma\left[\frac{1}{2}\eta_j\frac{\partial\chi_k}{\partial\eta_j} + (v+\sigma_k)|\eta|^2\chi_k\right], \quad j=1,2; \quad \eta=\eta_1+i\eta_2 \tag{23.1.7}$$

The general solution can be found at once [200, 201],

$$\begin{aligned}&\chi_k(\eta,t)=\exp[-(v+\sigma_k)|\eta|^2 q]\chi_k(p^{1/2}\eta,0), \quad \text{where}\\ &p=\exp(-\tau), \quad q=1-\exp(-\tau), \quad \tau=\gamma t.\end{aligned} \tag{23.1.8}$$

Equation (23.1.8) for $\chi_0(\eta,t)$ has also been obtained in [205]; there the starting point was not (23.1.7) but a model of a damped oscillator [202] with an approximated integration of the Heisenberg equations for the operator $\hat{a}(t)$. Clearly, the model is equivalent to the kinetic equation (23.1.1).

Now let us describe the oscillator state using a quasiprobability distribution $W_k(\alpha)$. It is related to the characteristic functions $\chi_k(\eta)$ by the Fourier transformation

$$W_k(\alpha) = \frac{1}{\pi^2}\int \chi_k(\eta)\exp(\bar{\eta}\alpha - \eta\bar{\alpha})\,d^2\eta. \tag{23.1.9}$$

Evidently, $W_N(\alpha)$ coincides with the weight function in the integral representation of the density matrix

$$\varrho = \int P(\alpha)|\alpha\rangle\langle\alpha|\,d^2\alpha, \qquad W_N(\alpha)\equiv P(\alpha). \tag{23.1.10}$$

Furthermore, $W_0(\alpha)\equiv W(\alpha)$ is the Wigner distribution [212] (the density in the phase space) and, finally, $W_A(\alpha)=\pi^{-1}Q(\alpha)$, where $Q(\alpha)=\langle\alpha|\varrho|\alpha\rangle$. (A more detailed discussion of properties of functions $W_k(\alpha)$ can be found in Sect. 1.7).

As shown in [200, 201], the quasiprobability distributions $W_k(\alpha)$ satisfy the Fokker-Planck equation

$$\begin{aligned}&\frac{\partial W_k}{\partial t} = -\frac{\partial}{\partial\alpha_i}(A_i W_k) + \frac{1}{2}\frac{\partial^2}{\partial\alpha_i\partial\alpha_j}(B_{ij}W_k), \quad \text{where}\\ &A_i = -\tfrac{1}{2}\gamma\alpha_i, \quad B_{ij}=\tfrac{1}{2}\gamma(v+\sigma_k)\delta_{ij}, \quad \alpha=\alpha_1+i\alpha_2.\end{aligned} \tag{23.1.11}$$

The solution of (23.1.11) is [200, 213]

$$W_k(\alpha,t) = \int d^2\alpha' W_k(\alpha',0)\frac{1}{\pi q(v+\sigma_k)}\exp\left(\frac{|\alpha-\alpha' p^{1/2}|^2}{q(v+\sigma_k)}\right). \tag{23.1.12}$$

A few remarks on the time evolution of the quasiprobabilities $W_k(\alpha,t)$ are not out of place.

1) The complex amplitude in the α plane undergoes a damping proportional to $p^{1/2}$. Besides, for $v\neq 0$ fluctuations in the system increase due to its interaction with the thermostat.

2) As t goes to infinity,

$$W_k(\alpha, t) = \frac{1}{\pi(\nu+\sigma_k)} \exp\left(-\frac{|\alpha|^2}{\nu+\sigma_k}\right) \tag{23.1.13}$$

independently of the initial distribution.

3) For an arbitrary initial distribution $W_k(\alpha, 0)$, the expectation value $\langle\alpha_i\rangle$ and its dispersion $D_{ij} = \langle(\alpha_i - \langle\alpha_i\rangle)(\alpha_j - \langle\alpha_j\rangle)\rangle$ are changed as

$$\langle\alpha_i(t)\rangle = p^{1/2}\langle\alpha_i(0)\rangle \tag{23.1.14}$$
$$D_{ij}(t) = p(t) D_{ij}(0) + \tfrac{1}{2} q(t)(\nu+\sigma_k)\delta_{ij}.$$

Here $\langle\alpha_1 + \mathrm{i}\alpha_2\rangle$ is universal for all k and is equal to $\mathrm{tr}\,(\varrho\hat{a})$, while the dispersions D_{ij} are different for different k, and are simply related to each other,

$$D_{ij}^{(k)} = D_{ij}^{(N)} + \tfrac{1}{2}\sigma_k\delta_{ij}. \tag{23.1.15}$$

If the initial distribution $W_k(\alpha, 0)$ was Gaussian, it retains its functional form in the relaxation process, and its parameters are given by (23.1.14).

Let us reproduce some examples given in [201].

1) Suppose the oscillator was in thermodynamic equilibrium at $t=0$, with temperature T_0 different from the thermostat temperature T:

$$\varrho_{mn}(0) = (1-\xi_0)\xi_0^n\delta_{mn}, \qquad \xi_0 = \exp(-\hbar\omega/kT_0). \tag{23.1.16}$$

Then

$$W_k(\alpha, t) = \frac{1}{\pi(\mu+\sigma_k)} \exp\left(-\frac{|\alpha|^2}{\mu+\sigma_k}\right), \qquad \mu(t) = p(t)\nu_0 + q(t)\nu, \tag{23.1.17}$$

i.e., at any moment of time $W_k(\alpha, t)$ is Gaussian. Therefore, distributions of populations $W_n(t)$ retain Planck form at any time, and the oscillator temperature varies from T_0 to T, according to

$$\bar{n}(t) = p(t)\nu_0 + q(t)\nu. \tag{23.1.18}$$

Note that this result was obtained in another way be *Schwinger* [214].

2) The initial state was a superposition of a coherent state $|\alpha_0\rangle$ and the Planck state, with a given parameter $\nu = \nu_0$. Then

$$W_k(\alpha, t) = \frac{1}{\pi(\mu+\sigma_k)} \exp\left(-\frac{|\alpha - \sqrt{p}\,\alpha_0|^2}{\mu(t)+\sigma_k}\right), \tag{23.1.19}$$

$$\mu(t) = p(t)\nu_0 + q(t)\nu. \tag{23.1.20}$$

3) Suppose that at $t=0$ there was a superposition (following *Glauber* [7, 8]) of the Planck distribution and the coherent state averaged over its phase. Now

$$W_k(\alpha,t)=\frac{\mathrm{I}_0\left(\frac{2\sqrt{p}|\alpha\alpha_0|}{\mu+\sigma_k}\right)}{\pi(\mu+\sigma_k)}\exp\left(-\frac{|\alpha|^2+p(t)|\alpha_0|^2}{\mu+\sigma_k}\right). \tag{23.1.21}$$

where $\mathrm{I}_0(x)$ is the Bessel function.

4) There were exactly N quanta in the initial state $\varrho(0)=|N\rangle\langle N|$. Then

$$W_k(\alpha,t)=\frac{(qv-p+\sigma_k)^N}{\pi(qv+\sigma_k)^{N+1}}\exp\left(-\frac{|\alpha|^2}{qv+\sigma_k}\right)\mathrm{L}_N(x) \tag{23.1.22}$$

where $\mathrm{L}_N(x)$ is the Laguerre polynomial, and

$$x=\frac{p|\alpha|^2}{(qv+\sigma_k)(p-qv-\sigma_k)}. \tag{23.1.23}$$

23.1.3 Use of Operator Symbols

To obtain explicit formulae for the occupation functions $W_n(t)$, let us use the symbol of the density operator $\hat{\varrho}$,

$$R(\bar{\alpha},\beta)=\exp\left[\tfrac{1}{2}(|\alpha|^2+|\beta|^2)\right]\langle\alpha|\hat{\varrho}|\beta\rangle. \tag{23.1.24}$$

Now

$$R(z_1,\bar{z}_2;t)=\sum\varrho_{mn}(t)\frac{z_1^m\bar{z}_2^n}{(m!n!)^{1/2}}, \tag{23.1.25}$$

so $R(z_1,\bar{z}_2)$ is a generating function for the matrix elements $\varrho_{mn}(t)$. In terms of $R(z_1,\bar{z}_2)$ the kinetic equation (23.1.1) is [200, 201]

$$\frac{\partial R}{\partial t}=\gamma\left[(v+1)\frac{\partial^2R}{\partial z_1\partial\bar{z}_2}-\left(v+\frac{1}{2}\right)\left(z_1\frac{\partial R}{\partial z_1}+\bar{z}_2\frac{\partial R}{\partial\bar{z}_2}\right)+v(z_1\bar{z}_2-1)R\right]. \tag{23.1.26}$$

It is suitable to employ the function $R(z_1,\bar{z}_2)$ in a more complicated case where the relaxation occurs in the presence of an external force [200, 201]. Then the density matrix satisfies

$$\frac{\partial\varrho}{\partial t}=-\mathrm{i}[V,\varrho]-\frac{\gamma}{2}[(v+1)(a^+a\varrho-2a\varrho a^++\varrho a^+a)$$

$$+v(aa^+\varrho-2a^+\varrho a+\varrho aa^+)],\quad\text{where} \tag{23.1.27}$$

$$V=-f(t)x=-(2\omega)^{-1/2}[\bar{f}(t)\mathrm{e}^{-\mathrm{i}\omega t}a+f(t)\mathrm{e}^{\mathrm{i}\omega t}a^+], \tag{23.1.28}$$

and (23.1.26) is extended to

$$\frac{\partial R}{\partial t}=\gamma\left[(\nu+1)\frac{\partial^2 R}{\partial z_1\partial \bar{z}_2}-\left(\nu+\frac{1}{2}\right)\left(z_1\frac{\partial R}{\partial z_1}+\bar{z}_2\frac{\partial R}{\partial \bar{z}_2}\right)+\nu(z_1\bar{z}_2-1)R\right]$$

$$-\frac{\mathrm{i}}{\sqrt{2\omega_0}}\left[f(t)\,\mathrm{e}^{\mathrm{i}\omega_0 t}\left(\frac{\partial}{\partial \bar{z}_2}-z_1\right)-\bar{f}(t)\,\mathrm{e}^{-\mathrm{i}\omega t}\left(\frac{\partial}{\partial z_1}-\bar{z}_2\right)\right]R. \qquad (23.1.29)$$

It can be shown that for arbitrary initial conditions (23.1.29) has the following solution

$$R(z_1,\bar{z}_2;t)=\int d\mu(\zeta_1)\,d\mu(\zeta_2)G(z_1,z_2;\zeta_1,\zeta_2|t)R(\zeta_1,\bar{\zeta}_2;0), \qquad (23.1.30)$$

where the integration measure is

$$d\mu(\zeta)=\pi^{-1}\exp(-|\zeta|^2)\,d^2\zeta. \qquad (23.1.31)$$

The Green's function is given by

$$G(z_1,z_2;\zeta_1,\zeta_2|t)=\frac{1}{1+q\nu}\exp\left\{\frac{F(z_1,z_2;\zeta_1,\zeta_2|t)}{1+q\nu}\right\}\quad\text{with} \qquad (23.1.32)$$

$$F=q\nu z_1\bar{z}_2+q(\nu+1)\bar{\zeta}_1\zeta_2+p^{1/2}(z_1\bar{\zeta}_1+\bar{z}_2\zeta_2-\bar{v}\bar{\zeta}_1-v\zeta_2)$$

$$+vz_1+\bar{v}\bar{z}_2-|v|^2. \qquad (23.1.33)$$

The notations used are

$$p=\exp(-\gamma t),\qquad q=1-\exp(-\gamma t), \qquad (23.1.34)$$

$$v(t)=\frac{\mathrm{i}}{\sqrt{2\omega_0}}\int_0^t f(t')\exp\left[-\frac{\gamma}{2}(\mathrm{t}-\mathrm{t}')+\mathrm{i}\omega_0 t'\right]dt'$$

($v(t)$ is the complex amplitude for the forced classical oscillator excited by the force $f(t)$).

In principle, (23.1.30) is the solution of the evolution problem for an arbitrary initial state. To get the populations of different levels, $W_n(t)$, and nondiagonal elements of the density matrix $\varrho_{mn}(t)$, one has to expand the function in powers of z_1 and z_2.

Note that the evolution of the density matrix $\hat{\varrho}(t)$, given by (23.1.30), can be described as a result of three consequent transformations of the initial matrix $\hat{\varrho}(0)$.

1) One transformation involves a relaxation at zero thermostat temperature and $f(t)\equiv 0$. This process corresponds to contraction of the complex amplitude in the α plane by a factor of $p^{-1/2}$. The corresponding Green's function is obtained from (23.1.32, 33) with $v=\nu=0$.

2) Another transformation involves an increase in the Gaussian fluctuations of the amplitude (at $f(t)\equiv 0$ and not taking the damping into account). The Green's function which corresponds to this transformation is obtained from (23.1.32) under the substitution $p\to 1$, $q\to 0$, $qv\to v[1-\exp(-\gamma t)]$, $v\to 0$.

3) The third is a unitary transformation

$$\hat{\varrho}\to\hat{\varrho}'=D(v)\hat{\varrho}D^{-1}(v), \qquad D(v)=\exp(va^{+}-\bar{v}a), \tag{23.1.35}$$

i.e. a shift of the complex amplitude by a vector $v(t)$. The transformation is formally identical to substituting for $\varrho(t)$ the oscillator with no damping under the external force [the damping factor γ enters $D(v)$ only via $v(t)$, (23.1.34)].

Note that transformations (1) and (2) are nonunitary, a fact specific to any relaxation process.

We now turn to some examples.

1) The initial symbol is

$$R(z_1,\bar{z}_2;0)=\frac{1}{1+\nu_0}\exp\left(\frac{\nu_0 z_1\bar{z}_2+\alpha z_1+\bar{\alpha}\bar{z}_2-|\alpha|^2}{1+\nu_0}\right), \tag{23.1.36}$$

where $\nu_0\geq 0$, α is an arbitrary complex number. This density matrix corresponds to the characteristic function

$$\chi_N(\eta)=\exp(-\nu_0|\eta|^2-\bar{\alpha}\eta-\alpha\bar{\eta}), \tag{23.1.37}$$

i.e., to a superposition of the coherent state $|\alpha\rangle$ and the Gaussian noise, which is the Planck distribution with the parameter

$$\nu_0=[\exp(\hbar\omega_0/kT_0)-1]^{-1}.$$

As shown by *Glauber* [7, 8], a superposition of two states corresponds to a sum of electromagnetic fields generated by independent sources. So the example given in (23.1.36) is relevant to the case common in quantum optics, where an input field consists of a coherent signal with a random thermal noise superposed. Substituting (23.1.36) in (23.1.30) yields

$$R(z_1,\bar{z}_2;t)=\frac{1}{1+\mu}\exp\left(\frac{\mu z_1\bar{z}_2+\beta z_1+\bar{\beta}\bar{z}_2-|\beta|^2}{1+\mu}\right), \tag{23.1.38}$$

where

$$\mu=p\nu_0+q\nu, \qquad \beta=\beta(t)=p^{1/2}\alpha+v(t). \tag{23.1.39}$$

Comparing (23.1.38 with 36) reveals the invariance of the state's form during its relaxation process, while only the parameters β and μ are varied. Here $\mu(t)$ is the mean energy of the oscillator which undergoes relaxation in the absence of an external force, and $\beta(t)$ is the fluctuation amplitude of the classical oscillator with the initial condition $\beta(0)=\alpha$.

Explicit expressions for populations stem from (23.1.38),

$$W_n(t)=\frac{\mu^n}{(1+\mu)^{n+1}}\exp\left(-\frac{|\beta|^2}{1+\mu}\right)\mathrm{L}_n\left(-\frac{|\beta|^2}{1+\mu}\right). \tag{23.1.40}$$

Here $\mathrm{L}_n(x)$ is the standard Laguerre polynomial. (Note that $\mathrm{L}_n(x)>0$ at $x\leq 0$.)

2) Setting $\alpha=0$ gives the formulae describing relaxation of the Planck distribution (with a parameter v_0 and an initial temperature T_0). Here $\beta(t)\equiv v(t)$. If we neglect the relaxation now, $\mu(t)=v_0$ and $\gamma=0$ must be inserted in (23.1.34) for $v(t)$. After these simplifications, (23.1.40) transforms into a formula obtained by *Schwinger* [214].

3) In a particular case $v_0=0$, the formulae of Sect. 23.1.1 describe the relaxation of the coherent state $|\alpha\rangle$ (here $\mu=qv$). An especially simple result is obtained for zero thermostat temperature ($v=0$): $\hat{\varrho}(t)=|\beta(t)\rangle\langle\beta(t)|$, i.e, the oscillator is in a coherent state $|\beta(t)\rangle$ at any time moment. This is the only case where the relaxation conserves a pure state, in spite of interacting with a dissipative subsystem (the case in view has vacuum fluctuations of an electromagnetic field). This fact indicates once more a quasiclassical nature of coherent states.

4) With $\alpha=v_0=0$ the oscillator is in ground state, $\hat{\varrho}(0)=|0\rangle\langle 0|$. Equations (23.1.16, 17) are still valid for the matrix elements, where one has to set $\mu=vq$, $\beta(t)=v(t)$. The Poisson distributions are for $W_n(t)$ at $v=0$,

$$W_n(t)=\mathrm{e}^{-\lambda}\frac{\lambda^n}{n!},\qquad \lambda=\lambda(t)=|v(t)|^2. \tag{23.1.41}$$

Feynman [56] obtained this formula for an oscillator without damping. It is remarkable that for $v=0$, taking damping into account does not change the form of the distribution (23.1.41) and is manifested only in the value of the amplitude $v(t)$. This is not the case for $v\neq 0$.

5) If the oscillator has N quanta in its initial state, $\hat{\varrho}(0)=|N\rangle\langle N|$, the problem is more cumbersome to treat. Calculating integral (23.1.30) [200, 201] leads to

$$R(z_1,z_2;t)=\frac{[q(1+v)]^N}{(1+qv)^{N+1}}\mathrm{L}_N\left(-\frac{p(z_1-\bar{v})(\bar{z}_2-v)}{q(1+v)(1+qv)}\right)$$
$$\times\exp\left(\frac{qvz_1\bar{z}_2+vz_1+\bar{v}\bar{z}_2-|v|^2}{1+qv}\right), \tag{23.1.42}$$

In the general case, the expression for $W_n(t)$ derived from (23.1.42) is rather complicated. Let us look at some simple examples.

a) Setting $z_1=z_2=0$ in (23.1.42) gives the ground-level population

$$W_0(t)=\frac{[q(1+v)]^N}{(1+qv)^{N+1}}\exp\left(-\frac{|v|^2}{1+qv}\right)\mathrm{L}_N\left(-\frac{p|v|^2}{q(1+v)(1+qv)}\right). \tag{23.1.43}$$

b) For $v=0$ (zero thermostat temperature) the expression for $W_n(t)$ is simplified essentially [200, 201]:

$$W_n(t)=e^{-\lambda}\sum_{k=0}^{N}\frac{n_<!N!}{n_>!k!(N-k)!}p^k q^{N-k}\lambda^m\times|L_{n_<}^m(\lambda)|^2, \text{ where} \quad (23.1.44)$$

$$n_<=\min(k,n),\quad n_>=\max(k,n),\quad m=|k-n|,\quad \lambda=|v(t)|^2.$$

c) The generating function for the populations $W_n(t)$ at $v=0$ is

$$G_N(z,t)=\sum W_n(t)z^n=(1-p\zeta)^N e^{-\lambda\zeta}\,L_N\left(-\frac{p\lambda\zeta^2}{1-p\zeta}\right),\quad \zeta=1-z. \quad (23.1.45)$$

23.2 Relaxation of a Spinning Particle to Thermodynamic Equilibrium in the Presence of a Magnetic Field

Let us consider a particle with spin s and magnetic moment $\mu=gs$ under the action of a homogeneous magnetic field H. The system has $(2s+1)$ equidistant energy levels, $E_m=\hbar\omega_0 m$, $\omega_0=-gH$. For definiteness, we shall suppose that the gyromagnetic ratio is negative, $g<0$; then $m=-s$ at the lowest level and $m=+s$ at the highest level. If the system is in contact with the thermostat at temperature T, its state is described in terms of the density matrix $\varrho(t)$, and the relaxation due to the magnetic dipole radiation proceeds according to an equation proposed in [215, 216]

$$\begin{aligned}\dot{\varrho}=-\tfrac{1}{2}\gamma[(v+1)(S_+S_-\varrho-2S_-\varrho S_++\varrho S_+S_-)\\ +v(S_-S_+\varrho-2S_+\varrho S_-+\varrho S_-S_+)],\end{aligned} \quad (23.2.1)$$

where γ is a constant describing the system-thermostat interaction, and v is given by the Planck formula

$$v=[\exp(\hbar\omega_0/kT)-1]^{-1}, \quad (23.2.2)$$

$S_\pm=S_x\pm iS_y$ (the z axis is along the magnetic field H). This equation is adequate also for relaxation in a system of N identical two-level atoms confined in a region whose size is less than the wavelength (the *Dicke* problem [150]). As every atom can be in one of two states, it is suitable to introduce the "energy spin" operators [150], which are just the usual Pauli matrices $\sigma_\pm$, σ_z. The atoms do not emit their spontaneous radiation independently, since they are interacting because of the radiation field. It can be shown [150] that the total "spin" $\left(s=\frac{1}{2}\sum_{i=1}^{N}\sigma_i\right)$ is conserved in the relaxation process. In other words, the states with different values of s decay independently and have different specific lifetimes. In this interpretation $m=(n_+ - n_-)/2$, where $n_\pm$ is the upper (lower) level occupation number, and $|m| \leq s \leq N/2$.

Another physical example is depolarization of positive muons in condensed media. This phenomenon has been considered in [217–219]. As the magnetic moment of μ^+ is small compared with that of an electron, the direct interaction of the μ^+ spin with the medium can be neglected. The hyperfine splitting in a muonium is $\hbar\omega_0 \sim 0.1$ K, so that at room temperature $v \gg 1$, $(v+1) \approx v$. Taking this dact into account, we write the following equation for the density matrix [215, 216]

$$\dot{\varrho} = -\mathrm{i}[V, \varrho] + W(\hat{R}\varrho + \varrho\hat{R} - S_-\varrho S_+ - S_+\varrho S_-), \tag{23.2.3}$$

where $V = \frac{1}{4}\hbar\omega_0 \boldsymbol{S}\boldsymbol{S}_\mu - g_e \boldsymbol{S}\boldsymbol{H}$, $\boldsymbol{S}$ is the electron spin, $\boldsymbol{S}_\mu$ the muon spin, $R = (S_+S_- + S_-S_+)/2$, $W = v\gamma$.

Kinetic equation (23.2.3) describes an effect of a medium with infinite temperature. A direct comparison shows that the equation for the muonium density matrix used in [217–219] does actually coincide with that in (23.2.3), though written in another form. The constant W was interpreted there as the probability of muon spin flip.

Note an important property of (23.2.1): if at the initial time moment $t=0$ the density matrix satisfied three fundamental requirements,

A) $\mathrm{tr}\{\varrho\} = 1$ (normalization),
B) $\varrho^+ = \varrho$ (hermiticity),
C) $\langle\psi|\varrho|\psi\rangle \geq 0$ for any state vector $|\psi\rangle$ (nonnegativitivity),

then they are satisfied at an arbitrary time moment, $t > 0$. Only the third requirement (C) is not trivial here; the corresponding proof can be found in [215].

Now we turn to the simplest case where the thermostat is at zero temperature ($v=0$). Equation (23.2.1) is reduced to

$$\dot{\varrho} = \tfrac{1}{2}\gamma\{[S_-, \varrho S_+] + [S_-\varrho, S_+]\}. \tag{23.2.4}$$

The density matrix is determined by its symbol $P(\boldsymbol{n}) = P(\theta, \varphi)$,

$$\varrho = \int d\mu_s(\boldsymbol{n})\, P(\boldsymbol{n}) |\boldsymbol{n}\rangle\langle\boldsymbol{n}|, \tag{23.2.5}$$

where the integration measure is $d\mu_s(n) = (4\pi)^{-1}(2s+1)\, d\Omega(\boldsymbol{n})$, and $\{|\boldsymbol{n}\rangle\}$ are the coherent states for the three-dimensional rotation group. For simplicity, suppose also, that $P(\boldsymbol{n}) = P(\theta)$, i.e., the density matrix is independent of the azimuthal angle. Then using the decomposition of unity

$$\int d\mu_s(\boldsymbol{n}) |\boldsymbol{n}\rangle\langle\boldsymbol{n}| = \hat{1},$$

one gets an equation for the function $P(\theta;t)$, which is of the type obtained and investigated in [148, 149]

$$\frac{\partial}{\partial t}[\sin\theta\, P(\theta;t)] = \frac{\partial}{\partial\theta}\{[s\sin\theta + \tfrac{1}{2}\sin\theta/(1+\cos\theta)]\sin\theta\, P(\theta;t)\}$$

$$+\frac{\partial^2}{\partial\theta^2}[\tfrac{1}{2}(1-\cos\theta)\sin\theta\, P(\theta,t)]. \tag{23.2.6}$$

This is just the Fokker-Planck equation on the unit sphere $S^2 = \{\boldsymbol{n}: \boldsymbol{n}^2 = 1\}$ for a function $f(\theta;t) = \sin\theta\, P(\theta,t)$. The first term on the rhs of (23.2.6) is a displacement responsible for motion of the distribution as a whole; furthermore the distribution expands due to the diffusion coefficient $D(\theta) = (1-\cos\theta)/2$, which has its maximum at $\theta = 0$ and vanishes at $\theta = \pi$. The combined effect of the displacement and diffusion is an expansion of the distribution on the sphere and a motion of its maximum to the pole $\theta = 0$. As t goes to infinity, the density matrix tends to $\varrho = |s, -s\rangle\langle s, -s|$. The position of the distribution maximum θ_{max} satisfies the differential equation

$$\frac{d}{dt}\theta_{max} = -\gamma s \sin\theta_{max}, \tag{23.2.7}$$

which has the solution

$$\tan\tfrac{1}{2}\theta_{max}(t) = \tan\tfrac{1}{2}\theta_{max}(0)\exp(-\gamma s t). \tag{23.2.8}$$

Suppose now that the angular momentum is very large, $s \gg 1$. Then in the first approximation (23.2.6) for $f(\theta,t)$ is

$$\partial f/\partial t = \gamma\frac{\partial}{\partial\theta}(s\sin\theta f). \tag{23.2.9}$$

This equation is easily integrated by means of characteristics, with an initial condition $f(\theta,0) = f_0(\theta)$:

$$f(\theta,t) = (\cosh\gamma s t - \cos\theta\sinh\gamma s t)^{-1}. \tag{23.2.10}$$

$$f_0(2\arctan\exp(\gamma s t)\tan(\tfrac{1}{2}\theta)).$$

One can verify that the function $f(\theta,t)$ is normalized correctly for all times. For a particular initial condition

$$f_0 = \sin\theta\,\delta(\cos\theta - \cos\theta_0),$$

one has

$$f(\theta,t) = \sin\theta\,\delta\left(\cos\theta - \frac{\cos\theta_0\cosh\gamma s t + \sinh\gamma s t}{\cos\theta_0\sinh\gamma s t + \cosh\gamma s t}\right).$$

Turning back to the exact equation (23.2.6), the variables can be changed as suggested in [149],

$$f(\theta,t)=2^{2s+1}\sin\theta(1-\cos\theta)^{-2(s+1)}h(z,t), \qquad z=\cot^2(\tfrac{1}{2}\theta). \tag{23.2.11}$$

The equation is transformed to

$$\partial h/\partial t=[1+2(s+1)z]\partial h/\partial z+z(1+z)\partial^2 h/\partial z^2. \tag{23.2.12}$$

This equation with an initial condition $h(z,0)=h_0(z)$ has been solved in [149]:

$$\begin{aligned} h(z,t)=2\int_0^\infty d\sigma\sigma &\begin{Bmatrix}\tanh\pi\sigma\\ \coth\pi\sigma\end{Bmatrix} F(s+\tfrac{1}{2}+\mathrm{i}\sigma, s+\tfrac{1}{2}-\mathrm{i}\sigma;1;-z)\\ &\times\exp\{-[\sigma^2+(s+\tfrac{1}{2})^2]t\}\\ &\times\int_0^\infty F(s+\tfrac{1}{2}+\mathrm{i}\sigma;s+\tfrac{1}{2}-\mathrm{i}\sigma;1;-z')h_0(z')(1+z')^2dz', \end{aligned}$$

where $F(a,b,c;x)$ is the standard hypergeometric function; tanh in braces stands for integer s, and coth stands for half-integer s.

For special initial data, there is a solution with separable time dependence, $h(z,t)=\Phi(z)\exp(-\lambda t)$:

$$\Phi_\pm(z,\lambda)=(1+z)^{-s-\frac{1}{2}\mp\varrho}F(s+\tfrac{1}{2}\pm\varrho,-s+\tfrac{1}{2}\pm\varrho,1\pm2\varrho;(1+z)^{-1}),$$

where $\varrho=[(s+\frac{1}{2})^2-\lambda]^{1/2}$. Note that the square root in the definition of ϱ can be made unique by cutting the complex λ plane from $(s+\frac{1}{2})^2$ to $+\infty$ along the positive real axis and requiring that ϱ be positive for real positive $(s+\frac{1}{2})^2-\lambda$. Consequently, $\mathrm{Re}\{\varrho\}>0$, if $\mathrm{Im}\{\lambda\}\neq0$.

24. Landau Diamagnetism

In this chapter, following [220], we use coherent states to obtain Landau diamagnetism [221] for a free electron gas.

We shall start from the retarded Green's function for the harmonic oscillator in CS representation. The units to be used are $\hbar=1$, $k=1$ (k is the Boltzmann constant), and throughout the chapter the oscillator energy is referred to the ground level $E_0=\hbar\omega/2$, unless indicated otherwise.

By definition, the retarded Green's function is

$$G(\alpha, t|\beta, t')=\langle\alpha(t)|\beta(t')\rangle=\langle\alpha|e^{-iH\tau}|\beta\rangle, \tag{24.1}$$

$\tau=t-t'>0$, $H=\omega a^+a$ is the oscillator Hamiltonian. Since $\alpha(t)=\alpha\exp(-i\omega t)$,

$$G(\alpha, t|\beta, t')=\exp\left(\bar{\alpha}\beta e^{i\omega\tau}-\frac{|\alpha|^2+|\beta|^2}{2}\right). \tag{24.2}$$

Note that (24.2) is much simpler than the well-known expression for the Green's function in coordinate representation.

Equation (24.2) leads directly to an expression for another Green's function

$$G_E(\alpha, \beta)=i\langle\alpha|(E-H)^{-1}|\beta\rangle=\int_0^\infty e^{iE\tau}G(\alpha, \tau|\beta, 0)$$

$$=\frac{i}{E}\exp\left(-\frac{(|\alpha|^2+|\beta|^2)}{2}\right)\Phi\left(-\frac{E}{\omega}, 1-\frac{E}{\omega}\middle|\bar{\alpha}\beta\right). \tag{24.3}$$

which is also much simpler than the corresponding Green's function in coordinate representation. Here $\Phi(a, b|z)$ is the confluent hypergeometric function.

The hypergeometric function in (24.3) has poles at $E=n\omega$, $n=1, 2, 3, \ldots$ which determine the oscillator energy levels. Actually,

$$G_E(\alpha, \beta)=\sum_{n=0}^{\infty}\overline{\psi_n(\alpha)}\,\psi_n(\beta)\,\frac{1}{E-n\omega}, \quad \text{where} \tag{24.4}$$

$$\psi_n(\alpha)=\frac{\alpha^n}{\sqrt{n!}}\,e^{-|\alpha|^2/2}. \tag{24.5}$$

The temperature Green's function

$$\langle\alpha|e^{-H/T}|\beta\rangle=\exp\left(\bar{\alpha}\beta e^{-\omega/T}-\frac{(|\alpha|^2+|\beta|^2)}{2}\right) \tag{24.6}$$

is obtained from (24.2) by substitution $\tau\to i/T$, hence giving the partition function for the oscillator

$$Z=\int\frac{d^2\alpha}{\pi}\langle\alpha|e^{-H/T}|\alpha\rangle=\frac{1}{1-e^{-\omega/T}}. \tag{24.7}$$

One application of these formulae relates to the magnetism of free electrons. Here, $\omega=e\mathscr{H}/mc$ is the electron cyclotron frequency in magnetic field $\mathscr{H}$. Let us consider a unit volume of electron gas in a homogeneous magnetic field $\mathscr{H}$ directed along the z axis. To calculate the partition function it is appropriate to use here a representation (α, p_y, p_z) for the wave functions. The result is

$$Z_{el}=\frac{m\omega}{2\pi}\left(\frac{2\pi}{mT}\right)^{1/2}(1-e^{-\omega/T})^{-1}. \tag{24.8}$$

The first factor in (24.8) is due to the degeneracy of the Landau levels in p_y, the second and the third factors result from integration over p_z and α. Hence we obtain an expression for the free energy, adding the ground-state level $E_0=\omega/2$,

$$F=\frac{\omega}{2}-T\ln Z_{el}. \tag{24.9}$$

The magnetization is

$$M=-\frac{\partial F}{\partial\mathscr{H}}=-\frac{e}{mc}\left(\frac{1}{2}-\frac{T}{\omega}+\frac{1}{e^{\omega/T}-1}\right). \tag{24.10}$$

For small ω, M is proportional to the magnetic field $\mathscr{H}$, and one gets the familiar expression for the magnetic susceptibility

$$\chi=-\frac{1}{12T}\left(\frac{e}{mc}\right)^2. \tag{24.11}$$

Up to now we have worked within classical statistics. For a strong degeneracy, where the de Haas-van Alfen effect takes place, one should bear the Fermi-Dirac statistics of electrons in mind. It is appropriate to write the basic formula of quantum statistics

$$\Omega_Q=-T\sum_a\ln\{1+\exp[(\mu-E_a)/T\} \tag{24.12}$$

by means of a contour integral

$$\Omega_{\mathrm{Q}}=\int_{L}\Omega_{\mathrm{cl}}(\beta)\,\frac{\mathrm{e}^{\mu\beta}\,\mathrm{d}\beta}{2\mathrm{i}\beta\sin(\pi\beta T)}, \tag{24.13}$$

as suggested by *Rumer* [222]. Here

$$\Omega_{\mathrm{cl}}(\beta)=-T\sum_{a}\mathrm{e}^{-\beta E_a}=-T\,\frac{m\omega}{2\pi}\sqrt{\frac{m}{2\pi\beta}}\,\mathrm{e}^{-\frac{\beta\omega}{2}}\,(1-\mathrm{e}^{-\beta\omega})^{-1} \tag{24.14}$$

is the classical partition function and the contour L is a vertical straight line in the β plane intersecting the real axis on the interval $(0, 1/T)$.

In the presence of single-particle excitations in the system, function $\Omega_{\mathrm{cl}}(\beta)$ has poles on the imaginary β axis. The contribution from such a pole $\beta=\mathrm{i}n/\omega$ to Ω_{Q} contains an oscillating factor $\exp(\mathrm{i}n/\omega)$. The integrand in (24.13) has branch points at $\beta=0$ and $\beta=\infty$ [so that one should make a cut $(-\infty<\beta\leq 0)$] and poles at the points

$$\beta_s=\pm\frac{2\pi\mathrm{i}s}{\omega}. \tag{24.15}$$

For high temperature one can close the contour L in the right half-plane, thereby getting the conventional formula for Ω_{Q}.

For weak fields, where $\mu>\omega/2$, the contour L can be closed to the left, giving the pole contribution,

$$\Omega_{\mathrm{osc}}=-T\left(\frac{m\omega}{2\pi}\right)\sum_{s=-\infty}^{\infty}\sqrt{\frac{m}{2\pi}}\left(\frac{\mathrm{e}^{\beta\left(\mu-\frac{\omega}{2}\right)}}{2\mathrm{i}\beta^{3/2}\sin(\pi\beta T)}\,\mathrm{e}^{-\frac{2\pi\mathrm{i}}{\beta\omega}}\right)\Bigg|_{\beta=2\pi\mathrm{i}s/\omega}$$

$$=\frac{T}{4\pi^2}\sum_{s=1}^{\infty}(-1)^s\left(\frac{m\omega}{s}\right)^{3/2}\frac{\cos\left[(2\pi s\mu/\omega)-\frac{\pi}{4}\right]}{\sinh(2\pi^2 sT/\omega)}. \tag{24.16}$$

This expression varies periodically with the field intensity $\mathscr{H}$ (the de Haas-van Alfen effect), that is an effect of analytical properties of Green's functions.

The electron spin has been neglected in (24.14). To take it into account one should add a factor $\exp(-\beta g\mu_{\mathrm{B}}\sigma\mathscr{H})$ due to the spin energy in the magnetic field (μ_{B} is the Bohr magneton, $\sigma=\pm 1/2$), and sum up over spin projections. Consequently, a factor $2\cosh(g\beta\omega/4)$ appears in (24.14), and a factor $2\cos(\pi g s/2)$ in (24.16).

For very weak fields $\omega\ll\mu$, and all the terms in the sum in (24.16) oscillate fast, so that $\Omega_{\mathrm{osc}}\to 0$, leaving only the left cut contribution which leads to Landau diamagnetism. As $\omega\to 0$,

$$\Omega_{\mathrm{cl}}(\beta)\approx -T\left(\frac{m}{2\pi\beta}\right)^{3/2}\left(1-\frac{\beta^2\omega^2/4}{3!}+\dots\right), \tag{24.17}$$

hence (at $T \to 0$)

$$\Omega_Q \approx \Omega_0 + \int \frac{e^{\beta H}\, d\beta}{2i\beta \sin(\pi\beta T)} \frac{T\mu_B^2 \mathscr{H}^2}{6} \left(\frac{m}{2\pi\beta}\right)^{3/2} \beta^2$$

$$= \Omega_0 + \frac{\mu_B^2 \mathscr{H}^2}{6} \left(\frac{m}{2\pi}\right)^{3/2} \int_L \frac{e^{\mu\beta}\, d\beta}{2\pi i \beta^{3/2}}. \tag{24.18}$$

The contour integral equals $\sqrt{\mu}\,\Gamma(\frac{3}{2})$, and

$$\Omega_Q \approx \Omega_0 + \frac{\mu_B^2 \mathscr{H}^2}{3} \left(\frac{m}{2\pi}\right)^{3/2} \sqrt{\frac{\mu}{\pi}} = \Omega_0 + \frac{\mu_B^2 \mathscr{H}^2}{8\mu}. \tag{24.19}$$

The second derivative over $\mathscr{H}$ gives

$$\chi = -\frac{\mu_B^2}{4\mu}. \tag{24.20}$$

Account for the electron spin leads to an extra factor of 2 in χ, and to addition of the Pauli paramagnetism; the resulting expression is

$$\begin{gathered} \chi = \chi_{\text{dia}} + \chi_{\text{para}} \\ \chi_{\text{dia}} = -\frac{\mu_B^2}{2\mu}, \qquad \chi_{\text{para}} = \frac{3\mu_B^2}{2\mu}. \end{gathered} \tag{24.21}$$

25. The Heisenberg-Euler Lagrangian

In this chapter, following [223] and using coherent states, we derive nonlinear corrections to the electromagnetic field Lagrangian, the Heisenberg-Euler correction [224].

Let us consider a spinless relativistic particle in constant electric $\boldsymbol{\mathscr{E}}$ and magnetic $\boldsymbol{\mathscr{H}}$ field. Suppose that $(\boldsymbol{\mathscr{E}}\boldsymbol{\mathscr{H}}) \neq 0$. Then a reference frame exists, where

$$\boldsymbol{\mathscr{E}}=(0,0,\mathscr{E}), \qquad \boldsymbol{\mathscr{H}}=(0,0,\mathscr{H}). \tag{25.1}$$

The particle state is characterized by a generalized momentum

$$\Pi_\mu = p_\mu - eA_\mu, \quad [\Pi_\mu, \Pi_\nu] = \mathrm{i}eF_{\mu\nu}, \quad \mu,\nu=0,1,2,3, \tag{25.2}$$

where A_μ is the electromagnetic field potential, and $F_{\mu\nu}$ is the electromagnetic field tensor. In our case,

$$[\Pi_1, \Pi_2] = \mathrm{i}e\mathscr{H}, \qquad [\Pi_0, \Pi_3] = \mathrm{i}e\mathscr{E}. \tag{25.3}$$

It follows from (25.3) that the operators Π_μ can be expressed in terms of boson operators

$$\begin{aligned} \Pi_1 &= \sqrt{\frac{e\mathscr{H}}{2}}\,(a+a^+), \qquad \Pi_2 = -\mathrm{i}\sqrt{\frac{e\mathscr{H}}{2}}\,(a-a^+), \\ \Pi_3 &= -\mathrm{i}\sqrt{\frac{e\mathscr{E}}{2}}\,(b-b^+), \qquad \Pi_0 = \sqrt{\frac{e\mathscr{E}}{2}}\,(b+b^+). \end{aligned} \tag{25.4}$$

To find nonlinear corrections to the electromagnetic field Lagrangian it is convenient to use the proper time method which reduces the problem to that with the Hamiltonian

$$H = \mu^2 - \Pi_\mu^2 = \mu^2 + e\mathscr{H}(a^+a + aa^+) - e\mathscr{E}(bb + b^+b^+), \tag{25.5}$$

where μ is the particle mass.

The resulting electromagnetic field action functional acquires a correction

$$W' = -\mathrm{i}\,\mathrm{tr}\,\{\ln G\}, \tag{25.6}$$

where $G=(\mu^2-\Pi_\mu^2)^{-1}$ is the Green's function for the scalar particle in the external electromagnetic field. In our case this quantity is given by

$$W' = -\mathrm{i}\,\mathrm{tr}\,\{\ln[(\mu^2-\Pi^2)^{-1}]\} = -\mathrm{i}\,\mathrm{tr}\left\{\int_0^\infty \frac{ds}{s}\exp(-sH)\right\}$$

$$= -\mathrm{i}\int_0^\infty \frac{ds}{s}\,\mathrm{e}^{-\mu^2 s}\,\mathrm{tr}\,\{\exp[s\,e\mathscr{E}(bb+b^+b^+)$$

$$-s\,e\mathscr{H}(a^+a+aa^+)]\}. \tag{25.7}$$

The trace in (25.7) is calculated over all particle quantum numbers, so the degeneracy of the Landau levels should be taken into account. Therefore there are additional factors of $e\mathscr{H}L_xL_y/2\pi$ and $e\mathscr{E}L_zT/2\pi$, which are due to the degeneracy of transverse and longitudinal components. Here L_x, L_y, L_z are the sizes of the region where the external field is present, T is the time period during which the field is nonzero. The action W' turns out to be proportional to the four-volume $L_xL_yL_zT$, as it must be anticipated for a constant and homogeneous field. The coefficient at the four-volume is just the desired Lagrangian correction,

$$\mathscr{L}' = -\mathrm{i}\,\frac{e\mathscr{H}}{2\pi}\,\frac{e\mathscr{E}}{2\pi}\int_0^\infty \frac{ds}{s}\,\mathrm{e}^{-s\mu^2}\,\mathrm{tr}\,\{\exp[se\mathscr{E}(b^2+(b^+)^2)$$

$$-s\,e\mathscr{H}(aa^+ + a^+a)]\}. \tag{25.8}$$

The symbol tr{ } here stands for the trace over variables relevant to the operators a and b, and it can be calculated easily. The first factor is

$$\mathrm{tr}\,\{\exp[-se\mathscr{H}(a^+a+aa^+)]\} = \sum_{n=0}^{\infty} \exp[-se\mathscr{H}(2n+1)]$$

$$= [2\sinh se\mathscr{H}]^{-1}. \tag{25.9.}$$

To calculate the second factor we make a transformation diagonalizing the operator $(bb+b^+b^+)$

$$bb+b^+b^+ = \mathrm{i}\,U(b^+b+bb^+)U^{-1}$$

$$U=\exp\left[\frac{\mathrm{i}}{8}\pi(b^+b^+-bb)\right]. \tag{25.10}$$

Thus we obtain

$$\mathrm{tr}\,\{\exp[se\mathscr{E}(bb+b^+b^+)]\} = \mathrm{tr}\,\{U\exp[\mathrm{i}se\mathscr{E}(b^+b+bb^+)]U^{-1}\}$$

$$=\frac{\mathrm{i}}{2\sin(se\mathscr{E})}. \tag{25.11}$$

The final expression for $\mathscr{L}'$ takes the form

$$\mathscr{L}' = \frac{1}{16\pi^2} \int \frac{ds}{s} \frac{e\mathscr{H}}{\sinh(se\mathscr{H})} \frac{e\mathscr{E}}{\sin(se\mathscr{E})} \mathrm{e}^{-s\mu^2}. \tag{25.12}$$

This expression should be regularized by subtracting the first terms of the Taylor series in $\mathscr{E}^2$ and $\mathscr{H}^2$ from the integrand

$$\mathscr{L}' = \frac{1}{16\pi^2} \int \frac{ds}{s^3} \left[\frac{se\mathscr{H}}{\sinh(se\mathscr{H})} \frac{se\mathscr{E}}{\sin(se\mathscr{E})} - 1 - \frac{e^2 s^2}{6} (\mathscr{E}^2 - \mathscr{H}^2) \right] \mathrm{e}^{-s\mu^2}. \tag{25.13}$$

For weak fields the fourth-order correction in $\mathscr{E}$ and $\mathscr{H}$ is obtained

$$\mathscr{L}_{\mathrm{sc}}^{(4)} = \frac{e^4}{16\pi^2} \frac{7(\mathscr{E}^2 - \mathscr{H}^2)^2 + 4(\boldsymbol{\mathscr{E}}\boldsymbol{\mathscr{H}})^2}{360\mu^4}. \tag{25.14}$$

A correction to the electromagnetic field Lagrangian due to the interaction with spin-1/2 particles can be obtained analogously. One should take into account only two additional factors. First, because of the Fermi statistics the sign in (25.6) must be changed, second, a factor 1/2 appears in the square of the Green's function,

$$W_{1/2} = \mathrm{i}\,\mathrm{tr}\{\ln G\} = \mathrm{i}\,\mathrm{tr}\{\ln[m + \gamma\Pi]^{-1}\} = \frac{\mathrm{i}}{2}\,\mathrm{tr}\{\ln[m^2 - (\gamma\Pi)^2]^{-1}\}, \tag{25.15}$$

where m is the particle mass.

The quantity $(\gamma\Pi)^2 = \Pi^2 + e\sigma_{\mu\nu}F_{\mu\nu}$ contains a term with $\sigma_{\mu\nu}F_{\mu\nu}$ which commutes with the operator Π^2. Therefore the spin projection trace gives a factor $2\cosh(se\mathscr{H})$ for the magnetic part and a factor $2\cos(se\mathscr{E})$ for the electric part.

Hence we get an expression for the Heisenberg-Euler Lagrangian [224],

$$\mathscr{L}'_{1/2} = -\frac{1}{8\pi^2} \int_0^\infty \frac{ds}{s^3} \left[\frac{se\mathscr{H}}{\tanh se\boldsymbol{\mathscr{H}}} \frac{se\mathscr{E}}{\tan se\boldsymbol{\mathscr{E}}} - 1 + \frac{e^2 s^2}{3} (\mathscr{E}^2 - \mathscr{H}^2) \right] \mathrm{e}^{-sm^2}. \tag{25.16}$$

For weak fields, the familiar expression is reconstructed,

$$\mathscr{L}_{1/2}^{(4)} = \frac{e^4}{8\pi^2} \frac{(\mathscr{E}^2 - \mathscr{H}^2)^2 + 7(\boldsymbol{\mathscr{E}}\boldsymbol{\mathscr{H}})^2}{45m^4}. \tag{25.17}$$

Note that the integrands in (25.13, 16) have poles at the points $s_n = \pi n/e\mathscr{E}$, so the Lagrangian has an imaginary part, indicating a vacuum instability with respect to the particle pair production.

26. Synchrotron Radiation

Synchrotron radiation is quantum radiation of a charged particle in a homogeneous magnetic field, or, which is the same, a radiative decay of the Landau levels. So this effect can be described by calculating the mass operator for the particle in a homogeneous magnetic field [225, 226]. As shown in [223], which we follow in this chapter, using coherent states simplifies the calculations substantially.

As in Chap. 25, we consider a relativistic charged particle in a homogeneous magnetic field. Here the longitudinal components of the momentum are conserved, and they can be replaced by their eigenvalues,

$$\Pi_{\|}=(\Pi_0,\Pi_z)=(E,0). \tag{26.1}$$

The transverse components are expressed in terms of boson operators, as in Chap. 25,

$$\Pi_1=\sqrt{\frac{e\mathscr{H}}{2}}\,(a+a^+),\qquad \Pi_2=-\mathrm{i}\sqrt{\frac{e\mathscr{H}}{2}}\,(a-a^+). \tag{26.2}$$

The particle mass operator in the magnetic field is given by

$$M=\frac{e^2}{(2\pi)^4}\int\frac{d^4k}{k^2}\int\limits_0^\infty ds(2\Pi-k)\exp\left[\mathrm{i}s(\Pi-k)^2-\mathrm{i}s\mu^2\right](2\Pi-k). \tag{26.3}$$

The exponential operator factor must be transposed to the extreme right. As the operators Π_1 and Π_2 do not commute with this operator, they are transformed to

$$\begin{aligned}\Pi_1\to\Pi_1(s)&=(\Pi_1-k_1)\cos 2se\mathscr{H}-(\Pi_2-k_2)\sin 2se\mathscr{H}+k_1,\\ \Pi_2\to\Pi_2(s)&=(\Pi_2-k_2)\cos 2se\mathscr{H}+(\Pi_1-k_1)\sin 2se\mathscr{H}+k_2.\end{aligned} \tag{26.4}$$

Restricting ourselves to the leading terms in (k/E), we obtain

$$M=\frac{4e^2}{(2\pi)^4}\int\frac{d^4k}{k^2}\int\limits_0^\infty ds\Pi\Pi(s)\exp\left[\mathrm{i}s(\Pi-k)^2-\mathrm{i}s\mu^2\right] \tag{26.5}$$

$$\Pi\Pi(s)=\Pi^2+\Pi_\perp^2(1-\cos 2se\mathscr{H})+\mathrm{i}e\mathscr{H}\sin 2se\mathscr{H}.$$

On the mass shell,

$$\Pi^2=\mu^2, \qquad \Pi_\perp^2=E^2-\mu^2. \tag{26.6}$$

Further, for a weak magnetic field and a relativistic particle

$$e\mathscr{H} \ll \mu^2 \ll E^2, \tag{26.7}$$

then

$$M=\frac{4e^2\mu^2}{(2\pi)^4}\int\frac{d^4k}{k^2}\int_0^\infty ds\left(1+2\frac{E^2}{\mu^2}\sin^2 se\mathscr{H}\right)\exp\left[\mathrm{i}s(\Pi-k)^2-\mathrm{i}s\mu^2\right]. \tag{26.8}$$

As the operator M is just an exponential of a quadratic form of the operators a and a^+, it is suitable to represent the operator by its symbol

$$M(\alpha)=\langle\alpha|M|\alpha\rangle,$$

where $|\alpha\rangle$ is the standard Glauber CS

$$|\alpha\rangle=\exp\left(-\frac{|\alpha|^2}{2}\right)\exp(\alpha a^+)|0\rangle.$$

Using formulae from Sect. 1.2, and retaining only linear terms in the exponent, then

$$\langle\beta|\exp\left[\mathrm{i}s(\Pi-k)^2-\mathrm{i}s\mu^2\right]|\beta\rangle=\exp(-2\mathrm{i}sk_0E+2\mathrm{i}\boldsymbol{N}\boldsymbol{k}), \tag{26.9}$$

where the vector $\boldsymbol{N}$ has transverse components only,

$$\boldsymbol{N}=\frac{|\beta|}{\sqrt{2e\mathscr{H}}}(\sin 2se\mathscr{H}, 1-\cos 2se\mathscr{H}, 0). \tag{26.10}$$

Since

$$|\beta|^2=\langle\beta|a^+a|\beta\rangle=(E^2-\mu^2-e\mathscr{H})/2e\mathscr{H}, \quad \text{then} \tag{26.11}$$

$$|\boldsymbol{N}|=\frac{(E^2-\mu^2)^{1/2}}{e\mathscr{H}}\sin(se\mathscr{H}), \quad \text{and} \tag{26.12}$$

$$M(\beta)=\langle\beta|M|\beta\rangle=\frac{4e^2\mu^2}{(2\pi)^4}\int\frac{d^4k}{k^2}\int_0^\infty ds\left(1+2\frac{E^2}{\mu^2}\sin^2 se\mathscr{H}\right) \times\exp\left[2\mathrm{i}(\boldsymbol{N}\boldsymbol{k}-sEk_0)\right]. \tag{26.13}$$

Furthermore, in view of the optical theorem, the decay probability W_β of state $|\beta\rangle$ is proportional to the imaginary part of the mass operator,

$$W_\beta = -\frac{1}{E}\,\mathrm{Im}\,\{W(\beta)\}, \tag{26.14}$$

so that

$$W_\beta = -\frac{4e^2\mu^2}{E}\int\frac{d\boldsymbol{k}}{2\omega(2\pi)^3}\int_0^\infty ds\left(1+\frac{E^2}{\mu^2}\sin^2 se\mathscr{H}\right)$$
$$\times\cos(2\boldsymbol{N}\boldsymbol{k}-2sE\omega). \tag{26.15}$$

Integration over the angles of the photon emission leads to

$$\int d\Omega\cos(2\boldsymbol{N}\boldsymbol{k}-2sE\omega)=\frac{\pi}{N\omega}\,[\sin(2\omega Es+2N\omega)$$
$$-\sin(2\omega Es-2N\omega)]. \tag{26.16}$$

Note that in the case of a weak electromagnetic field, the main contribution is due to the region $se\mathscr{H}\sim\mu/E\ll 1$ in the integral over s. Expanding over this parameter gives

$$\int d\Omega\cos(2\boldsymbol{N}\boldsymbol{k}-2Es\omega)=\frac{\pi}{Es}\left[\sin(4\omega Es)-\sin\omega Es\left(\frac{\mu^2}{E^2}+\frac{(se\mathscr{H})^2}{3}\right)\right]. \tag{26.17}$$

The spectral distribution of the radiation is

$$dw=\frac{e^2}{4\pi^2}\left(\frac{\mu}{E}\right)^2 d\omega\int_0^\infty\frac{ds}{s}\left[1+2\,\frac{E^2}{\mu^2}\,(se\mathscr{H})^2\right]$$
$$\times\left[\sin\omega Es\left(\frac{\mu^2}{E^2}+\frac{s^2e^2\mathscr{H}^2}{3}\right)-\sin 4\omega Es\right]\quad\text{or} \tag{26.18}$$

$$dw=\frac{2}{\pi}\left(\frac{\mu}{E}\right)^2 d\omega\left[\int_0^\infty\frac{dx}{x}\,(1+2x^2)\,\sin\xi\left(x+\frac{x^3}{3}\right)-\frac{\pi}{2}\right], \tag{26.19}$$

where

$$\xi=(\omega\mu^3)/(e\mathscr{H}E^2).$$

This formula was originally obtained by *Schwinger* [226], but in a more complicated way.

Retaining the terms of order ω, one can get corrections of order $\chi=eE\mathscr{H}/\mu^3$ which are of quantum nature.

27. Classical and Quantal Entropy

A "classical entropy" of quantum-mechanical systems, as introduced in [227], has a natural definition in the framework of the CS method. Some inequalities for this quantity are easily proved by this method.

The classical entropy of a quantum system, S^{cl}, is a useful concept proposed by *Wehrl* [227], who also investigated some of its properties (a review was given in [228]). Wehrl conjectured, in particular, that $S^{\text{cl}} \geq 1$. This statement was proved later by *Lieb* [229].

Let us start from the simplest quantum system with one degree of freedom; let p and q be the canonical momentum and coordinate operators. As is known, the system state is described by means of the density matrix, that is a nonnegative operator with unit trace. The corresponding quantal entropy of the system is

$$S^{\text{qu}}(\hat{\varrho}) = -\operatorname{tr}\{\hat{\varrho} \ln \hat{\varrho}\}, \tag{27.1}$$

while the classical entropy for a phase space distribution $\varrho(p, q)$ is

$$S(\varrho) = -\int \frac{dp\, dq}{2\pi\hbar}\, \varrho \ln \varrho. \tag{27.2}$$

One would suppose that the $\hbar \to 0$ limit of the quantal entropy is just the classical entropy. This is not the case, however, since the quantal entropy is always ≥ 0, while the classical entropy can be negative, or even $-\infty$. As shown in [228], this is due to the fact that the classical phase space distribution can be confined to an arbitrary narrow domain with an area less than $2\pi\hbar$. Moreover, such a distribution is incompatible with the uncertainty principle and cannot be realized in a real physical system. At best, one would deal with a quantal state characterized by the least possible uncertainty $\Delta p \Delta q$ localized near a phase space point (p, q), i.e., the coherent state $|\alpha\rangle$, $\alpha = (2\hbar)^{-1/2}(q + ip)$.

By definition, the classical distribution is

$$\varrho(\alpha) = \langle \alpha | \hat{\varrho} | \alpha \rangle, \tag{27.3}$$

and classical entropy is

$$S^{\text{cl}} = S(\varrho(\alpha)). \tag{27.4}$$

The function $\varrho(\alpha)$ satisfies the following conditions:

i) $0 \leq \varrho(\alpha) \leq 1$,
ii) $\varrho(\alpha)$ is a continuous function,
iii) $\varrho(\alpha) \to 0$ as $|\alpha| \to \infty$.

Hence it can be concluded that

i) in agreement with the uncertainty principle, the distribution $\varrho(p, q) \equiv \varrho(\alpha)$ cannot be confined to a domain of an area less than $2\pi\hbar$;
ii) an important inequality holds,

$$S^{\mathrm{cl}} \geq S^{\mathrm{qu}}. \tag{27.5}$$

Actually, as the function $s(x) = -x \ln x$ is concave for positive x, then

$$S(\langle\alpha|\hat{\varrho}|\alpha\rangle) \geq \langle\alpha|S(\hat{\varrho})|\alpha\rangle, \quad \text{so} \tag{27.6}$$

$$S^{\mathrm{cl}} = \int \frac{d^2\alpha}{\pi} S(\varrho(\alpha)) \geq \int \frac{d^2\alpha}{\pi} \langle\alpha|S(\varrho)|\alpha\rangle = \mathrm{tr}\{S(\varrho)\} = S^{\mathrm{qu}}(\varrho). \tag{27.7}$$

Furthermore, because the function $\varrho(\alpha)$ is continuous, the equality $S^{\mathrm{cl}} = S^{\mathrm{qu}}$ would imply $S(\langle\alpha|\hat{\varrho}|\alpha\rangle) = \langle\alpha|S(\hat{\varrho})|\alpha\rangle$ for all α, and as the function $s(x)$ is strongly concave one would conclude that every $|\alpha\rangle$ must be an eigenvector of $\hat{\varrho}$, an impossible requirement. Note that the minimum of S^{qu} is zero (it attains the minimum at any pure state), while the minimum of S^{cl} is never zero.

Another important property of the entropy S^{cl} is that it is a monotonous function. Let $\hat{\varrho}_{12}$ be the density matrix for a quantum system in the Hilbert space $L^2(R) \otimes L^2(R)$, and $|\alpha_1, \alpha_2\rangle = |\alpha_1\rangle \otimes |\alpha_2\rangle$. By definition,

$$\varrho_{12}^{\mathrm{cl}}(\alpha_1, \alpha_2) = \langle\alpha_1, \alpha_2|\hat{\varrho}_{12}|\alpha_1, \alpha_2\rangle, \tag{27.8}$$

and the symbol $\varrho_{12}^{\mathrm{cl}}(\alpha_1)$ corresponds to the density matrix $\hat{\varrho}_1 = \mathrm{tr}_2\, \hat{\varrho}_{12}$. *Wehrl* [227] proved an inequality

$$S_{12}^{\mathrm{cl}} \equiv S(\varrho_{12}^{\mathrm{cl}}(\alpha_1, \alpha_2)) \geq S(\varrho_1^{\mathrm{cl}}) = S_1^{\mathrm{cl}}. \tag{27.9}$$

Note that this entropy property, which is desirable from the physical point of view, is not in general fulfilled for quantal entropy.

Wehrl [227] also advanced the following hypothesis.

Hypothesis. The minimum of S^{cl} is 1 (and is independent of $\hbar$). This minimum is attained only for density matrices of the type $\hat{\varrho} = |\alpha\rangle\langle\alpha|$, where $|\alpha\rangle$ is the usual Glauber coherent state. *Lieb* proved this statement in [229], so it will not be reproduced here. Here we present only the following lemma.

Lemma. If a density matrix $\hat{\varrho}$ provides a minimum of S^{cl}, then $\hat{\varrho} = \sum \lambda_i |\psi_i\rangle\langle\psi_i|$.

Proof. Let $\hat{\varrho}=\sum \lambda_i|\psi_i\rangle\langle\psi_i|$, where $\lambda_i>0$ and $\sum \lambda_i=1$. Clearly, $\varrho^{\mathrm{cl}}(z)=\sum \lambda_i\varrho_i(z)$, and $\varrho_i(z)=\langle z|P_{\psi_i}|z\rangle$, where $P_{\psi_i}=|\psi_i\rangle\langle\psi_i|$. As the function $s(x)$ is concave, $S(\varrho^{\mathrm{cl}}(z))\geq\sum \lambda_i S(\varrho_i(z))$, and the equality is attained, if and only if $\varrho_i(z)=\varrho_j(z)$ almost everywhere for all i and j. Suppose ϱ_i is the projection to a state vector $\psi_i\in L^2(R)$. Let us consider a function $f_i(w)=\int \psi_i(x)\exp(-\frac{1}{2}x^2+wx)d^2x$, $w=q+ip\in\mathbb{C}$. Evidently, $f_i(w)$ is an entire function of w, and if $\varrho_i(z)=\varrho_j(z)$ almost everywhere, then $|f_i(w)|=|f_j(w)|$ for all w, so that $f_i(w)=f_j(w)\exp[i\theta(w)]$, where $\theta(w)$ is real and analytical at the points at which $f_i\neq 0$. Consequently, $\theta(w)=\mathrm{const}$. As the Fourier transform is unique, $\psi_i=\beta\psi_j$, where $|\beta|=1$ almost everywhere, so $P_{\psi_i}=P_{\psi_j}$, in contradiction to the initial assumption.

Extending Wehrl's hypothesis, *Lieb* [229] advanced a stronger hypothesis for spin CS.

Let $\{|\boldsymbol{n}\rangle\}$ be the set of spin CS. Then the spin density matrix corresponds to a function $\varrho^{\mathrm{cl}}(\boldsymbol{n})=\langle\boldsymbol{n}|\hat{\varrho}|\boldsymbol{n}\rangle$, and

$$S^{\mathrm{cl}}(\hat{\varrho})=S(\varrho^{\mathrm{cl}}(\boldsymbol{n})),\quad \text{where} \tag{27.10}$$

$$\begin{aligned} S(f)&=-\int f(\boldsymbol{n})\ln f(\boldsymbol{n})\,d\mu_j(\boldsymbol{n}),\\ d\mu_j(\boldsymbol{n})&=(4\pi)^{-1}(2j+1)\sin\theta\, d\theta\, d\varphi. \end{aligned} \tag{27.11}$$

Even in this case S^{cl} is monotonous, and $S^{\mathrm{cl}}\geq S^{\mathrm{qu}}$. It can be easily verified that since $\langle\boldsymbol{n}'|P_{\boldsymbol{n}}|\boldsymbol{n}'\rangle=(\cos\frac{1}{2}\theta)^{4j}$, where θ is the angle between $\boldsymbol{n}$ and $\boldsymbol{n}'$,

$$S^{\mathrm{cl}}(P_{\boldsymbol{n}})=2j/(2j+1). \tag{27.12}$$

Hypothesis [229].

$$S^{\mathrm{cl}}(\varrho^{\mathrm{qu}})\geq 2j/(2j+1). \tag{27.13}$$

The validity of this statement for $j=1/2$ can be proved easily [229], but no proof is known for arbitrary j.

Another inequality for entropy, written in terms of one-particle distributions, was obtained by *Thirring* [230].

Let $\hat{\varrho}$ be the density matrix for a system of bosons or fermions. A one-particle density matrix is associated with $\hat{\varrho}$,

$$\hat{\varrho}_1=\sum_{m,m'}|m'\rangle\langle m|\,\mathrm{tr}(\hat{\varrho}a_m^+,a_m), \tag{27.14}$$

and one can define a one-particle phase-space distribution, $\varrho(q,k)=\mathrm{tr}(\varrho a_{(q,k)}^+a_{(q,k)})$. Here the vectors $|m\rangle\in L^2(R)$ are elements of an orthonormalized basis, and $|q,k\rangle\sim g(x-q)\exp(ikx)$, with a function $g\in L^2(R)$, is a complete set of state vectors,

$$\int |q,k\rangle\langle q,k|d^3q\,d^3k/(2\pi)^3=\hat{1}.$$

Let $a_f = \int d^3x a(x) f^*(x)$ and $a_f^* = \int d^3x a^+(x) f(x)$ be the annihilation-creation operators for the state given by function $f(x)$. Then the entropy of the whole system is bounded by some one-particle expressions.

Theorem.

$$S(\varrho) \equiv -\mathrm{tr}\,(\hat{\varrho}\ln\hat{\varrho}) \leq -\mathrm{tr}\,[\hat{\varrho}_1 \ln\hat{\varrho}_1 \pm (1\mp\hat{\varrho}_1)\ln(1\mp\hat{\varrho}_1)]$$
$$\equiv S_1(\varrho_1) \leq -\int [\varrho\ln\varrho \pm (1\mp\varrho)\ln(1\mp\varrho)]\, d^3q\, d^3k/(2\pi)^3. \qquad (27.15)$$

The function in the integrand is $\varrho(q,k)$. Here the symbol tr stands for the trace of operators in space $L^2(R^3)$, and the alternative signs correspond to Fermi (Bose) particles.

Remarks. 1) The quantity $N = \mathrm{tr}\{\varrho_1\}$ is the average particle number. As $n_m \leq 1$ for fermions, operator inequalities $0 \leq \varrho_1 \leq 1/N$ arise. Respectively, $\int \varrho(q,k)\, d^3q\, d^3k/(2\pi)^3 = N$, and $0 \leq \varrho(q,k) \leq 1/N$. Thus both contributions to S_1 are positive for fermions.

2) The theorem asserts that the product states have the maximal entropy compatible with the given one-particle distribution. Thus absence of correlations leads to maximal chaos. The quantity S_1 is not given just by the one-particle density matrix, it differs from $-\mathrm{tr}\{\varrho_1 \ln\varrho_1\}$, but holes described by the density matrix $(1-\varrho)$ also contribute to it.

3) It is a consequence of Klein's inequality that the density matrix $\varrho_1 = \{\exp[-\beta(h-\mu)] \pm 1\}^{-1}$ maximizes S_1 for fixed $\mathrm{tr}\{\varrho_1 h\}$ and $\mathrm{tr}\{\varrho_1\}$. Similarly, for fixed integrals

$$\int h(q,k)\varrho(q,k)\, d^3q\, d^3k/(2\pi)^3 \quad \text{and} \quad \int \varrho(q,k)\, d^3q\, d^3k/(2\pi)^3,$$

the densities

$$\varrho_{F,B}(q,k) = (\exp\{-\beta[h(q,k)-\mu]\} \pm 1)^{-1}$$

maximize the integral for S.

These inequalities have been used in [230] to construct a simple derivation of the Hartree and Thomas-Fermi theory.

Appendix A
Proof of Completeness for Certain CS Subsystems

It was proven in Sect. 1.4 that the CS subsystem $\{|\alpha_k\rangle|$ is complete if and only if there is no entire function $\psi(\alpha)$ such that $\psi(\alpha_k)=0$ and $\int|\psi(\alpha)|^2 \exp(-|\alpha|^2)d^2\alpha < \infty$.

The relevance of Example i) in Sect. 1.4 was evident. Let us prove the statement of Example ii): system $\{|\alpha_k\rangle\}$ is complete if the point set $\{\alpha_k\}$ does not contain the origin $\alpha=0$, and

$$\sum_k |\alpha_k|^{-2-\varepsilon}=\infty \tag{A.1}$$

at some positive $\varepsilon>0$. This statement is a consequence of some general theorems concerning a relation between the order of an entire function and the distribution of its zeros in the complex plane.

Recall that the order λ for an entire analytical function $f(z)$ is the number

$$\lambda=\overline{\lim_{r\to\infty}}\frac{\ln[\ln M(r)]}{\ln r}, \tag{A.2}$$

where $M(r)$ is the maximum of modulus of the function $f(z)$ on the circle $|z|=r$, $\overline{\lim}=\lim\sup$ is the upper limit. Respectively, the number

$$\mu=\overline{\lim_{r\to\infty}}\frac{\ln M(r)}{r^{\lambda}} \tag{A.3}$$

is called the type of the function $f(z)$.

Let $\{\alpha_k\}$ be a sequence of zeros of the entire function $f(z)$. To characterize the sequence $\{\alpha_k\}$, the so-called convergence index λ_1 is introduced. The definition is that for arbitrary small positive numbers ε and δ the series $\sum_k|\alpha_k|^{-\lambda_1-\varepsilon}$ is converging, while the series $\sum_k|\alpha_k|^{-\lambda_1+\delta}$ is diverging. The series $\sum|\alpha_k|^{-\lambda_1}$ may be either converging or diverging.

From the theory of entire functions [231], it is known that the order of an entire function with zeros at some points $\{\alpha_k\}$ is no less than the convergence index of the sequence $\{\alpha_k\}$: $\lambda\geq\lambda_1$.

Condition (A.1) means that $\lambda_1>2$. Therefore the order λ for the function $f(z)$ with zeros at the points $\{\alpha_k\}$ is above 2: $\lambda\geq\lambda_1>2$. Clearly, here

$\int |f|^2 \exp(-|z|^2) d^2z = \infty$, so the CS system $\{|\alpha_k\rangle\}$ is complete. Correspondingly, at $\lambda_1 < 2$, as well as at $\lambda_1 = 2$, the CS system is incomplete if the series $\sum |\alpha_k|^{-2}$ is converging.

If, however, $\lambda_1 = 2$ and the series $\sum |\alpha_k|^2$ is diverging, then to find out whether the subsystem is complete, one must have more specific information on the distribution of zeros α_k in the complex α plane [231]. Here it is not difficult to construct an entire function of $\lambda = 2$ with zeros at the points α_k. According to Theorem 9.1.1 in the book by *Boas* [232], the type μ for such a function $f(z)$ must satisfy the inequalities

$$\mu \geq \frac{1}{2} \varliminf_{r \to \infty} \frac{N(r)}{r^2} \tag{A.4}$$

$$\mu \geq \frac{1}{2e} \varlimsup \frac{N(r)}{r^2}, \tag{A.5}$$

where $N(r)$ is the number of points of the set $\{\alpha_k\}$ within the circle $|z| \leq r$, $\varliminf = \lim\inf$ is the lower limit, $\varlimsup = \lim\sup$ is the upper limit. Hence, the CS system is complete at $\lambda_1 = 2$ and under any one of these conditions:

$$\varliminf_{r \to \infty} \frac{N(r)}{r^2} > 1, \tag{A.6}$$

$$\varlimsup_{r \to \infty} \frac{N(r)}{r^2} > e. \tag{A.7}$$

Let us consider now Example iii), where the points form a regular lattice in the α plane:

$$\{\alpha_k\} = \{\alpha_{mn} = m\omega_1 + n\omega_2\}, \tag{A.8}$$

where the lattice spacings ω_1 and ω_2 are linearly independent, Im $\{\omega_2 \bar{\omega}_1\} \neq 0$, and m and n are arbitrary integers.

Note, firstly, that for a regular lattice with cells of an area S

$$\varliminf_{r \to \infty} \frac{N(r)}{r^2} = \varlimsup_{r \to \infty} \frac{N(r)}{r^2} = \frac{\pi}{S}. \tag{A.8}$$

Now condition (A.6) leads directly to the first part of the theorem in Sect. 1.4.

Next consider the lattice with $S > \pi$. To prove the second part of the theorem, it is appropriate to construct an entire function belonging to the space $\mathscr{F}$ and vanishing at the lattice points α_{mn}. The last condition is fulfilled by the Weierstrass σ function [47]:

$$\sigma(\alpha) = \alpha \prod_{m,n}{}' \left(1 - \frac{\alpha}{\alpha_{mn}}\right) \exp\left[\frac{\alpha}{\alpha_{mn}} + \frac{1}{2}\left(\frac{\alpha}{\alpha_{mn}}\right)^2\right]. \tag{A.9}$$

The product here is over all integers excluding the factor with $m=n=0$. It is well known [47] that the order of the σ function is $\lambda=2$. It may also be shown that such a number ν exists, that the type μ of the function $\tilde{\sigma}(\alpha)=\exp(-\nu\alpha^2)\sigma(\alpha)$ is the minimal possible at the given zeros $\{\alpha_{mn}\}$ and is given by

$$\mu=\frac{\pi}{2S}. \tag{A.10}$$

It was found that this function satisfies

$$|\tilde{\sigma}(\alpha)|^2=\varrho(\alpha,\bar{\alpha})\exp(2\mu|\alpha|^2), \tag{A.11}$$

where $\varrho(\alpha,\bar{\alpha})$ is a doubly periodic function with the periods ω_1 and ω_2. It is not difficult to see that at $S>\pi$ the integral $I=\int|\psi|^2\exp(-|z|^2)d^2z$ is convergent. In this case the function $\tilde{\sigma}(\alpha)$ determines a state vector $|\tilde{\psi}\rangle\in\mathscr{H}$ orthogonal to all the states $|\alpha_{mn}\rangle$ and therefore the system $\{|\alpha_{mn}\rangle\}$ is not complete. Thus the second part of the theorem is proven. The relation (A.11) also shows that at $S=\pi$ integral I is divergent for any entire function, having zeros at the points α_{mn}. So the CS system $\{|\alpha_{mn}\rangle\}$ is complete at $S=\pi$.

Note further that if a state $|\alpha_{m_0n_0}\rangle$ is removed from the system, we get a function $\tilde{\sigma}_1(\alpha)=(\alpha-\alpha_{m_0n_0})^{-1}\tilde{\sigma}(\alpha)$ instead of $\tilde{\sigma}(\alpha)$. So in this situation the completeness problem is reduced to investigating convergence for the integral [38]

$$I_{m_0n_0}=\int\frac{\varrho(\alpha,\bar{\alpha})}{|\alpha-\alpha_{m_0n_0}|^2}\,d^2\alpha. \tag{A.12}$$

Note, first of all, that since the function $\varrho(\alpha,\bar{\alpha})$ is periodic, integral $I_{m_0n_0}$ is reduced to I_{00}:

$$I_{m_0n_0}=I_{00}=\int\frac{d^2\alpha}{|\alpha|^2}\,\varrho(\alpha,\bar{\alpha}). \tag{A.13}$$

Here ϱ is a nonnegative double-periodic function, all zeros of which are at the lattice points, $\alpha_{mn}=m\omega_1+n\omega_2$. Let us estimate the integral I_{00}. To this end, we calculate the integral in (A.13) over the domain that is a union of nonoverlapping disks of a radius r_0, with centers at $\gamma_{m_0n_0}$, coinciding with centers of the lattice cells $\varrho(\alpha)\neq 0$, $\alpha\in D_{mn}$:

$$I_{00}>\sum_{m,n}{}' J_{mn}, \qquad J_{mn}=\int_{D_{mn}}\frac{d^2\alpha}{|\alpha|^2}\,\varrho(\alpha,\bar{\alpha}),$$
$$D_{mn}=\{\alpha:|\alpha-\gamma_{mn}|<r_0\}. \tag{A.14}$$

Evidently,

$$J_{mn} > \frac{\varrho_0 \pi r_0^2}{(|\gamma_{mn}| + r_0)^2} > \frac{\pi r_0^2 \varrho_0}{4|\gamma_{mn}|^2}, \tag{A.15}$$

where ϱ_0 is the minimal value of the function $\varrho(\alpha, \bar{\alpha})$ in the domain D_{mn}. Hence

$$I_{00} > \frac{\pi r_0^2 \varrho_0}{4} \sum_{m,n} \frac{1}{|\gamma_{mn}|^2} \tag{A.16}$$

and $I_{m_0 n_0} = I_{00} = \infty$. Thus the CS system $\{|\alpha_{mn}\rangle\}$ remains complete after the state is removed.

Suppose that two states, say $|\alpha_{m_1 n_1}\rangle$ and $|\alpha_{m_2 n_2}\rangle$, are removed from the CS system. Again we have to investigate the integral

$$\int \frac{\varrho(\alpha, \bar{\alpha}) d^2\alpha}{|\alpha - \alpha_{m_1 n_1}|^2 |\alpha - \alpha_{m_2 n_2}|^2}. \tag{A.17}$$

It is not difficult to see that the integral is converging, so the system becomes incomplete if any two states are removed from it. This concludes the proof of the theorem.

Appendix B
Matrix Elements of the Operator $D(\gamma)$

First of all, we deduce (1.6.36) for the matrix elements of the operator $D(\gamma)$. The generating function is given by (1.6.34)

$$G(\bar{\alpha},\beta)=\sum_{m,n} D_{mn}\frac{\bar{\alpha}^m}{\sqrt{m!}}\frac{\beta^n}{\sqrt{n!}}$$

$$=\exp(-|\gamma|^2/2)\exp(\bar{\alpha}\beta+\bar{\alpha}\gamma-\beta\bar{\gamma}). \tag{B.1}$$

Expanding in powers of $\bar{\alpha}$ and β yields

$$G=\exp(-\tfrac{1}{2}|\gamma|^2)\sum_{n_1n_2n_3}\frac{\bar{\alpha}^{n_1+n_2}\beta^{n_1+n_3}\gamma^{n_2}(-\bar{\gamma})^{n_3}}{n_1!n_2!n_3!}$$

$$=\exp(-\tfrac{1}{2}|\gamma|^2)\sum_{m,n,n_1}\frac{\gamma^{m-n_1}(-\bar{\gamma})^{n-n_1}}{n_1!(m-n_1)!(n-n_1)!}\bar{\alpha}^m\beta^n \tag{B.2}$$

so that

$$D_{mn}=\sqrt{m!n!}\sum_{n_1=0}^{\min(m,n)}\frac{\gamma^{m-n_1}(-\bar{\gamma})^{n-n_1}}{n_1!(m-n_1)!(n-n_1)!}\exp\left(-\frac{|\gamma|^2}{2}\right). \tag{B.3}$$

Consider first the case $m=n+k\geq n$. Then

$$D_{mn}=\sqrt{\frac{n!}{m!}}\exp\left(-\frac{1}{2}|\gamma|^2\right)\gamma^k\sum_{n_1=0}^{n}\frac{(-\gamma\bar{\gamma})^{n-n_1}(n+k)!}{(n-n_1)!n_1!(n+k-n_1)!}. \tag{B.4}$$

Looking at an expression for the Laguerre polynomials (see, e.g., the definition in [233])

$$L_n^k(x)=\sum_{s=0}^{n}\frac{(-x)^s}{s!}\frac{\Gamma(n+k+1)}{(n-s)!\Gamma(k+s+1)}, \tag{B.5}$$

one gets at once

$$D_{mn}=\sqrt{\frac{n!}{m!}}\exp\left(-\frac{1}{2}|\gamma|^2\right)\gamma^{m-n}L_n^{m-n}(|\gamma|^2). \tag{B.6}$$

A similar formula holds for $m \leq n$:

$$D_{mn} = \sqrt{\frac{m!}{n!}} \exp\left(-\frac{1}{2}|\gamma|^2\right)(-\bar{\gamma})^{n-m} L_m^{n-m}(|\gamma|^2). \tag{B.7}$$

Note that (B.5) for the Laguerre polynomials is meaningful also for negative k. The corresponding symmetry relation is

$$\frac{1}{m!} L_{m+k}^{-k}(x) = \frac{1}{(m+k)!}(-x)^k L_m^k(x). \tag{B.8}$$

Thus (B.6, 7) are equivalent. Note an inequality

$$|e^{-r/2} r^{k/2} L_n^k(r)| \leq \sqrt{\frac{(n+k)!}{n!}} \quad \text{at} \quad (n+k) > 0 \tag{B.9}$$

resulting from the unitarity $|D_{mn}| \leq 1$.

Next we write the matrix element $\langle m|D(\gamma)|n\rangle$ in the Fock-Bargmann representation. Starting from the antinormal form of the operator, $D(\gamma) = \exp(\frac{1}{2}|\gamma|^2)\exp(-\bar{\gamma}a)\exp(\gamma a^+)$, we get the following integral representation for the matrix element

$$\langle m|D(\gamma)|n\rangle = \frac{1}{\sqrt{m!n!}} e^{\frac{|\gamma|^2}{2}} \int d\mu(z) e^{-|z|^2} e^{\gamma\bar{z} - \bar{\gamma}z} \bar{z}^m z^n. \tag{B.10}$$

Comparison with equality (B.6) leads to an integral representation for the Laguerre polynomials:

$$\gamma^k L_n^k(|\gamma|^2) = \frac{1}{n!} e^{|\gamma|^2} \int d\mu(z) e^{\gamma\bar{z} - \bar{\gamma}z} z^k |z|^{2n} e^{-|z|^2}. \tag{B.11}$$

Introducing the polar coordinates in the integral, $z = \varrho \exp(i\varphi)$, and integrating over φ, then

$$L_n^k(x) = \frac{x^{-k/2} e^x}{n!} \int_0^\infty dt\, e^{-t} t^{n+k/2} J_k(2\sqrt{tx}). \tag{B.12}$$

Multiplying both sides of (B.11) by t^n, summing over n, and calculating the integral, one obtains the generating function for the Laguerre polynomials

$$\sum_n t^n L_n^k(|\gamma|^2) = \gamma^{-k} e^{|\gamma|^2} \int d\mu(z) e^{-|z|^2 + t|z|^2} e^{\gamma\bar{z} - \bar{\gamma}z} \bar{z}^k \tag{B.13}$$

$$\sum_n t^n L_n^k(x) = (1-t)^{-(k+1)} e^{-tx/(1-t)}, \quad |t| < 1, \quad k \geq 0. \tag{B.14}$$

Another variant of the generating function is obtained by setting $k=m-n$ in (B.11), and then calculating the sum over n with the factors t^n, and the integral over z. The result is

$$\sum_n t^n L_n^{m-n}(x)=(1+t)^m e^{-tx}. \tag{B.15}$$

Still another integral representation for the Laguerre polynomials is obtained by using the Fock-Bargmann representation and (1.3.25) describing the action of the operator $D(\gamma)$ in this representation,

$$\gamma^k L_n^k(|\gamma|^2)=\int d\mu(z)\, e^{-|z|^2} e^{\gamma z}\bar{z}^{n+k}\frac{(z-\bar{\gamma})^n}{n!}. \tag{B.16}$$

This integral representation may be also rewritten as

$$D_{n+k,n}(\gamma)=\int d\mu(z)\,\bar{\varphi}_{n+k}(z)\,\varphi_n(z-\bar{\gamma})\, e^{(\gamma z-\bar{\gamma}\bar{z})/2}, \tag{B.17}$$

where the following notation is used

$$\varphi_n(z)=\frac{z^n}{\sqrt{n!}}\, e^{-\frac{|z|^2}{2}}=\frac{\varrho^n}{\sqrt{n!}}\, e^{-\varrho^2/2}\, e^{in\varphi}. \tag{B.18}$$

It is remarkable that $\varphi_n(z)$ is the eigenfunction of the Hamiltonian operator for the two-dimensional harmonic oscillator.

Let us consider the case of $n \gg 1$. In this asymptotic the function $\varphi_n(z)$ has a sharp maximum, determined by the condition

$$\varrho=\sqrt{n}. \tag{B.19}$$

In other words, the function $\varphi_n(z)$ is nonzero mainly near the circle of the radius $\varrho=\sqrt{n}$. The corresponding geometrical structure on the complex z plane is shown in Fig. B.1 for $n \gg 1$ and $m \gg 1$. Two situations should be studied separately:

i) $\gamma>\sqrt{n}$. The circles $|\alpha|=\sqrt{m}$ and $|\alpha-\bar{\gamma}|=\sqrt{n}$ intersect, so there is no exponential smallness in $D_{mn}(\gamma)$ provided that

$$|\gamma|-\sqrt{n}<\sqrt{m}<|\gamma|+\sqrt{n}; \tag{B.20}$$

ii) $|\gamma|<\sqrt{n}$. Now $D_{mn}(\gamma)$ is not an exponentially small quantity if

$$\sqrt{n}-|\gamma|<\sqrt{m}<|\gamma|+\sqrt{n}. \tag{B.21}$$

Both conditions are combined to give

$$|\sqrt{n}-|\gamma||<\sqrt{m}<\sqrt{n}+|\gamma|, \tag{B.22}$$

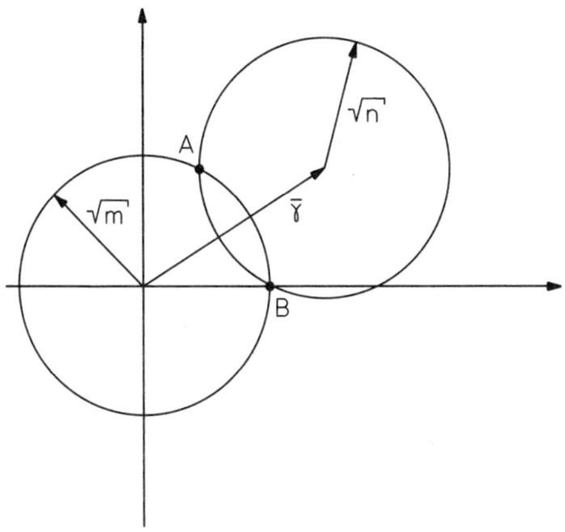

Fig. B. 1. At $m, n \gg 1$ integral (B. 17) is contributed mainly by vicinities of points A and B in the complex z plane

meaning that the quantities $\sqrt{m}$, $\sqrt{n}$ and $|\gamma|$ must satisfy the inequalities as sides of a triangle. Under these conditions, the $m \gg 1$, $n \gg 1$ asymptotic for $W_{mn} = |D_{mn}(\gamma)|^2$, averaged over the oscillations, is given by

$$\bar{W}_{mn} = \frac{1}{2\pi} \int dp\, dq\, \delta\left(\frac{p^2+q^2}{2} - m\right) \delta\left(\frac{(p-p_1)^2 + (q_1-q)^2}{2} - n\right)$$
$$\alpha = (q + \mathrm{i}p)/\sqrt{2}, \quad \alpha - \bar{\gamma} = (q_1 + \mathrm{i}p_1)/\sqrt{2}. \tag{B.23}$$

Finally, the matrix element $D_{mn}(\gamma)$ may be calculated also by coordinate representation. Hence we get the integral

$$\int dx\, \mathrm{e}^{-x^2 \mp \mathrm{i}\sqrt{2}rx} H_m(x) H_n(x)$$
$$= \sqrt{\pi}\, 2^{(m+n)/2} m!\, (\mathrm{i}r)^{n-m} \exp\left(\frac{r^2}{2}\right) L_m^{n-m}(r^2). \tag{B.24}$$

In conclusion we present some other useful formulae. The matrix element in view may be transformed to

$$\langle m|D(\alpha)|n\rangle = \frac{1}{\sqrt{m!\,n!}} \langle 0|a^m D(\alpha)(a^+)^n|0\rangle$$
$$= \frac{1}{\sqrt{m!\,n!}} \exp\left(\frac{|\alpha|^2}{2}\right) \langle 0|a^m \mathrm{e}^{-\bar{\alpha}a} \mathrm{e}^{\alpha a^+} (a^+)^n|0\rangle$$
$$= \frac{1}{\sqrt{m!\,n!}} \exp\left(\frac{|\alpha|^2}{2}\right) \left(-\frac{\partial}{\partial\bar{\alpha}}\right)^m \left(\frac{\partial}{\partial\alpha}\right)^n \exp(-|\alpha|^2). \tag{B.25}$$

The generating function for the matrix elements squared is obtained using (B.10),

$$F(u,v;\gamma)=\sum_{m,n} u^m v^n |\langle m|D(\gamma)|n\rangle|^2$$
$$=e^{|\gamma|^2}\int e^{-(|\alpha|^2+|\beta|^2)} e^{\alpha\gamma-\alpha\gamma} e^{\beta\gamma-\beta\gamma} e^{u\alpha\beta} e^{v\beta\alpha} d\mu(\alpha)\, d\mu(\beta). \tag{B.26}$$

This integral is easily calculated, yielding

$$F(u,v;\gamma)=(1-uv)^{-1}\exp(|\gamma|^2)\exp\left(-\frac{|\gamma|^2(1-u)(1-v)}{1-uv}\right). \tag{B.27}$$

Appendix C
Jacobians of Group Transformations for Classical Domains

Let us consider the first type of domain. Any element of this domain is a matrix with p rows and q columns which satisfies the condition

$$\mathrm{I}^{(p)} - zz^+ > 0.$$

The group G acts in the following way

$$z \to z_1 = z \cdot g = (A'z + C')(B'z + D')^{-1}.$$

It can be rewritten as

$$z_1(B'z + D') = (A'z + C').$$

Differentiating this equality and substituting $z = 0$ gives

$$dz_1 \cdot D' = [A' - C'(D')^{-1}B']\,dz \tag{C.1}$$

$$J_g(0) = [\det(A - BD^{-1}C)]^q [\det D]^p. \tag{C.2}$$

Since

$$\det[A - BD^{-1}C] = \det\begin{bmatrix} A & B \\ C & D \end{bmatrix}(\det D)^{-1} \tag{C.3}$$

and $\det\begin{bmatrix} A & B \\ C & D \end{bmatrix} = 1$, then $J_g(0) = [\det D]^{-(p+q)}$. (C.4)

It can be easily seen that the transformation g_z translating z to 0 is

$$g_z = \begin{bmatrix} A_1 & 0 \\ 0 & D_1 \end{bmatrix}\begin{bmatrix} \mathrm{I}^{(p)} & \bar{z}_1 \\ z_1' & \mathrm{I}^{(q)} \end{bmatrix}, \quad \text{where} \tag{C.5}$$

$$A_1^+ A_1 = (\mathrm{I} - \bar{z}_1 z_1')^{-1}, \qquad D_1^+ D_1 = (\mathrm{I} - z_1'\bar{z}_1)^{-1}. \tag{C.6}$$

Finally

$$J_g(z) = [\det(B'z + D')]^{-(p+q)}. \tag{C.7}$$

The calculations for the domains of the second and third types are analogous yielding (12.2.15, 16).

The case of the fourth type of domain should be considered separately. The domain D_{IV} consists of p-dimensional vectors $z=(z_1, \ldots, z_p)$ satisfying the conditions

$$1+|zz'|^2-2\,\bar{z}z'>0, \qquad |zz'|<1.$$

The group action in this domain is

$$\left\{\left[\left(\frac{1}{2}(zz'+1),\ \frac{\mathrm{i}}{2}(zz'-1)\right)A'+zB'\right]\begin{pmatrix}1\\ \mathrm{i}\end{pmatrix}\right\} z_1$$
$$=\left\{\left(\frac{1}{2}(zz'+1),\ \frac{\mathrm{i}}{2}(zz'-1)\right)C'+zD'\right\}, \tag{C.8}$$

where A, B, C and D are real 2×2, $2\times p$, $p\times 2$ and $p\times p$ matrices, respectively, which satisfy the conditions

$$\det\begin{bmatrix}A & B\\ C & D\end{bmatrix}=1, \quad AA'-BB'=\mathrm{I}^{(2)}, \qquad DD'-CC'=\mathrm{I}^{(p)}, \qquad AC'=BD'.$$

Differentiating (C.8) and substituting $z=0$ gives

$$Kdz_{1i}=\left(D_{ij}-\frac{1}{2K}\,c_ib_j\right)dz_j, \quad \text{where} \tag{C.9}$$

$$K=\frac{1}{2}(1,\mathrm{i})\,A\begin{pmatrix}1\\ -\mathrm{i}\end{pmatrix}, \qquad c_i=C_{i\alpha}\begin{pmatrix}1\\ -\mathrm{i}\end{pmatrix}_{\alpha}, \qquad b_j=(1,\mathrm{i})_\beta B_{\beta j}. \tag{C.10}$$

Therefore the desired expression for the Jacobian is

$$J_g(0)=\frac{\det D}{2K^{p+1}}\left[(1,\mathrm{i})\,A\begin{pmatrix}1\\ -\mathrm{i}\end{pmatrix}-(1,\mathrm{i})\,BD^{-1}C\begin{pmatrix}1\\ -\mathrm{i}\end{pmatrix}\right]. \tag{C.11}$$

Using the identity

$$\det(\delta_{kl}-a_kb_l)=1-(a,b),$$

(C.11) is transformed to

$$J_g(0)=K^{-p}\det D\,\frac{(1,\mathrm{i})(A')^{-1}\begin{pmatrix}1\\ \mathrm{i}\end{pmatrix}}{(1,\mathrm{i})(A)\begin{pmatrix}1\\ -\mathrm{i}\end{pmatrix}}. \tag{C.12}$$

For any 2×2 matrix

$$\langle j, v'|\hat{P}_{lm}|j, v\rangle = \sqrt{\frac{2l+1}{4\pi}}\, \langle j, v'; l, m|j, v\rangle \langle j, -j; l, 0|j, -j\rangle, \tag{4.3.42}$$

$$\begin{aligned} &\|f\|_k < \infty, \\ &\|f\|_k^2 = \frac{2k-1}{\pi} \int_D |f(z)|^2 (1-|z|^2)^{2k-2} d^2z, \qquad D = \{z : |z| < 1\}. \end{aligned} \tag{5.1.18}$$

$$f_n(z) = z^n = \mathrm{e}^{\mathrm{i}n\theta}. \tag{5.2.66}$$

$$u_n^\lambda(\zeta) = \frac{1}{2\pi}\int_0^{2\pi} \mathrm{e}^{-\mathrm{i}n\theta}(1-|\zeta|^2)^{1/2-\mathrm{i}\lambda}|1-\zeta z|^{-1+2\mathrm{i}\lambda}d\theta$$

$$\mathscr{I}_g(z)=[\mathscr{I}_{g_z}(0)]^{-1}\,\mathscr{I}_{g_zg}(0). \tag{12.2.10}$$

$$\mathscr{I}_g(z)=|\mathscr{I}_{g_z}(0)|^{-1}\,\mathscr{I}_{g_zg}(0). \tag{12.2.13}$$

$$\begin{aligned}
&1.\ \ SL(p,\mathbb{C})\times\mathbb{R}^{+} && \text{for } D^{\mathrm{I}};\\
&2.\ \ SL(p,\mathbb{R})\times\mathbb{R}^{+} && \text{for } D^{\mathrm{II}};\\
&3.\ \ SU^{*}(2p,\mathbb{R})\times\mathbb{R}^{+} && \text{for } D^{\mathrm{III}};
\end{aligned} \tag{13.6.9}$$

$$(f, \Phi_{\bar{\zeta}}) = f(\zeta) \tag{16.5.12}$$

$$[a_{ij}]=\begin{pmatrix} \sqrt{\varrho} & -\sqrt{1-\varrho} \\ -\sqrt{1-\varrho} & -\sqrt{\varrho} \end{pmatrix}, \qquad \begin{aligned} y_1&=\sqrt{\nu(1-\varrho)}\,\mathrm{e}^{\mathrm{i}\varphi} \\ y_2&=-\sqrt{\nu}\,(\mathrm{e}^{-\mathrm{i}\varphi}-\sqrt{\varrho}\,\mathrm{e}^{\mathrm{i}\varphi}) \end{aligned} \tag{18.4.14}$$

$$\begin{aligned} &W_{00}=\sqrt{1-\varrho}\exp\left[-\nu(1-\sqrt{\varrho}\cos 2\varphi)\right], \\ &W_{10}=\nu(1-\varrho)\,W_{00}, \quad W_{01}=\nu(1-2\sqrt{\varrho}\cos 2\varphi+\nu^2\varrho)\,W_{00}, \\ &W_{11}=(1-\varrho)\left[(1-\nu)^2+2\nu(1-\nu)\sqrt{\varrho}\cos 2\varphi+\nu^2\varrho\right]W_{00}. \end{aligned} \tag{18.4.16}$$

$$|\det \tilde{D}| = |\det(D^+ D + \mathrm{e}^{-2i\theta} B^+ B)|$$
$$= |\det D|^2 |\det(\mathrm{I} + \mathrm{e}^{-2i\theta}\zeta^+\zeta)|. \tag{19.3.66}$$

$$a^{\mu}_{n}=\frac{1}{\sqrt{2}}\,[\chi^{\mu}_{1}(n)-\mathrm{i}\chi^{\mu}_{2}(n)],\qquad (a^{\mu}_{n})^{+}=\frac{1}{\sqrt{2}}\,[\chi^{\mu}_{1}(n)+\mathrm{i}\chi^{\mu}_{2}(n)],$$
$$\{a^{\mu}_{n},\ a^{\nu}_{m}\}=0,\qquad \{a^{\mu}_{n},\ (a^{\nu}_{m})^{+}\}=\hbar\delta_{\mu\nu}\delta_{mn},\qquad a^{\mu}_{n}|0\rangle=0. \tag{22.1.12}$$

$$A_{mn}=\tfrac{1}{4}\,[(a^{\mu}_{m})^{+}a^{\mu}_{n}-a^{\mu}_{n}(a^{\mu}_{m})^{+}],\qquad B_{mn}=\tfrac{1}{2}\,a^{\mu}_{m}a^{\mu}_{n},$$
$$B^{+}_{mn}=\tfrac{1}{2}\,(a^{\mu}_{n})^{+}(a^{\mu}_{m})^{+}\quad\text{or}\quad B^{*}_{mn}=-B^{+}_{nm}=\tfrac{1}{2}\,(a^{\mu}_{m})^{+}(a^{\mu}_{n})^{+}. \tag{22.1.13}$$

$$R(z_1,\bar{z}_2;0)=\frac{1}{1+v_0}\exp\left(\frac{v_0 z_1\bar{z}_2+\alpha z_1+\bar{\alpha}z_2-|\alpha|^2}{1+v_0}\right), \tag{23.1.36}$$

$$\nu=[\exp(\hbar\omega_0/kT)-1]^{-1}, \tag{23.2.2}$$

$$=[2\sinh se\mathscr{H}]^{-1}. \qquad (25.9.)$$

$$\Pi_{\parallel}=(\Pi_0,\Pi_z)=(E,0). \tag{26.1}$$

$$K=\frac{1}{2}(1,i)\,A\begin{pmatrix}1\\ -i\end{pmatrix}, \qquad c_i=C_{i\alpha}\begin{pmatrix}1\\ -i\end{pmatrix}_{\alpha}, \qquad b_j=(1,i)_\beta B_{\beta j}. \tag{C.10}$$

$$J_g(0)=\frac{\det D}{2K^{p+1}}\left[(1,i)\,A\begin{pmatrix}1\\ -i\end{pmatrix}-(1,i)\,BD^{-1}C\begin{pmatrix}1\\ -i\end{pmatrix}\right]. \tag{C.11}$$

$$J_g(0)=K^{-p}\det D\,\frac{(1,i)(A')^{-1}\begin{pmatrix}1\\ i\end{pmatrix}}{(1,i)(A)\begin{pmatrix}1\\ -i\end{pmatrix}}. \tag{C.12}$$

$$\frac{(1,\mathrm{i})(A')^{-1}\begin{pmatrix}1\\ \mathrm{i}\end{pmatrix}}{(1,\mathrm{i})\,A\begin{pmatrix}1\\ -\mathrm{i}\end{pmatrix}}=\frac{1}{\det A},\quad \text{therefore} \tag{C.13}$$

$$J_g(0)=\frac{1}{K^p}\,\frac{\det D}{\det A}. \tag{C.14}$$

In our case then

$$\det\begin{vmatrix}A & B\\ C & D\end{vmatrix}=\det D\det[A-BD^{-1}C]$$

$$=\det D\det(A')^{-1}=\frac{\det D}{\det A},\quad \text{hence} \tag{C.15}$$

$$J_g(0)=\frac{1}{K^p}=\left[\frac{1}{2}(1,\,-\mathrm{i})\,A'\begin{pmatrix}1\\ \mathrm{i}\end{pmatrix}\right]^{-p}. \tag{C.16}$$

The group transformations translate 0 into the point $z=(2K)^{-1}(1,-\mathrm{i})C'$. Calculating the value of $1+|zz'|^2-2\bar{z}z'$ we get a useful identity

$$(1+|zz'|^2-2\bar{z}z')=|K|^{-2}, \tag{C.17}$$

and, immediately, we get (12.2.17) for $\varrho(z,\bar{z})$. The expression for $J_g(z)$ results from a similar calculation:

$$J_g(z)=\left\{\left[\left(\frac{1}{2}(zz'+1),\ \frac{\mathrm{i}}{2}(zz'-1)\right)A'+zB'\right]\begin{pmatrix}1\\ \mathrm{i}\end{pmatrix}\right\}^{-p}. \tag{C.18}$$

Addendum
Further Applications of the CS Method

Abundant information on various applications of the CS method is contained in a volume by *Klauder* and *Skagerstam* [A1]. An introductory review presents an exposition of properties of coherent states; it is followed by a vast collection of reprints from original papers related to the CS method.

It is not out of place to mention here a number of works in addition to those cited in Chap. 11, which cover the following applications of the CS method:

1. A description of soft-photon clouds around charged particles and an elimination of infrared divergences from quantum field theory [A2, A3]. A similar problem was considered in quantum gravity [A4].
2. Application of the CS method to the theory of magnetism [A5–A9].
3. Analysis of the measurement process [A10–A14]. In particular, a limit, due to quantum effects, on the sensitivity of a gravitational wave detector is considered.
4. Application of the CS method to thermodynamics, in particular to the calculation of partition functions [A15, A16].
5. Coherent states in nuclear physics [A17–A20].
6. Coherent states and path integrals [A1, A21].
7. Application of the CS method to quantum field theory and elementary particle physics [A22–A31] (and a lot of other works).
8. Construction of multi-dimensional wave functions by means of classical trajectories [A32] and description of collision processes [A33] in chemical physics.
9. Biological applications. Description of long-range forces between human blood cells [A34] and the long-range phase coherence in bacteriorhodopsin macromolecules [A35].
10. Resonance radiation field interaction with two level atoms in the framework of the Dicke model. The calculation of the phase transition into the superradiative state is simplified considerably by means of the CS method [A36, A37].
11. An approximate description of the interaction of a charged particle with a strong quantized radiation field. A derivation of the Schrödinger equation for the charged particle in a classical radiation field has been performed [A38] by means of a CS subsystem related to von Neumann's lattice (cf. Sect. 1.5).
12. Some applications of the CS method to the description of superfluidity of a weakly non-ideal Bose gas [A39, A40].

13. Coherent states in the theory of radars [A41].

14. Description of the quantum Hall effect [A42].

15. Fermion coherent states in nuclear physics (time-dependent Hartree-Fock theory) [A43].

16. A problem in non-equilibrium statistical physics: description of spin relaxation [A44].

17. Application to plasma physics, in particular to the Vlasov equation [A45].

18. An approach to Landau diamagnetism [A46].

19. Description of the Nielsen-Olesen vortices in quantum field theory using the CS method [A47].

20. Theory of a system of N identical subsystems, where N goes to infinity (the $1/N$-expansion) [A48].

In conclusion, I should mention some works dealing with overcomplete systems of states, different from those treated in the present book. A "continuous representation" of quantum states has been considered by *Klauder* [3, 27, 28] and *Berezin* [A49], based upon the resolution of unity and some properties of continuity. In general, systems of states arising in this approach are not necessarily related to a group structure, and their properties can be quite different from those of the generalized coherent states exposed here. A particular case of the continuous representation is the system proposed by *Barut* and *Girardello* [16], which is not subject of a natural action of the related Lie group $SU(1,1)$. Another system of this type has been considered recently [A50]. A generalization of coherent states for quantum motion in general potentials has been proposed, and uses a method which is an analytic complement to the group theory point of view [A51] (and references therein).

References

1. Schrödinger E.: Naturwissenschaften **14**, 664 (1926)
2. von Neumann J.: *Mathematische Grundlagen der Quantenmechanik* (Springer, Berlin, 1932)
3. Klauder J.R.: Ann. of Phys. **11**, 123 (1960)
4. Bargmann V.: Commun. Pure & Appl. Math. **14**, 187 (1961)
5. Segal I.E.: *Mathematical Problems of Relativistic Physics* (Am. Math. Soc., Providence, R.I., 1963)
6. Berezin F.A.: *The Method of Second Quantization* (Academic, New York, 1966)
7. Glauber R.J.: Phys. Rev. **130**, 2529 (1963)
8. Glauber R.J.: Phys. Rev. **131**, 2766 (1963)
9. Klauder J.R., Sudarshan E.C.G.: *Fundamentals of Quantum Optics* (Benjamin, New York, 1968)
10. Carruthers P., Nieto M.: Rev. Mod. Phys. **40**, 411 (1968)
11. Perelomov A.M.: Preprint ITEP No. 46 (Moscow, 1974) [in Russian]
12. Perelomov A.M.: Sov. Phys. Usp. **20**(9), 703 (1977)
13. Malkin I.A., Manko V.I.: *Dynamical Symmetries and Coherent States of Quantum Systems* (Nauka, Moscow, 1979) [in Russian]
14. Weyl H.: *Gruppentheorie und Quantenmechanik* (Hirzel, Leipzig, 1928)
15. Perelomov A.M.: Commun. Math. Phys. **26**, 22 (1972)
16. Barut A.O., Girardello L.: Commun. Math. Phys. **21**, 41 (1971)
17. Radcliffe J.M.: J. Phys. **A4**, 313 (1971)
18. Perelomov A.M.: *Dynamical Symmetries in Quantum Physics*, D. Sc. Thesis (ITEP, Moscow, 1973) [in Russian]
19. Kirillov A.A.: *Elements of the Theory of Representations* (Springer, Berlin, Heidelberg, New York, 1976)
20. Kostant B.: In *Lectures in Modern Analysis and Applications III*, ed. by C.T. Taam, Lecture Notes Math., Vol. 170 (Springer, Berlin, Heidelberg, New York, 1970), p. 87
21. Kostant B.: In *Group Representations in Mathematics and Physics*, ed. by V. Bargmann, Lecture Notes Phys., Vol. 6 (Springer, Berlin, Heidelberg, New York, 1970) p. 237
22. Berezin F.A.: Sov. Math. Izv. **38**, 1116 (1974, **39**, 363 (1975)
23. Cartier P.: In *Proc. Symp. Pure Math.*, Vol. 9, Algebraic Groups and Discontinuous Subgroups (Am. Math. Soc., Providence, R.I. 1966) p. 361–383
24. Glauber R.J.: Phys. Rev. **84**, 395 (1951)
25. Stone M.: Proc. Nat. Acad. Sci. USA **16**, 172 (1930)
26. von Neumann J.: Math. Ann. **104**, 570 (1931)
27. Klauder J.R.: J. Math. Phys **4**, 1055 (1963); **5**, 177 (1964)
28. Klauder J.R., Mc Kenna J.: J. Math . Phys. **5**, 878 (1964)
29. Berezin F.A.: Mathemat. Sbornik 1971, **86**, 578 (1971) [in Russian]
30. Aronszajn A.: Trans. Amer. Math. Soc. **68**, 337 (1950)
31. Bergmann S.: The kernel functions and conformal mapping. Math. Surv. No. 5, Amer. Math. Soc. No. 4 (1950)
32. Fock V.A.: Z. Phys. **49**, 339 (1928)
33. Bargmann V.: Rev. Mod. Phys. **34**, 829 (1962)
34. Bargmann V.: Commun. Pure Appl. Math. **20**, 1 (1967)

35. Itzykson C.: Commun. Math. Phys. **4**, 92 (1967)
36. Bargmann V.: In *Analytic Methods in Mathematical Physics*, Vol. 74 (Gordon and Breach, New York 1970) p. 1
37. Terentyev M.V.: Ann. Phys. **74**, 1 (1972)
38. Perelomov A.M.: Theor. Math. Phys. **6**, 213 (1971) [in Russian]
39. Bargmann V., Butera P., Girardello L., Klauder J.R.: Reps. Math. Phys. **2**, 221 (1971)
40. Bacry H., Grossmann A., Zak J.: Phys. Rev. **B12**, 1118 (1975)
41. Janssen A.: "Weighted Wigner Distributions Vanishing on Lattices", CALTECH preprint (1979)
42. Janssen A.: J. Math. Phys. **23**, 720 (1982)
43. Boon M., Zak J., Zucker J.: J. Math. Phys. **24**, 316 (1983)
44. Gårding L.: Proc. Nat. Acad. Sci. USA **33**, 331 (1947)
45. Chern S.-S.: *Complex Manifolds without Potential* (Van Nostrand, Princeton 1967)
46. Siegel C.L.: *Analytic Functions of Several Complex Variables*. Princeton, N.J. (1962)
47. Bellman R.: *A Brief Introduction to Theta Functions* (Holt, Rinehart and Winston, New York 1961)
48. Gel'fand I.M., Graev M.I., Pyatetsky-Shapiro I.I.: *Representation Theory and Automorphic Functions* (Nauka, Moscow 1966) [in Russian]
49. Dubrovin B.A., Novikov S.P.: ZhETF **79**, 1006 (1980) [in Russian]
50. Sudarshan E.C.G.: Phys. Rev. Lett. **10**, 277 (1963)
51. Berezin F.A.: Sov. Phys. Usp. **132**, 497 (1980)
52. Golden S.: Phys. Rev. **B137**, 1127 (1965)
53. Thompson C.J.: J. Math. Phys. **6**, 1812 (1965)
54. Leschke H., Stolze J., Moraweck M.: J. Phys. A **13**, 1311 (1980)
55. Schwinger J.: Phys. Rev. **91**, 728 (1953)
56. Feynman R.: Phys. Rev. **84**, 108 (1951)
57. Weyl H.: Z. Phys. **46**, 1 (1928)
58. Stratonovich R.L.: *Selected Problems of Fluctuation Theory in Radio Engineering* (Sovetskoye Radio, Moscow 1961) [in Russian]
59. Louisell W.H.: *Radiation and Noise in Quantum Electronics* (McGraw-Hill, New York 1965)
60. Holliday D., Glassgold A.E.: Phys. Rev. **A139**, 1717 (1965)
61. Mollow B.R.: Phys. Rev. **162**, 1256 (1967)
62. Zeldovich B.Ya., Perelomov A.M., Popov V.S.: preprints ITEP-612, 618 (1968) [in Russian]
63. Feller W.: *An Introduction to Probability Theory and Its Applications*, Vol. 2 (Wiley, New York 1966)
64. Schwarz A.S.: Dokl. AN SSSR **173**, 793 (1967) [in Russian]
65. Gutkin E.: Funk. anal. i ego Pril. **9**(3), 89 (1975) [in Russian]
66. Arnold V.I.: *Mathematical Methods of Classical Mechanics* (Springer, New York 1978)
67. Perelomov A.M.: Yad. Fiz. **29**, 1688 (1979) [in Russian]
68. Delbourgo R.: J. Phys. **A10**, 1837 (1977)
69. Delbourgo R., Fox J.R.: J. Phys. **A10**, 1233 (1977)
70. Zhelobenko D.P.: *Compact Lie Groups and Their Representations. Translation of Mathematical Monographs*, Vol. 40 (Am. Math. Soc., Providence, R.I. 1973)
71. Bacry H.: Phys. Rev. **A18**, 617 (1978)
72. Monastyrsky M.I., Perelomov A.M.: Dokl. AN SSSR **207**, 1303 (1972) [in Russian]
73. Monastyrsky M.I., Perelomov A.M.: Reps. Math. Phys. **6**, 1 (1974)
74. Rossi H., Vergne M.: J. Funct. Anal. **13**, 324 (1973)
75. Perelomov A.M.: Funkt. Analys i Ego Prilozh. **6**(4) 47 (1972) [in Russian]
76. Arecchi F.T., Courtens E., Gilmore R., Thomas H.: Phys. Rev. **A6**, 2211 (1972)
77. Gelfand I.M., Minlos R.A., Shapiro Z.Ya.: *Representations of Three-Dimensional Rotation Group and the Lorentz Group* (Fitzmatgiz, Moscow 1958) [in Russian]
78. Vilenkin N.Ya.: *Special Functions and Theory of Group Representations* (Nauka, Moscow 1965) [in Russian]

79. Barut A., Raczka R.: *Theory of Group Representations and Applications* (Polish Scientific Publishers, Warszawa 1977)
80. Onofri E.: J. Math. Phys. **16**, 1087 (1975)
81. Helgason S.: *Differential Geometry, Lie Groups and Symmetric Spaces* (Academic, New York 1978)
82. Lieb E.H.: Commun. Math. Phys. **32**, 327 (1973)
83. Bargmann V.: Ann. Math. **48**, 568 (1947)
84. Gelfand I.M., Graev M.I., Vilenkin N.Ya.: *Integral Geometry and Related Problems of Representation Theory* (Fizmatgiz, Moscow 1962) [in Russian]
85. Perelomov A.M., Popov V.S.: ZhETF **56**, 1375 (1969) [in Russian]
86. Molchanov V.F.: Funk. Anal. i Ego Prilož. **1**(2) 1 (1967)
87. Ford L.R.: *Automorphic Functions* (McGraw-Hill, New York 1929)
88. Naimark M.A.: *Linear Representations of Lorentz Group* (Fizmatgiz, Moscow 1958) [in Russian]
89. Shapiro I.S.: Dokl. AN SSSR **106**, 647 (1956) [in Russian]
90. Perelomov A.M., Popov V.S.: ZhETF **50**, 179 (1966) [in Russian]
91. Perelomov A.M., Popov V.S.: ZhETF **54**, 1799 (1968) [in Russian]
92. Weil A.: Acta Math. **113**, 143 (1964)
93. Hua L.-K.: *Harmonic Analysis of Functions of Several Complex Variables in Classical Domains* (Am. Math. Soc., Providence, R.I. 1963)
94. Berezin F.A.: Commun. Math. Phys. **63**, 131 (1978)
95. Kirillov A.A.: Usp. Mat. Nauk **17**(4), 57 (1962)
96. Moore C.C., Wolf J.A.: Trans. Am. Math. Soc. **185**, 445 (1973)
97. Brezin J.: In *Partial Differential Equations and Related Topics*, ed. by J.A. Goldstein, Lecture Notes Math., Vol. 446 (Springer, Berlin, Heidelberg, New York 1975) p. 13
98. Moscowici H.: Commun. Math. Phys. **54**, 63 (1977)
99. Borel A.: Proc. Nat. Acad. Sci. USA **40**, 1147 (1954)
100. Gilmore, R.: Ann. Phys. **74**, 391 (1972)
101. Harish-Chandra: Amer. J. Math. **77**, 743 (1955)
102. Harish-Chandra: Amer. J. Math. **78**, 564 (1956)
103. Harish-Chandra: Acta Math. **116**, 1 (1966)
104. Cartan E.: Abh. Math. Sem. Univ. Hamburg **11**, 116 (1935)
105. Siegel C.L.: Ann. Math. **44**, 674 (1943)
106. Hano I.: Amer. J. Math. **79**, 885 (1957)
107. Perelomov A.M.: Commun. Math. Phys. **44**, 197 (1975)
108. Monastyrsky M.I., Perelomov A.M.: Ann. Inst. H. Poincaré **23**, 23 (1975)
109. Mackey G.W.: Bull. Amer. Math. Soc. **69**, 628 (1963)
110. Kostant B.: Bull. Amer. Math. Soc. **75**, 627 (1969)
111. Gel'fand I.M., Graev M.I.: Trudy Moscow Matem. Ob-wa **8**, 321 (1959)
112. Helgason S.: Adv. Math. **5**, 1 (1970)
113. Freudental H.: *Oktaven, Ausnahmegruppen und Oktavengeometry* (University of Utrecht, Utrecht 1951)
114. Harish-Chandra: Am. J. Math. **80**, 241 (1958)
115. Gindikin S.G., Karpelevič F.I.: Dokl. Akad. Nauk SSSR **145**, 252 (1962)
116. Landau L.D., Lifshitz E.M.: *Quantum Mechanics* (Fizmatgiz, Moscow 1963)
117. Moore C.: Am. J. Math. **86**, 201, 358 (1964)
118. Perelomov A.M.: Funk. anal. i ego Pril. **7**, 3, 57 (1973) [in Russian]
119. Gelfand I.M., Graev M.I., Pyatetsky-Shapiro I.I.: *Representation Theory and Automorphic Functions* (Nauka, Moscow 1966) [in Russian]
120. Ford L.R.: *Automorphic Functions* (McGraw-Hill Book Co., New York 1929)
121. Lehner J.: *Discontinuous Groups and Automorphic Functions* (Am. Math. Soc., Providence, R.I. 1964)
122. Hedlund E.: Duke Math. J. **2**, 530 (1936)
123. Boas R.P.: Ann. Math. **47**, 21 (1946)
124. Margulis G.A.: Funk. analyz i ego Pril. **3**(4) 89 (1969) [in Russian]

125. Moore C.C.: Am. J. Math. **88**, 154 (1966)
126. Siegel C.L.: Ann. Math. **46**, 708 (1945)
127. Poincaré H.: Acta Math. **1**, 193 (1882)
128. Borel A.: Topology **2**, 111 (1963)
129. Hirzebruch F.: *Topological Methods in Algebraic Geometry* (Springer, New York 1966)
130. Berezin F.A.: "General Concept of Quantization". Preprint ITP-74-20E, Kiev (1974); Commun. Math. Phys. **40**, 153 (1975)
131. Berezin F.A.: Dokl. Akad. Nauk SSSR **241**, 15 (1978)
132. Berezin F.A.: Funk. Anal. i Ego Prilož. **1**(2), 1 (1967)
133. Bogolyubov N.N.: Izv. AN SSSR, Ser. Phys. **11**, 77 (1947)
134. Bissichis W.H., Foldy L.L.: Phys. Rev. **A133**, 935 (1964)
135. Solomon A.I.: J. Math. Phys. **12**, 390 (1971)
136. Anderson P.W.: Phys. Rev. **112**, 1900 (1958)
137. Chung V.: Phys. Rev. **B140**, 1110 (1965)
138. Kibble T.W.: J. Math. Phys. **9**, 315 (1968)
139. Kibble T.W.: Phys. Rev. **173**, 1527; **174**, 1882; **175**, 1624 (1968)
140. Gochev I.G.: JINR preprint, E 17-84-253 (1984)
141. Shapiro I.S.: Pis'ma ZhETF **21**, 624 (1975); ZhETF **70**, 2050 (1976)
142. Cahill K.: Phys. Lett. **B53**, 174 (1974)
143. Vinciarelli P.: Phys. Lett. **B59**, 380 (1975)
144. Rühl W.: Commun. Math. Phys. **83**, 455 (1982)
145. Dubrovin B.A., Novikov S.P.: ZhETF **79**, 1006 (1980)
146. Flammand G.: J. Math. Phys. **7**, 1924 (1966)
147. Demkov Yu.N., Ostrovsky V.N., Solov'ev E.A.: ZhETF **66**, 125 (1974)
148. Narducci L.M., Bowden C.M., Coulter C.A.: Lett. Nuovo Cim. **8**, 57 (1973)
149. Narducci L.M., Bowden C.M., Coulter C.A.: Phys. Rev. **A9**, 829, 999 (1974)
150. Dicke R.H.: Phys. Rev. **93**, 99 (1954)
151. Hepp K., Lieb E.H.: Ann. Phys. **76**, 360 (1973); Phys. Rev. **A8**, 2517 (1973)
152. Gilmore R.: Phys. Lett. **A55**, 459 (1976)
153. Gilmore R., Bowden C.M.: J. Math. Phys. **17**, 1617 (1976)
154. Gilmore R.: J. Math. Phys. **15**, 2090 (1974)
155. D'Ariano G., Rasetti M., Vadacchino M.: J. Phys. **A18**, 1295 (1985)
156. D'Ariano G.M., Rasetti M.G.: Phys. Lett. **A107**, 291 (1985)
157. Rasetti M.: Intern. J. Theor. Phys. **14**, 1 (1975)
158. Filippov G.F., Vasilevsky V.S. Chopovsky L.L.: Particles and Nuclei **15**, 1338 (1984)
159. Nieto M., Simmons L.M.Jr., Gutschick V.P.: Phys. Rev. **D23**, 927 (1981)
160. Martin J.L.: Proc. Roy. Soc. **A251**, 543 (1959)
161. Montonen C.: Nuovo Cim. **A19**, 69 (1974)
162. Ohniki Y., Kashiwa T.: Progr. Theor. Phys. **60**, 548 (1978)
163. Bars I., Günaydin M.: Commun. Math. Phys. **91**, 31 (1983)
164. Zeldovič Ya.B.: ZhETF **51**, 1492 (1966)
165. Ritus V.I.: ZhETF **51**, 1544 (1966)
166. Zeldovič Ya.B.: Usp. Fiz. Nauk **95**, 209 (1968)
167. Zeldovič Ya.B., Pitaevsky L.P., Popov V.S., Starobinsky A.A.: Usp. Fiz. Nauk **105**, 780 (1971)
168. Husimi K.: Progr. Theor. Phys. **9**, 381 (1953)
169. Dykhne A.M.: ZhETF **38**, 570 (1960)
170. Lewis H.R., Riesenfeld W.B.: J. Math. Phys. **10**, 1458 (1969)
171. Perelomov A.M., Popov V.S.: ZhETF **56**, 1375 (1969)
172. Perelomov A.M., Popov V.S.: ZhETF **57**, 1684 (1969)
173. Perelomov A.M., Popov V.S.: Teor. Mat. Fiz. **1**, 360 (1969)
174. Baz' A.I., Zeldovič Ya.B., Perelomov A.M.: *Scattering, Reactions and Decays in Nonrelativistic Quantum Mechanics* (Nauka, Moscow 1971) [in Russian]
175. Landau L.D., Lifschitz E.M.: *Mechanics* (Pergamon, New York 1976), §27

176. Erdelyi A. et al.: *Higher Transcendental Functions*, Vol. 2 (McGraw-Hill Book Co., New York 1953)
177. Perelomov A.M.: Teor. Mat. Fiz. **6**, 213 (1971)
178. Calogero F.: J. Math. Phys. **12**, 419 (1971)
179. Majorana E.: Nuovo Cim. **9**, 335 (1932)
180. Grib A.A., Mamaev S.G., Mostepaneneko V.M.: *Quantum Effects in Intensive External Fields* (Atomizdat, Moscow 1980) [in Russian]
181. Perelomov A.M.: Phys. Lett. **A39**, 165 (1972); Teor. Mat. Fiz. **16**, 303 (1973)
182. Popov V.S.: ZhETF **62**, 1248 (1972)
183. Parker L.: Phys. Rev. **183**, 1057 (1969)
184. Parker L.: Phys. Rev. **D3**, 346 (1971)
185. Zeldovič Ya.B., Starobinsky A.A.: ZhETF **61**, 2161 (1971)
186. Harish-Chandra: Am. J. Math. **78**, 1 (1956)
187. Perelomov A.M.: Phys. Lett. **A39**, 353 (1972)
188. Marinov M.S., Popov V.S.: Yad. Fiz. **15**, 1271 (1972)
189. Marinov M.S., Popov V.S.: Fortschr. Phys. **25**, 373 (1977)
190. Perelomov A.M.: Teor. Mat. Fiz. **19**, 83 (1974)
191. Narozny N.B., Nikishov A.I.: Yad. Fiz. **11**, 1072 (1970)
192. Rabi I.: Phys. Rev. **51**, 652 (1937)
193. Bellissard J., Holtz R.: J. Math. Phys. **15**, 1275 (1974)
194. Simon B.: Commun. Math. Phys. **71**, 247 (1980)
195. Fuller W., Lenard A.: Commun. Math. Phys. **69**, 99 (1979)
196. Papanicolaou N.: Ann. Phys. **136**, 210 (1981)
197. Gross P., Neveu A.: Phys. Rev. **D10**, 3235 (1974)
198. Neveu A., Papanicolaou N.: Commun. Math. Phys. **58**, 31 (1978)
199. Zakharov V.E., Mikhailov A.V.: Commun. Math. Phys. **74**, 21 (1980)
200. Perelomov A.M., Popov V.S., Zeldovič B.Ya.: Preprints ITEP, Nos. 612, 618 (1968) [in Russian]
201. Perelomov A.M., Popov V.S., Zeldovič B.Ya.: ZhETF **55**, 589 (1968); **57**, 196 (1969)
202. Senitzky I.R.: Phys. Rev. **119**, 670 (1960)
203. Gordon J.P., Walker L.R., Louisell W.H.: Phys. Rev. **130**, 806 (1963)
204. Shen Y.R.: Phys. Rev. **155**, 921 (1967)
205. Louisell W.H.: *Radiation and Noise in Quantum Electronics* (McGraw-Hill, New York 1965)
206. Shelepin L.A.: ZhETF **54**, 1463 (1968)
207. Makhviladze T.M., Shelepin L.A.: Tr. FIAN **70**, 120 (1973)
208. Lizin I.M., Makhviladze T.M., Shelepin L.A.: Tr. FIAN **87**, 21 (1976)
209. Karasev V.P., Shelepin L.A.: Tr. FIAN **124**, 49 (1980)
210. Lax M.: Phys. Rev. **145**, 110 (1960)
211. Landau L.D.: Z. Phys. **45**, 430 (1927), [see also: Collected Papers of L.D. Landau (Pergamon, New York 1965)]
212. Wigner E.P.: Phys. Rev. **40**, 749 (1932)
213. Holliday D., Glassgold A.E.: Phys. Rev. **A139**, 1717 (1965)
214. Schwinger J.: J. Math. Phys. **2**, 407 (1961)
215. Belavin A.A., Perelomov A.M., Popov V.S., Zeldovich B.Ya.: Preprint ITEP No. 622 (1968) [in Russian]
216. Belavin A.A., Perelomov A.M., Popov V.S., Zeldovich B.Ya.: ZhETF **56**, 264 (1969)
217. Perelomov A.M.: ZhETF **40**, 1418 (1961)
218. Nosov V.G., Yakovleva T.V.: ZhETF **43**, 1751 (1963)
219. Ivanter I.G., Smilga V.P.: ZhETF **54**, 559 (1968)
220. Granovsky Ya.I., Dimashko Yu.A.: Ukr. Phys. J. **19**, 1456 (1974)
221. Landau L.D.: Z. Phys. **64**, 629 (1930)
222. Rumer Yu.B.: ZhETF **18**, 1081 (1948)
223. Granovsky Ya.I., Dimashko Yu.A.: ZhETF **68**, 1991 (1975)
224. Heisenberg W., Euler H.: Z. Phys. **98**, 714 (1936)

225. Ritus V.I.: ZhETF **57**, 2176 (1969)
226. Schwinger J.: Phys. Rev. **D7**, 1696 (1973)
227. Wehrl A.: Reps. Math. Phys. **16**, 353 (1979)
228. Wehrl A.: Rev. Mod. Phys. **50**, 221 (1978)
229. Lieb E.H.: Commun. Math. Phys. **62**, 35 (1978)
230. Thirring W.: Lett. Math. Phys. **4**, 67 (1980)
231. Levin B.Ya.: Distribution of Zeroes of Entire Functions (GITTL, Moscow, Leningrad 1956) [in Russian]
232. Boas R.P.: *Entire Functions* (Academic, New York 1954)
233. Gradstein I.S., Ryzhik I.M.: *Tables of Integrals, Sums, Series and Products* (Fizmatgiz, Moscow 1963) [in Russian]

References to Addendum

A1. Klauder, J.R., Skagerstam, B.-S.: *Coherent States. Applications in Physics and Mathematical Physics* (World Scientific Publishing, Singapore 1985)
A2. Greco, M., Palumbo, F., Pancheri-Srivastava, G., Srivastava, Y.: Phys. Lett. **B77**, 282 (1978)
A3. Kibble, T.: In: *Mathematical Methods in Theoretical Physics*, ed by K.T. Mahanthappa and W.E. Brittin (Gordon and Breach, New York 1969) p. 387
A4. Gomatam, J.: Phys. Rev. **D3**, 1292 (1971)
A5. Granovsky, Ya.I., Zhedanov, A.S.: JETP Lett. **41**, 312 (1985)
A6. Jevicki, A., Papanicolaou, N.: Ann. Phys. **120**, 107 (1979)
A7. Mead, L., Papanicolaou, N.: Phys. Rev. **B28**, 1633 (1983)
A8. Shankar, R.: Phys. Rev. Lett. **45**, 1088 (1980)
A9. Vuillermot, P.: Commun. Math. Phys. **76**, 1 (1980)
A10. Braginsky, V.B., Vorontsov, Yu.I., Khalili, F.Ya.: JETP **73**, 1340 (1977)
A11. Caves, C., Thorne, K., Drever, R., Sandberg, V., Zimmermann, M.: Rev. Mod. Phys. **52**, 341 (1980)
A12. Hollenhorst, J.: Phys. Rev. **D19**, 1669 (1979)
A13. Moncrief, V.: Ann. Phys. **114**, 201 (1978)
A14. Unruh, W.: Phys. Rev. **D17**, 1180 (1978)
A15. Gilmore, R.: In: *Symmetries in Science*, ed. by B. Gruber and R.S. Millman (Plenum, New York 1980) p. 105
A16. Rasetti, M.: Int. J. Theor. Phys. **14**, 1 (1975)
A17. Janssen, D.: Yadern. Fiz. USSR **25**, 897 (1977)
A18. Fowler, G., Stelte, N., Weiner, R.: Nucl. Phys. **A319**, 349 (1979)
A19. Gilmore, R., Feng, D.: Phys. Lett. **B76**, 26 (1978)
A20. Gilmore, R., Feng, D.: Nucl. Phys. **A301**, 189 (1978)
A21. Hammer, C., Schrauner, J., de Facio, B.: Phys. Rev. **D18**, 373 (1978)
A22. Botke, J., Scalapino, D., Sugar, R.: Phys. Rev. **D9**, 813 (1974)
A23. Carruthers, P., Shih, C.: Phys. Lett. **B127**, 242 (1983)
A24. Carruthers, P. In: Proc. XIVth Intern. Symposium on Multiparticle Dynamics (World Scientific Publisher, Singapore 1984) p. 825
A25. Curci, G., Greco, M., Srivastava, Y.: Phys. Rev. Lett. **43**, 834 (1979)
A26. Eriksson, K.-E., Skagerstam, B.-S.: Phys. Rev. **D18**, 3858 (1978)
A27. Eriksson, K.-E., Mukunda, N., Skagerstam, B.-S.: Phys. Rev. **D24**, 2615 (1981)
A28. Horn, D., Silver, R.: Ann. Phys. (NY) **66**, 509 (1971)
A29. Ni, G.-J., Wang, Y.-P.: Phys. Rev. **D27**, 969 (1983)
A30. Rayne, F.: Nucl. Phys. **B159**, 244 (1979)
A31. Skagerstam, B.-S.: Zs. f. Phys. **C24**, 97 (1985)
A32. Davies, M., Heller, E.: J. Chem. Phys. **75**, 3919 (1981)

A33. Shi, S.: J. Chem. Phys. **79**, 1343 (1983)
A34. Paul, R.: Phys. Lett. **A96**, 263 (1983)
A35. Dunne, L., Clark, A., Braudas, E.: Chem. Phys. Lett. **97**, 573 (1983)
A36. Narducci, L., Bowden, C., Bluemel, V., Garrazana, G., Tuft, R.: Phys. Rev. **A11**, 973 (1975)
A37. Wang, Y., Hioe, F.: Phys. Rev. **A7**, 831 (1973)
A38. Toyoda, T., Wildermuth, K.: Phys. Rev. **D22**, 2391 (1980)
A39. Cummings, F., Johnston, R.: Phys. Rev. **151**, 105 (1966)
A40. Langer, J.: Phys. Rev. **167**, 183 (1968)
A41. Klauder, J.R.: Syst. Techn. J. **34**, 909 (1960)
A42. Klauder, J.R.: Phys. Rev. **A29**, 2036 (1984)
A43. Suzuki, T.: Nucl. Phys. **A398,** 557 (1983)
A44. Takahashi, Y., Shibata, F.: J. Phys. Soc. Jpn. **38**, 656 (1975)
A45. Ghosh, G., Dutta-Roy, B., Bhaumik, D.: Plasma Phys. **19**, 1051 (1977)
A46. Feldman, A., Kahn, A.: Phys. Rev. **B1**, 4584 (1970)
A47. Pottinger, D.: Phys. Lett. **B78**, 476 (1978)
A48. Yaffe, L.: Rev. Mod. Phys. **54**, 407 (1982)
A49. Berezin, F.A.: Usp. Fiz. Nauk **132**, 497 (1980)
A50. Skagerstam, B.-S.: J. Phys. **A18**, 1 (1985)
A51. Nieto, M.M.: "Coherent States with Classical Motion; from an Analytic Method Complementary to Group Theory" In: *Group Theoretical Methods in Physics*, Vol. 2, ed. by M.A. Markov (Nauka, Moscow 1983) p. 174

Subject Index